D0781226

RELATIVITY
for Scientists and Engineers

Ray Skinner

Professor of Physics,
University of Saskatchewan

Dover Publications, Inc.
New York

TO MY PARENTS

Published in Canada by General Publishing Company, Ltd., 30 Lesmill Road, Don Mills, Toronto, Ontario.
Published in the United Kingdom by Constable and Company, Ltd., 10 Orange Street, London WC2H 7EG.

This Dover edition, first published in 1982, is an unabridged republication of the work originally published by Blaisdell Publishing Company, Waltham, Massachusetts in 1969 under the title *Relativity*. Typographical errors have been corrected in the present edition, and a revised Answers to Problems has been prepared by Professor Skinner.

International Standard Book Number: 0-486-64215-1
Library of Congress Catalog Card Number: 81-65709

Manufactured in the United States of America
Dover Publications, Inc.
180 Varick Street
New York, N.Y. 10014

Note to the Dover Edition

This edition differs from the original in that typographical errors have been corrected and a new solutions section contains hints and answers, many corrected, for almost all of the problems.

The original goal of a five-volume series covering all of the major branches of physics was never realized because of a change in ownership of the original publishing house.

Preface to the First Edition

This is the second of a planned series of five volumes designed for use in under-graduate physics courses by students majoring in science or engineering. The aim of the series is to present a thorough discussion of the basic concepts of physics from a modern point of view. The first volume, *Mechanics*, examines newtonian mechanics; subsequent books in the series will include volumes on electromagnetism, thermal physics, and quantum mechanics.

The volumes in this series are intended for use in an introductory course of four or more semesters. Since the subject matter is discussed in greater depth than that usually found in introductory-level textbooks, each volume may be useful for a second course or an intermediate-level course in the specific subject covered by the individual volumes.

The theories of relativity have a fascination that captures the interest of even those students who are without much knowledge of physics. Furthermore, many of the basic concepts may be taught without the use of the calculus or difficult mathematical analyses. For these reasons, a short course on relativity can provide an introduction to physics that is attractive to students. Appropriate material may be chosen from the present text for a short introduction to the kinematics of special relativity, which is suitable for students having little background in physics beyond some familiarity with newtonian kinematics. The book may be used in various other ways moreover, even as a text for a one-semester, intermediate-level course in modern physics.

This book is divided into three chapters, which cover the kinematics and dynamics of special relativity and include an introduction to the ideas of general relativity; each chapter covers one major subject. The introduction to each chapter contains a review of the relevant newtonian mechanics, although this material need not be covered in depth. In addition, much of the chapter on kinematics and parts of the chapter on the dynamics of special relativity do not require any knowledge of the calculus, although such knowledge would be helpful.

The present text is self-contained, but desirable additional background for this text may be obtained from a study of newtonian mechanics as presented in *Mechanics*. Although that text is not prerequisite to the present book, a number of topics discussed in this text are dealt with in greater detail there; these topics include vectors, Newton's laws, scattering, and wave motion.

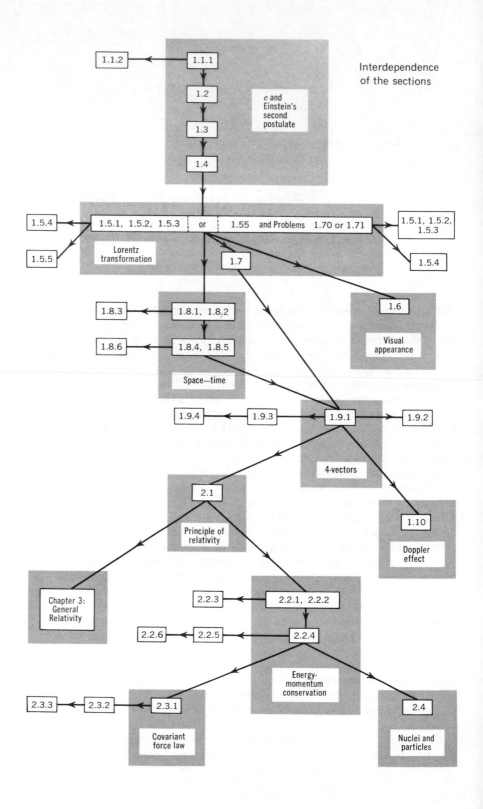

Interdependence of the sections

The chapter on the kinematics of special relativity includes a discussion of space-time geometry and of the group-theoretical basis for the importance of 4-vectors. These discussions culminate, in the chapter concerning the dynamics, in the development of the covariant formulation of the conservation laws and the equation of motion. Many applications of special relativity in nuclear and particle physics are presented in a section containing a description of the development of those subjects and the properties of the particles.

The third chapter provides an introduction to general relativity. The relevance of Riemannian geometry to a theory of gravitation and the first-order metric of the principle of equivalence are discussed in some detail, but Einstein's equations, the Schwarzschild solution, and its consequences are described only briefly. Nevertheless, the chapter contains more material on general relativity than many physics students have encountered in the past—the students' shortcoming in this field is unfortunate in view of the recent resurgence of interest in both the theoretical and experimental aspects of general relativity.

The chapters are divided into subsections, each of which contains material that usually can be covered in one lecture. However, there are a number of sections on given concepts that may be left for assigned reading or that may be correlated for one discussion. The possibilities for this approach are diagrammed on the flow chart. Almost all subsections are followed by problems, and each chapter is followed by an additional set of problems.

References to the corresponding material in a few selected texts are given at the beginning of most subsections; these references may be used for supplementary reading or as a source of additional problems. The book contains numerous references to books and articles, and these may be used to supplement the discussion in the text or as assigned reading. There are also a few advanced references at the end of each chapter to be used as a basis for more advanced discussions.

The order of presentation of the material need not follow the text exactly, although this might be helpful. The flow chart shows the grouping of subject matter according to section topics, and indicates the relationship of sections to one another. Each section is dependent only on the material listed along the lines that flow into the box marked with the number of the section. Those sections from which there are no outgoing lines, or whose topics do not lead into material selected for study, may be omitted. This is particularly true for those sections that lie outside the shaded regions and that provide additional discussions of the relevant topics.

I wish to thank all those who offered suggestions for the improvement of this book and those who encouraged me in this task; they include J. C. Bergstrom, M. H. Brennan, V. K. Gupta, R. N. H. Haslam, L. Katz, H. N. Rundle, and G. J. Sofko. A particular expression of thanks is due my good friend and former colleague on the faculty here, G. J. D. Taylor, for many interesting discussions. In addition, I owe a debt of gratitude to E. L. Tomusiak, who read the manuscript and made many worthwhile suggestions.

I am indebted to Professor R. W. Fuller, who made valuable comments on an initial draft. I am grateful for the thoughtful criticism of the consulting editors, Professors M. J. Klein and B. F. Stearns, who read through a number of drafts and always provided valuable advice and numerous suggestions for improvement.

Full responsibility for any errors, however, rests on my shoulders. I will always welcome any corrections and suggestions.

My wife Lilianne deserves special commendation. She typed all drafts of the manuscript and provided encouragement, humor, and patience throughout the project.

RAY SKINNER

Saskatoon, Canada

Contents

2

Special Relativity Theory: Introductory Dynamics   133

3

General Theory of Relativity   283

The Greek Alphabet

Alpha	A	α	a′lfə
Beta	B	β	bā′tə or bē′tə
Gamma	Γ	γ	gam′ə
Delta	Δ	δ	del′tə
Epsilon	E	ε	ep′sə lon or ep sī′lən
Zeta	Z	ζ	zā′tə or zē′tə
Eta	H	η	ā′tə or ē′tə
Theta	Θ	θ, ϑ	thā′tə or thē′tə
Iota	I	ι	ī ō′tə
Kappa	K	κ	kap′ə
Lambda	Λ	λ	lam′də
Mu	M	μ	mū
Nu	N	ν	nü or nū
Xi	Ξ	ξ	sī or zī or ksē
Omicron	O	o	om′e kron or ō′me kron
Pi	Π	π	pī
Rho	P	ρ	rō
Sigma	Σ	σ, ς	sig′mə
Tau	T	τ	tô, tou
Upsilon	Υ	υ	ūp′sə lon
Phi	Φ	φ, φ	fī or fē
Chi	X	χ	kī
Psi	Ψ	ψ	sī or psē
Omega	Ω	ω	ō meg′ə or ō meg ə or o māg′ə

* Key to pronunciation: ace (ās), accuse (ə kūz′), even (ē′vən), ice (īs), old (ōld), order (ôr′dər), ounce (ouns), rule (rül).

Prefixes for Metric Units

Prefix	Meaning	Symbol
Atto-	10^{-18}	
Femto-	10^{-15}	f
Pico-	10^{-12}	p
Nano-	10^{-9}	n
Micro-	10^{-6}	μ
Milli-	10^{-3}	m
Centi-	10^{-2}	c
Deci-	10^{-1}	
Deca-	10	
Hecto-	10^2	
Kilo-	10^3	k
Mega-	10^6	M
Giga-	10^9	G (or B)
Tera-	10^{12}	T

Table of Values

$e = 2.71828\ldots$

$\pi = 3.14159\ldots$

Standard acceleration of free fall $= g_n = 9.80665$ m/sec^2

Speed of light in empty space $= c = (2.997925 \pm 0.000003) \times 10^8$ m/sec

1 elementary charge $=$ elem ch $= e = (1.60210 \pm 0.00007) \times 10^{-19}$ C

Planck's constant $= h = (6.6256 \pm 0.0005) \times 10^{-34}$ J$\cdot$sec

Permittivity of the vacuum $= \varepsilon_0 = (8.85418 \pm 0.00002) \times 10^{-12}$ C^2/N$\cdot$m^2

Gravitational constant $= G = (6.670 \pm 0.015) \times 10^{-11}$ N$\cdot$m^2/kg^2

Electron mass $= m_e = (9.1091 \pm 0.0004) \times 10^{-31}$ kg

Proton mass $= m_p = (1.67252 \pm 0.00008) \times 10^{-27}$ kg

Neutron mass $= m_n = (1.67482 \pm 0.00008) \times 10^{-27}$ kg

Mean radius of the earth $= 6.371 \times 10^6$ m

Mass of the earth $= 5.977 \times 10^{24}$ kg

Mean radius of the earth's orbit $= 1.495 \times 10^{11}$ m

Radius of the moon $= 1.738 \times 10^6$ m

Mass of the moon $= 7.35 \times 10^{22}$ kg

Mean radius of the moon's orbit $= 3.844 \times 10^8$ m

Mass of the sun $= 1.989 \times 10^{30}$ kg

Radius of the sun $= 6.960 \times 10^8$ m

Special Relativity Theory: Kinematics

Our notions of space and time are derived from experiences with relative positions, clocks, motions, etc. These notions are refined as our range of experience increases, as can be confirmed by anyone who has watched an infant reach for a distant object or waited for a young child to return "in just a moment," or by one who has become more and more familiar with motions of objects and the laws to which these motions are subject. An improvement in understanding of our everyday notions of space and time is not enough, though; like all our other ideas, these notions and the basis for them must be scrutinized continuously for hidden fallacies as the range of our experiences is expanded. If necessary, these notions must be modified with the insight provided by new experiences. We cannot maintain ideas that are contrary to experience or, in other words, that disagree with experiment. This book focuses on the modification of our usual views of space and time and the consequent changes in the laws of motion resulting from new information provided by experiments performed in the last half of the nineteenth century.

The basis for our familiar notions of space and time is provided by our knowledge of the common motions of everyday experience, such as the motions of a ball or a car. These motions are correctly described by a set of laws introduced by the great English physicist and mathematician Sir Isaac Newton (1642–1727). His laws of motion are also valid to a high degree of accuracy when applied to the motions of the planets in the Solar System, extending some 10^{13} m across space and involving periods of 10^{10} sec. On the other hand, Newton's laws were used[*] by the father of nuclear physics, Sir Ernest Rutherford (1871–1937), in his analysis of the experiments that led to the discovery of the atomic nucleus, an object measuring about 10^{-14} m across, which experienced collisions for about 10^{-20} sec in those experiments. The range of validity of Newton's laws is limited, however; the special theory of relativity is required for the description of some phenomena that are not correctly described by Newton's laws.

---

[*] That Newton's laws gave meaningful results in this case does not mean that they adequately describe phenomena at the nuclear level. They do not, in fact. See Section 2.4.

The special theory of relativity is based on concepts of space and time that differ from those applicable in newtonian mechanics (although the concepts of relativity theory are compatible with the newtonian concepts in their common range of validity). There is a natural division of subject matter in the special theory of relativity: (1) *kinematics*, which is the study of the concepts of space and time and the means of describing the motions of objects; (2) *dynamics*, which is the study of the regularities in these motions or the rules that govern them. The kinematics of special relativity form the subject matter of this chapter; the dynamics of special relativity are discussed in Chapter 2.

Newtonian mechanics represents a precise statement of our intuitive ideas about the common motions of everyday experience. Therefore, it is worthwhile at this point to review the kinematics of newtonian mechanics as a preliminary step in our study of the modifications in newtonian kinematics required by the special theory of relativity.

Newton maintained that time was absolute: that is, every observer can determine time intervals relative to a time standard or clock that depends in no way on that observer. It follows from this, for example, that if the time interval between two explosions is 94 sec according to your clock, the time interval between these two events is also 94 sec according to my clock (unless one of the clocks is not functioning properly). This idea—that the length of the time interval between two events is independent of who measures it—provided a cornerstone for acceptable explanations of observed physical phenomena until the turn of this century.

It was assumed also, until the beginning of this century, that space satisfied the axioms of Euclidean geometry. It follows, for example, that two straight rods of lengths A and B that form a right-angled V have their outer ends separated by the distance $\sqrt{A^2 + B^2}$, as given by the theorem of Pythagoras [Figure 1.1(a)], and that this is true relative to *all* observers. A similar formula results in three dimensions [Figure 1.1(b)], again with respect to all observers.

Newtonian mechanics is based on the assumption that we can specify the position of an object in this Euclidean space at any instant of (the absolute) time. The motion of an object is described by giving its position at the various instants of time during that motion. A real object is extended, occupying a volume in space—not only one point—and the description of its motions requires the specification of each point in that object at each instant of time. This description is not a simple matter in general because of the relative motions of the parts, and it would be difficult indeed to find in such complicated motions regularities that could be stated as simple rules that govern these motions. This complication is avoided in newtonian mechanics by the use of the concept of a *point particle*. A point particle is an idealized object that behaves in every way as a real object, except that it occupies no volume. Newton's laws of motion apply to the motions of point particles, and they describe the motions of real extended objects if these objects are treated as an assembly of interacting point particles. (The converse of this—real objects consist of point particles that satisfy Newton's laws—is not true.)

The position of a point particle P can be specified in a number of ways relative to a given point O fixed to a reference frame. We can set up one or another coordinate system relative to that reference frame, and we can specify the position by giving the coordinates of that point relative to the origin O (Figure 1.2). Different sets of coordinates describe the same displacement,

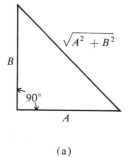

(a)

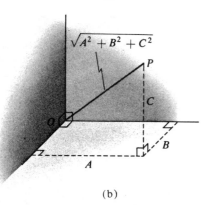

(b)

FIGURE 1.1 Consequences of the axioms of Euclidean geometry. (a) The theorem of Pythagoras. (b) $OP = \sqrt{A^2 + B^2 + C^2}$.

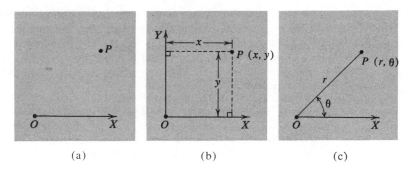

(a)   (b)   (c)

FIGURE 1.2 Coordinate systems appropriate for the description of motions on a plane. (a) Choice of origin O and a reference direction OX. (b) Rectangular or cartesian coordinates (x, y) of P. (c) Polar coordinates (r, θ) of P.

that from the origin to the position, and so the position may be designated by an entity, the position or *displacement vector*, that represents that displacement independent of the coordinate system used (Figure 1.3). The displacement vector determines the distance, the direction along the line between the position and the origin, and the sense—*from* the origin O *to* the position P—of that direction no matter what coordinate system is used to describe that vector. The description of a displacement vector depends on the choice of reference frame, and differs in this respect from a more familiar type of physical quantity, called a *scalar*, that is given by one number and its unit and that is independent of the reference frame. In this book a displacement vector is designated* by a boldface letter, such as **r**, or, if the end points are given, by use of the form (letter for the initial point) (letter for the final point), with an arrow over the two letters, such as $\overrightarrow{OP}$. The length of a displacement vector is a scalar and is called the magnitude of the vector. We denote the magnitude of a vector such as **r** by the corresponding letter in italic print, r, or by vertical bars around the vector's symbol such as $|\mathbf{r}|$ or $|\overrightarrow{OP}|$.

One displacement may be followed by another, and the combination of the corresponding vectors is called the vector sum. The combination process is called *vector addition*. Just as one displacement followed by another is equivalent to a third displacement, so the vector sum of two vectors is a third vector (Figures 1.4 and 1.5):

$$\overrightarrow{OP} + \overrightarrow{PQ} = \overrightarrow{OQ} \qquad \text{or} \qquad \mathbf{a} + \mathbf{b} = \mathbf{c}. \qquad (1.1)$$

The vector **b** is given by the process of vector subtraction of **a** from **c**:

$$\mathbf{b} = \mathbf{c} - \mathbf{a}. \qquad (1.2)$$

Any entity that can be specified by a magnitude, a direction, and a sense and satisfies a similar law of addition is also called a vector.

The motion of a point particle is specified by its position **r** at each instant of time t. We use a standard notation for functions to write (Figure 1.6)

$$\mathbf{r} = \mathbf{r}(t). \qquad (1.3)$$

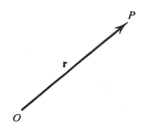

FIGURE 1.3 The displacement vector **r** from O to P.

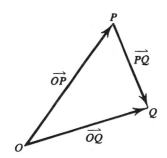

FIGURE 1.4 The addition of displacements.

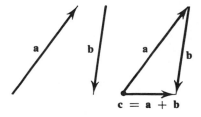

$$\mathbf{c} = \mathbf{a} + \mathbf{b}$$

FIGURE 1.5 The addition of the vectors **a** and **b**.

---

* You may find it convenient in handwritten material to use an arrow over the label to designate vectors that correspond to displacement vectors (as a reminder of their significance) and an underline to designate a generalization, introduced in Section 1.9, of these vectors.

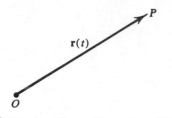

FIGURE 1.6 The position vector of P
relative to O at time t.

The motion also may be specified by the manner in which the position of the particle changes as time passes. Let $\Delta\mathbf{r}$ denote the displacement, the change in the position of the particle, in the time interval Δt (Figure 1.7). The *average velocity* $\bar{\mathbf{v}}$ of the particle between times t and $t + \Delta t$ is defined by

$$\bar{\mathbf{v}} = \frac{\mathbf{r}(t + \Delta t) - \mathbf{r}(t)}{\Delta t} = \frac{\Delta\mathbf{r}}{\Delta t}. \tag{1.4}$$

If the average velocity vectors are calculated for smaller and smaller values of Δt, the average velocities approach a vector $\mathbf{v}$ defined as the (instantaneous) *velocity* at the time t; the velocity is the limit, denoted by the expression "lim," of the average velocities, and is defined in mathematical notation by

$$\mathbf{v}(t) = \lim_{\Delta t \to 0} \frac{\mathbf{r}(t + \Delta t) - \mathbf{r}(t)}{\Delta t} = \lim_{\Delta t \to 0} \frac{\Delta\mathbf{r}}{\Delta t} = \frac{d\mathbf{r}}{dt}. \tag{1.5}$$

In much of the work in this chapter, the velocities under consideration will be constant vectors and so the respective average and instantaneous velocities will be equal. The magnitude v of the velocity $\mathbf{v}$ is called the *speed*.

The manner in which the velocity of a particle changes with time plays an important role in newtonian mechanics (see the introductory paragraphs of Chapter 2). The rate of change of the velocity with respect to the time is called the acceleration $\mathbf{a}$ and is defined by

$$\mathbf{a} = \lim_{\Delta t \to 0} \frac{\mathbf{v}(t + \Delta t) - \mathbf{v}(t)}{\Delta t} = \lim_{\Delta t \to 0} \frac{\Delta\mathbf{v}}{\Delta t}$$
$$= \frac{d\mathbf{v}}{dt} = \frac{d^2\mathbf{r}}{dt^2}. \tag{1.6}$$

The position vector of a particle specifies the site of the particle relative to a given frame of reference; thus it is of interest to determine the relation between the position vectors relative to two different frames. If the frames are at rest relative to one another, the relation involves only the vector $\mathbf{R}$ representing the displacement between the origins or reference points (Figure 1.8). The relation is given by the basic property of the addition of vectors:

$$\mathbf{r} = \mathbf{r}' + \mathbf{R}. \tag{1.7}$$

A similar relation for the instantaneous position vectors is true if the reference frames are in relative motion.

$$\mathbf{r}(t) = \mathbf{r}'(t) + \mathbf{R}(t), \tag{1.8}$$

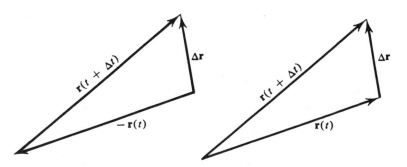

FIGURE 1.7   The definition of $\Delta\mathbf{r}$: $\Delta\mathbf{r} = \mathbf{r}(t + \Delta t) - \mathbf{r}(t)$.

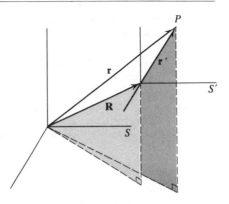

FIGURE 1.8   The relation between the reference systems S and S'.

where we have used the absolute character of time to label each vector by the same value of t. It follows directly from the definition of velocity and acceleration that

$$\mathbf{v}(t) = \mathbf{v}'(t) + \mathbf{V}(t) \tag{1.9}$$

and

$$\mathbf{a}(t) = \mathbf{a}'(t) + \mathbf{A}(t). \tag{1.10}$$

The description of the regularities in the motions of objects—the laws of mechanics—can be complex or simple, depending upon the reference frame used. Newton's laws of mechanics take on their simplest form relative to a frame of reference attached to the distant ("fixed") stars or to any other frame moving with a constant velocity relative to the fixed stars. Every such frame is called an *inertial frame of reference*, since a particle not experiencing any modification in its motion because of the presence of other objects in its environment exhibits only the property of inertia in such a motion; it moves with a constant velocity relative to each and every inertial reference frame.

The relation between the positions of an object relative to two inertial frames is given by the transformation law [Equations (1.8), (1.9), and (1.10)] for the case in which $\mathbf{V}(t) = \mathbf{V}$, a constant vector. This restricted transformation law is called the *galilean transformation law* after the Italian scientist Galileo Galilei (1564–1642), and for the particular case in which $\mathbf{R}$ is zero at $t = 0$ the law is given by

$$\mathbf{r}(t) = \mathbf{r}'(t) + \mathbf{V}t,$$
$$\mathbf{v}(t) = \mathbf{v}'(t) + \mathbf{V},$$
$$\mathbf{a}(t) = \mathbf{a}'(t). \tag{1.11}$$

Newton's laws of motion take on their most simple form and, indeed, the *same* form relative to all inertial frames related by the galilean transformation law. Therefore, according to Newton's laws, *it is impossible by a mechanical experiment to distinguish one inertial frame from another related by the galilean transformation law*. This is a statement of the *galilean principle of relativity*.

This concludes our review of the kinematics of newtonian mechanics. It can be seen from our discussion that this kinematics depends on the use of an absolute time, the same time for all observers (in the derivation of the galilean transformation) and on the properties of Euclidean geometry (in the representation of vectors by directed straight-line segments independent of the observer).

However, we shall see in this chapter that although these concepts are valid within the range of applicability of newtonian mechanics, they are not universally applicable.

Modifications in the pretwentieth century notions of space and time were found to be necessary as a result of experiments performed to determine the properties of light propagation. The propagation of light is a form of energy transmission that exhibits wave-like properties as demonstrated, for example, by the famous experiment (Figure 1.9) performed in the early part of the nineteenth century by the English physicist Thomas Young (1773–1829).

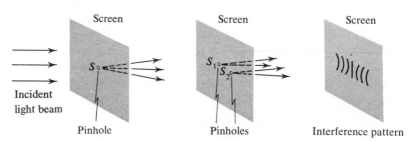

FIGURE 1.9   Young's experiment: The incident light that passes through S acts as a single source for light traveling along the two paths through S_1 and S_2, and gives an interference pattern on the final screen. The interference pattern can be explained on the basis of a wave theory of light; a stream of particles or corpuscles would produce an entirely different pattern according to our customary ideas of particle behavior.

Other waves such as sound waves, earthquake waves, and disturbances transmitted through solids, require a material medium for their transmission. Indeed, they are propagated by means of the interactions between neighboring elements; the motion of one part influences the behavior of the next part, and the disturbance is transmitted from one portion to the next, as shown in Figure 1.10. The properties of propagation of those types of waves, such as the speed of transmission, can be used to determine properties of the material medium through which they are propagated, and also can be used to gain information about the mechanism by which these waves are transmitted. Thus, the speed of sound in air (about 330 m/sec) is determined in part by the temperature of the air, and the speed of transmission of a disturbance in a solid is determined in part by the density of the solid.

Light propagates with a finite, but very great speed (Section 1.1). However, unlike the more familiar types of waves, light can be transmitted without the presence of a material medium. This is illustrated, for example, by the fact that starlight reaches us after traveling tremendous distances through space practically devoid of matter. This peculiarity of light transmission may also be demonstrated by an experiment that illustrates the fact that sound requires a material medium for its transmission (Figure 1.11). As the air is pumped out of the bell jar, the intensity of the sound of the ringing bell decreases to zero thus suggesting that sound requires a medium like air for its transmission, whereas the intensity of the light reflected from the bell does not decrease.

Since light does not require a material medium for its propagation, we can depict waves of light in the manner used to describe sound waves, for example, only if we postulate the existence of some medium that transmits the light waves.

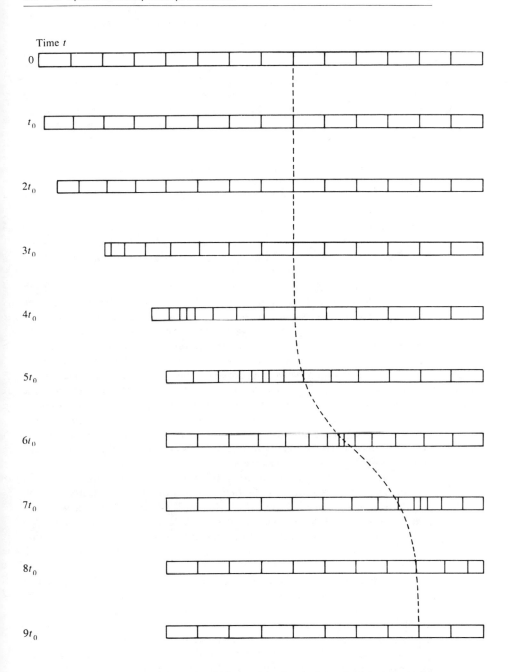

FIGURE 1.10 The manner in which a wave having the form of a short pulse is propagated along a solid rod. The rod is shown at rest in the top line. The vertical lines, etched in the rod, are equidistant when the enclosed regions are in equilibrium.

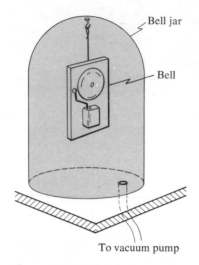

FIGURE 1.11   The bell can be seen even when there is no air in the jar.

Before the turn of this century, experimental techniques were sufficiently refined to the point where scientists could make measurements that would determine the reference frame in which this medium is at rest. All such experiments were unsuccessful in detecting that reference frame (Section 1.2). Nature appeared to be conspiring to prevent the detection of the medium that transmits light waves, and it remained for the great German–American theoretical physicist Albert Einstein (1879–1955) to point out that as a result, the assumption of the existence of such a medium was superfluous. Space itself, which we usually think of as a sort of "nothingness," has the property that it can transmit light even in the absence of matter.

The lack of success in determining the privileged reference frame in which the medium that transmits light waves is at rest, and other considerations, led Einstein to postulate that in fact there was no such preferred reference frame, and that the speed of light is the same relative to all inertial frames. These postulates of the *special theory of relativity** (Section 1.3) have been corroborated by many experiments since then and now have achieved the status of laws of nature. The kinematic consequences of the postulates of the special theory of relativity form the subject matter of Sections 1.4 to 1.10.

As a consequence of a study of the propagation of light, we gain further insight into the properties of space and time. We find that we must modify our notions on space and time as a result of experimental results (or new experiences, if you wish). These modifications appear very drastic indeed (Section 1.4), although the extraordinary consequences lie outside the range of our experiences of everyday life. We find that the division between space and time for one observer may not be the division between space and time for another. The concept of absolute time is demolished, and space and time become intermingled into one continuum, space-time. The valid transformation law (Section 1.5) relating the instant of time and the coordinates in space of an event relative to one inertial observer to the time and position of the same event relative to another inertial observer can be derived on the basis of Einstein's postulate of the equality of the speed of light for all inertial observers.

That the relation between observers is different than was believed prior to the turn of the century is shown by the fact that two observers in relative motion do not determine, in general, the same value for the length of an object nor the same value for the time interval between two events (Section 1.5). Even the visual appearance of an object is different for one observer than another (Section 1.6). Furthermore, velocity vectors combine in a manner different from that given by the galilean transformation law (Section 1.7).

The modifications in our notions of space and time required by experiment are formulated best in terms of occurrences or events and in terms of a geometry that does not satisfy the axioms of Euclidean geometry (Section 1.8). Events are the points of space-time and require four coordinates each for their specification. The analog of a position displacement in space is an *event displacement* and, just as a position displacement is the prototype of a vector in space, an event displacement is the prototype of a 4-vector in *space-time* (Section 1.9). These 4-component vectors and other similar entities are what we must use

---

* There exist many excellent books on the special theory of relativity written for the educated layman. See D. Bohm, *The Special Theory of Relativity*, W. A. Benjamin, New York, 1965; M. Born, *Einstein's Theory of Relativity*, rev. ed., Dover, New York, 1962; and the bibliography listed in *Special Relativity Theory, Selected Reprints*, American Institute of Physics, 1963.

henceforth in our formulation of the laws of physics, otherwise our statements of these laws will vary from observer to observer in circumstances involving relative speeds comparable to the speed of light.

A 4-component wave-propagation vector can be introduced to describe the propagation properties of a wave in space-time (Section 1.10).

In Chapter 2, we shall use the ideas developed in this chapter to amend the laws of mechanics to a form consistent with our new insight into space and time.

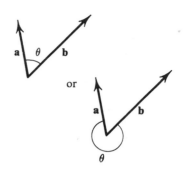

FIGURE 1.12 The angle θ between **a** and **b** may be defined in either of the ways shown.

Problem 1.1

The product of a scalar x and a vector **a**, $x\mathbf{a}$ is a vector of magnitude $|x|\,|\mathbf{a}|$ having the direction of **a** and the sense of **a** if x is positive, and having the opposite sense if x is negative.

(a) A vector **V** of magnitude V can always be written in the form $\mathbf{V} = V\hat{V}$, where $\hat{V}$ is called the unit vector in the direction of **V**. If **V** is a velocity, what are the units of $\hat{V}$?

(b) Let $\hat{x}$, $\hat{y}$, and $\hat{z}$ be unit vectors in the direction of the positive x, y, and z axes, respectively. Show that any vector **V** can be written in the form

$$\mathbf{V} = V_x\hat{x} + V_y\hat{y} + V_z\hat{z},$$

where V_x, V_y, and V_z are called the (rectangular) components of **V**. Do these components depend on the choice of coordinate axes?

(c) Show that two vectors **V** and **W** are equal if and only if their components are equal:

$$V_x = W_x, \qquad V_y = W_y, \qquad V_z = W_z.$$

(d) Use a vector diagram to show that $(\mathbf{V} + \mathbf{W}) + \mathbf{X} = \mathbf{V} + (\mathbf{W} + \mathbf{X})$.

(e) Show that the components of $\mathbf{V} + \mathbf{W}$ are $V_x + W_x$, $V_y + W_y$, $V_z + W_z$, and that the components of $\mathbf{V} - \mathbf{W}$ are $V_x - W_x$, $V_y - W_y$, $V_z - W_z$.

Problem 1.2

Two vectors can be multiplied together in three ways that are meaningful to physicists. One of these ways is described in this problem, another is described in Secton 1.9.4, and the third in Problem A1.17.

(a) The dot or *scalar product* of two vectors **a** and **b** is defined by $\mathbf{a} \cdot \mathbf{b} = ab \cos \theta$, where θ is the angle between **a** and **b** (Figure 1.12). Show that $\mathbf{a} \cdot \mathbf{b}$ is $|\mathbf{a}|$ times the component of **b** along the direction of **a**.

(b) Show from the result of (a) that

$$(\mathbf{a} + \mathbf{b}) \cdot \mathbf{c} = \mathbf{a} \cdot \mathbf{c} + \mathbf{b} \cdot \mathbf{c}.$$

(c) Show that $\hat{x} \cdot \hat{x} = 1$ and $\hat{x} \cdot \hat{y} = 0$. Find the values of the scalar products of all other combinations of unit vectors along the directions of the coordinate axes.

(d) Use the results of (b) and (c) to show that

$$\mathbf{a} \cdot \mathbf{b} = a_x b_x + a_y b_y + a_z b_z.$$

(e) Show that the number $a_x b_x + a_y b_y + a_z b_z$ formed from the components a_x, a_y, a_z and b_x, b_y, b_z of **a** and **b** respectively relative to one set of coordinate axes is independent of the choice of those axes.

Problem 1.3

A sinusoidal or harmonic plane wave traveling in the positive x direction is described by a wave function of the form $\mathbf{A} \sin (kx - \omega t)$ or by the complex form $\mathbf{A} e^{i(kx - \omega t)}$.

(a) The frequency ν is the number of complete oscillations that take place at one point x_0 per unit time. Show that $\omega = 2\pi \nu$. [The unit of frequency, cycle per second, is called the hertz (Hz) after the German physicist Heinrich Hertz (1857–1894).]
(b) A harmonic wave at any one instant t_0 consists of the repetition of an elementary unit in both directions. The size of the smallest such unit is called the wavelength λ (Figure 1.13). Show that $\lambda = 2\pi/k$.

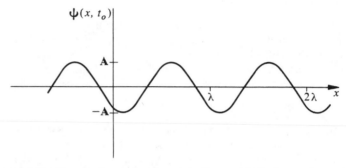

FIGURE 1.13   The waveform $\psi(x, t_0) = A \sin (kx - \omega t_0)$ of a harmonic wave at time t_0 consists of the repetition of an elementary unit of length λ.

(c) The period T of a harmonic wave is the time required for one cycle of the wave to occur at one point x_0. Show that $\nu T = 1$.
(d) The (phase) velocity $\mathbf{v}$ of the harmonic wave is given by $\mathbf{v} = v\hat{x}$, where v is the speed with which the waveform moves. Show that

$$\mathbf{A} \sin (kx - \omega t) = \mathbf{A} \sin [k(x + v \Delta t) - \omega(t + \Delta t)]$$

and, from this, that $v = \omega/k = \lambda \nu$.
(e) The propagation vector $\mathbf{k}$ of the harmonic wave is a vector of magnitude k directed along the direction of propagation of the wave. Show that $\mathbf{k} = k\hat{x}$ and that the wave function may be written as $\mathbf{A} \sin (\mathbf{k} \cdot \mathbf{r} - \omega t)$, where $\mathbf{r} = x\hat{x} + y\hat{y} + z\hat{z}$.
(f) The vector $\mathbf{A}$ determines whether the wave is longitudinal or transverse. If the wave is longitudinal, $\mathbf{A}$ lies along the direction of $\mathbf{k}$, and if transverse, along a direction perpendicular to $\mathbf{k}$. The direction of $\mathbf{A}$ in the latter case is called the direction of polarization of the transverse wave. Show that $\mathbf{A} \cdot \mathbf{k} = 0$ for a transverse wave.

Problem 1.4

Let the wave function $\mathbf{A} \sin (kx - \omega t)$ describe the displacement of the material at the position x at time t resulting from the transmission of a plane harmonic wave through the material. The displacement is referred to the reference frame S.

(a) An observer moves relative to the reference frame S with the velocity $V\hat{x}$. Use the galilean transformation law to show that the frequency ν' of the wave relative to that observer is given by $\nu' = \nu(1 - V/v)$, where ν and v are the frequency and speed of the wave relative to the reference frame S.

(b) A source of harmonic waves of frequency ν_s moves through the material with a velocity $V_s\hat{x}$ and generates the wave described relative to S by $A \sin (kx - \omega t)$. Use the fact that the wavelength λ is a scalar, and hence has the same value relative to the source and the reference frame S, to show that $\nu = \nu_s/(1 - V_s/v)$. These changes in frequency are called the *doppler effect* after the Austrian physicist Christian Johann Doppler (1803–1853).

Problem 1.5

Use the scalar product $\mathbf{v} \cdot \mathbf{v} = \mathbf{v}^2 = v^2$ to show that

$$v^2 = v'^2 + V^2 + 2v'V \cos \theta,$$

where θ is the angle between $\mathbf{v}'$ and $\mathbf{V}$. What are the maximum and minimum values of v for given values of v' and V?

Problem 1.6

Derive the galilean transformation law $\mathbf{v} = \mathbf{v}' + \mathbf{V}$ directly from $\mathbf{r}(t) = \mathbf{r}'(t) + \mathbf{R}(t)$ using the definition of velocity and the fact that $d\mathbf{R}/dt = \mathbf{V}$, a constant vector.

1.1   The Speed of Light

The speed of a wave in a material medium depends on the type of wave under consideration and on the properties and circumstances of that medium. Light exhibits wave properties, can be transmitted through various media, and even propagates through a vacuum. It might be expected, therefore, that the values of the speed of light in various media would provide some information on the mechanism of these waves.

1.1.1   *Measurements of the speed of light* *

Galileo appears to be the first scientist to have suggested a way to determine the speed of light even though some of his contemporaries believed that light was transmitted instantaneously. He suggested placing two observers some

---

* Kittel, Knight, and Ruderman, McGraw Hill, pp. 312–313; 318–322.
P.S.S.C. Physics (2nd ed.), D. C. Heath, Sec. 11–8, p. 196; Sec. 14–2, p. 242; Sec. 14–7, p. 249.
Resnick and Halliday (Part 2), John Wiley, Sec. 40–3, p. 998.
Note: References accompanying each subsection are listed in shortened footnote form. The following is a complete list of these standard references:
R. P. Feynman, R. B. Leighton, and M. Sands, *The Feynman Lectures on Physics* (vol. 1), Addison-Wesley, Reading, Mass., 1963.
C. Kacser, *Introduction to the Special Theory of Relativity*, Prentice-Hall, Englewood Cliffs, N.J., 1967.
C. Kittel, W. D. Knight, and M. A. Ruderman, *Mechanics, Berkeley Physics Course* (vol. 1), McGraw-Hill, N.Y., 1965.
P.S.S.C. Physics (2nd ed.), Physical Science Study Committee, D. C. Heath, Boston, Mass., 1965.
R. Resnick and D. Halliday, *Physics* (Part 2), John Wiley, N.Y., 1966.
R. Resnick, *Introduction to Special Relativity*, John Wiley, N.Y., 1968.
E. F. Taylor and J. A. Wheeler, *Spacetime Physics*, W. H. Freeman, San Francisco, 1966.

distance apart, each observer being equipped with a lamp that could be shut off quickly. The first observer shuts his lamp and, immediately upon seeing this, the second observer shuts his off. The first observer notes the time that elapses between the instant at which he shuts off his own lamp and the instant at which he sees the other observer shut his off. If this observed time interval is proportional to the separation distance of the two observers, then the speed of light is finite and can be determined from this experiment. However, a positive result cannot be obtained from Galileo's experiment over the distances available between terrestrial points.

The difficulty with Galileo's experiment, aside from the necessity of making measurements of very short time intervals, is that it takes a human observer a comparatively long time to react to a stimulus—that is, a finite time called the reaction time elapses between the time an observer sees the other's lamp being shut off and when the observer shuts off his own lamp. This human weakness can be overcome in the following way: Since the second observer acts merely as a reflector of the shutting off of the light from the first lamp, the second observer can be replaced by a mirror that reflects any light signal without delay. In addition, the measurement of the time interval can be performed by mechanical means in which the human reaction time plays no part. For example, an opaque sheet containing two holes moved across the light source allows the light out one hole and back into the source region through the other hole only for one particular speed, namely that for which the light goes through the first hole to the mirror and back in the exact length of time it takes the sheet to move the distance between the holes (Figure 1.14). This is the basis of the first terrestrial determination of the speed of light, performed in 1849 by the French physicist A. H. L. Fizeau (1819–1896) (Figure 1.15).

Notable measurements of the speed of light were performed by the American experimental physicist A. A. Michelson (1852–1931).* He used a technique

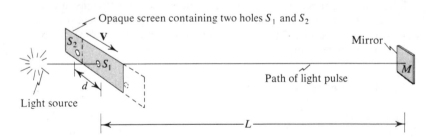

FIGURE 1.14 It takes the light pulse the time $2L/c$ to travel from S_1 to M and back through S_2 to the source, where c is the speed of light. During this time the screen moves a distance d at speed v. Therefore, in order that the light pass through S_2, it is necessary that $2L/c = d/v$ or $c = 2Lv/d$.

developed by the French physicist Léon Foucault (1819–1868) that involved the use of rotating mirrors in place of the toothed wheel of Fizeau's method (Figure 1.16) [1].†

---

* A biography of this scientist and a description of the state of science during his lifetime are contained in B. Jaffe, *Michelson and the Speed of Light*, Anchor Books, Garden City, N.Y., 1960.
† Throughout this book advanced references are indicated by bracketed numerals and are listed at the end of each chapter.

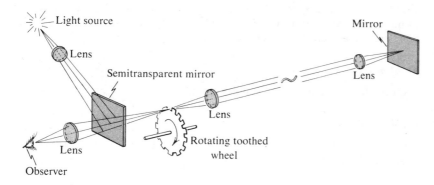

FIGURE 1.15 Fizeau's toothed-wheel apparatus for measuring the speed of light. The spaces between the teeth act as the holes of Figure 1.14.

The speed of light in a vacuum is denoted by the letter c.* The recommended value for c is

$$c = (2.997925 \pm 0.000003) \times 10^8 \text{ m/sec.} \qquad (1.12)$$

A vacuum transmits light, but light also propagates through matter such as glass. The speed of light c_m in a material medium can be determined through the techniques described above if part of the path of the light travels through a tube containing the medium. The speeds so measured are always less than c, and the ratio

$$\frac{c}{c_m} = n \qquad (1.13)$$

is called the *index of refraction* of the medium. The propagation of light in a material medium depends upon the interaction of the light with the matter, and this interaction differs for different wavelengths of the transmitted light. Therefore, the index of refraction depends (slightly) on the wavelength of the light as well as on the properties of the medium.

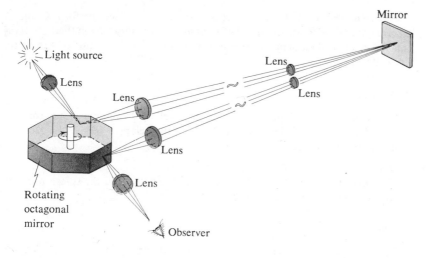

FIGURE 1.16 Michelson's rotating mirror apparatus.

---

* Although it is not a vector quantity, c is frequently called the velocity of light.

SUMMARY   The speed of light *in vacuo* is finite and has been measured to be $c = 2.998 \times 10^8$ m/sec. The speed of light in a material medium is less than c and is a characteristic of the medium.

Problem 1.7

Calculate the speed of light in miles per hour.

Problem 1.8

Calculate the time required for light to travel a distance equal to the circumference of the earth.

Problem 1.9

Calculate the time required for light to travel from the sun to the earth.

Problem 1.10

Calculate the time required for light to travel from the moon to the earth.

Problem 1.11

The semimajor axis of the orbit of the outer planet Pluto is 5.91×10^9 km. Calculate the time required for sunlight to cross the solar system.

Problem 1.12

Discuss the feasibility of measuring the speed of light by the apparatus shown in Figure 1.14 or by another simple piece of apparatus.

Problem 1.13

Two men, each of whom can react to a signal in 1/10 sec, perform Galileo's experiment and calculate the speed of light as 2×10^8 m/sec from their observations. Calculate their separation distance in terms of the radius of the earth, 6.4×10^6 m.

Problem 1.14

A man estimates the depth to the water in a well by dropping a stone from the mouth of the well and determining the time that elapses before the stone is seen to strike the water.

(a) Calculate the depth if this time is 0.25 sec.
(b) What fraction of that time is required for the light from the splash to reach the man at the top of the well?

Problem 1.15

Calculate the frequencies of the light with the following wavelengths (1 Å = 1 angstrom = 10^{-10} m).

Violet	4,500 Å	Yellow	5,700 to 5,900 Å
Blue	4,500 to 5,000 Å	Orange	5,900 to 6,100 Å
Green	5,000 to 5,700 Å	Red	6,100 Å

Problem 1.16

Fizeau used a toothed wheel with 720 teeth, and the distance between his toothed wheel and the mirror was 8,633 m. He observed the speed of rotation of the wheel when all the light was stopped by the teeth. Calculate that angular speed.

Problem 1.17

(a) Explain, as you would to a high school student, the principles of operation of Michelson's rotating mirror apparatus for measuring c.
(b) Michelson used a distance from a rotating octagonal mirror to a fixed mirror of 35 km. Calculate the angular speeds of the rotating mirror for which the light passes through the lens system as if the octagonal mirror were not rotating.

Problem 1.18

Calculate the speed of light in each of the media whose index of refraction is listed below:

Air (0°C, 760 mm Hg pressure)	1.0003	Ice	1.31
Water (20°C)	1.333	Amber	1.55
		Diamond	2.42

Problem 1.19

The index of refraction of a piece of crown glass has the following values at the given wavelengths:

| $\lambda =$ | 3.61×10^{-7} m | 4.86×10^{-7} m | 6.56×10^{-7} m | 12.0×10^{-7} m |
| $n =$ | 1.54 | 1.52 | 1.51 | 1.50 |

Calculate the speed of light for each of these wavelengths.

1.1.2 The aberration of starlight*

A direct manifestation of the finite speed of light is the phenomenon of aberration. This phenomenon was discovered by the English astronomer James Bradley (1693–1762).† *Aberration* is the systematic change in the relative position of a star as viewed from the earth during the earth's annual revolution

---

* Feynman, Leighton, and Sands (vol. 1), Addison-Wesley, Sec. 34–8, p. 34–10.
Kacser, Prentice-Hall, Sec. 4.6, p. 82.
Kittel, Knight, and Ruderman, McGraw-Hill, pp. 313–317.
Resnick, John Wiley, Sec. 1.7, p. 28.
† Bradley's investigations are described in A. B. Stewart, "The Discovery of Stellar Aberration," *Scientific American, 210*: 100, March 1964.

around the sun. A long narrow telescope attached to the earth must be inclined at an angle to the incoming starlight in order that the starlight reach the bottom of the telescope, since, during the (nonzero) time it takes the light to travel the length of the telescope, the earth has moved a small, but nevertheless nonzero, distance in its orbit (Figure 1.17). The angles involved are small, less than a minute of arc, but these are measurable and agree with the hypothesis that the earth·is circling in the inertial frame attached to the sun with a speed of $2\pi \times 1.5 \times 10^{11}$ m/yr.

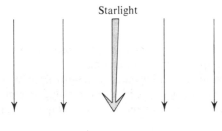

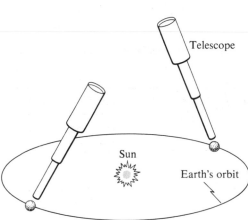

FIGURE 1.17   Stellar aberration.

SUMMARY   The finite speed with which light travels is reflected in the variations throughout the year of the apparent positions of stars.

Example 1.1

Q. Consider starlight striking the earth from the direction perpendicular to the plane of the earth's orbit (Figure 1.18).

    (a) A long narrow terrestrial telescope must be inclined at an angle to that direction in order that the starlight reach the bottom of the telescope. Calculate this angle.

    (b) Calculate the angular diameter of the apparent orbit of the star.

A. (a) Let Δt be the time it takes for the starlight to travel the length of the telescope (Figure 1.18). During this time, the telescope moves across the path of the starlight through a distance of $v\,\Delta t$, where v is the orbital speed of the earth,

$$v = 2\pi \times 1.49 \times 10^8 \text{ km/yr}$$
$$= 2.98 \times 10^4 \text{ m/sec.} \tag{1.14}$$

Therefore, the angle α at which the telescope is tilted is given by

$$\tan \alpha = \frac{v\,\Delta t}{c\,\Delta t} = \frac{v}{c} = \frac{2.98 \times 10^4}{3.00 \times 10^8} = 0.994 \times 10^{-4}. \tag{1.15}$$

Since $\tan \alpha \ll 1$,

$$\alpha = \tan \alpha = 0.994 \times 10^{-4} \text{ rad} = 20.5''. \tag{1.16}$$

(b) Relative to the earth, the telescope is tilted at an angle of 20.5 sec of arc with respect to the direction of the incoming starlight. The telescope must swing once about this direction each year (Figure 1.19), so the angular diameter of the apparent orbit is 2×20.5 sec = 41 sec of arc.

Problem 1.20

Discuss the phenomenon of aberration of starlight if the speed of light were as small as the speed of sound in air, about 330 m/sec. For example, would it be possible to see all stars situated above the horizon? What would the night sky look like?

Problem 1.21

A star lies on the plane of the earth's orbit. Take $t = 0$ to be that instant at which the earth's motion is directly toward the star (see Figure 1.20). At any given instant, a long narrow terrestrial telescope must be inclined at an angle to that direction in order that the starlight reach the bottom of the telescope. Calculate this angle as a function of the time t. Describe the apparent orbit of the star.

Problem 1.22

At some instant of time, the direction of the velocity $\mathbf{v}$ of the earth in its orbit around the sun makes an angle θ with the direction to a star, as shown in Figure 1.21. Let θ' be the angle with respect to $\mathbf{v}$ at which a long terrestrial telescope must be inclined in order that the starlight reach the bottom of the telescope.

(a) Show that

$$\tan \theta' = \frac{c \sin \theta}{c \cos \theta + v} = \frac{\sin \theta}{\cos \theta + v/c}.$$

(b) Show that for $|\theta' - \theta| \ll 1$

$$\tan \theta' = \tan \theta + (\theta' - \theta)\sec^2 \theta.$$

Hint: Use the relation $f(\theta') = f(\theta) + [df(\theta)/d\theta](\theta' - \theta)$, which you should justify.

(c) Show that for $v/c \ll 1$,

$$\tan \theta' = \tan \theta - \frac{v}{c}\sin \theta \sec^2 \theta.$$

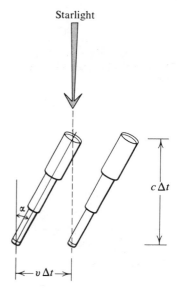

FIGURE 1.18 During the time Δt that it takes the starlight to travel the length of the telescope, the telescope moves a distance $v\,\Delta t$.

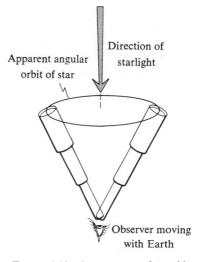

FIGURE 1.19 Apparent angular orbit of star.

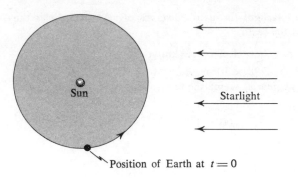

FIGURE 1.20 Direction of starlight from a star in the plane of the earth's orbit.

(d) Show that, for $v/c \ll 1$, we have $|\theta' - \theta| \ll 1$, and hence

$$\theta' - \theta = -\frac{v}{c}\sin\theta.$$

Hint: Use the approximations $\sin x \approx x$ and $\cos x \approx 1$, valid for small x. You should justify these approximations, at least with the use of a drawing of a triangle having one small angle.

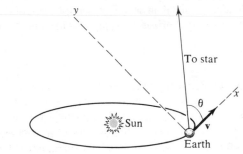

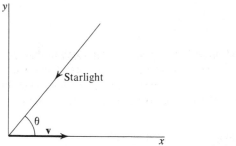

FIGURE 1.21 Direction of starlight.

1.2 The Michelson–Morley Experiment*

The experiments described in the preceding section show that light waves travel with the speed $c = 2.998 \times 10^8$ m/sec, although we did not specify the reference frame to which this speed is referred. Nor did we discuss in any

---

* See also R. S. Shankland, "The Michelson–Morley Experiment," *Scientific American*, *211*: 107, November 1964 and his article of the same title in the *American Journal of Physics*, *32*: 16 (1964).

detail the existence and properties of the medium that transmits light waves similar to the manner in which air carries sound waves to our ears. These problems were of concern to the physicists of the nineteenth century. They considered waves to be the oscillations of the material in a medium and, since light exhibits wavelike properties, they very naturally assumed the existence of a medium, called (*luminiferous*) *ether*,* that acted as a carrier of light waves. According to this view, light consists of oscillations in this medium, and c is the speed of the ether waves relative to that reference frame in which the ether is at rest.

The ether theory encountered insurmountable difficulties, aside from that that will be of concern to us later in this section. One such difficulty arose as follows: In order to explain the transmission of starlight, for example, it was necessary to assume that ether existed everywhere. Furthermore, ether must interact with matter, since the speed of the ether waves, light, is modified appreciably in the presence of matter (by the index of refraction factor). On the other hand, the planets pass through ether on their orbits in excellent agreement with Newton's law of gravitation if no account is taken of the presence of the ether. Thus, it was necessary to assume that there existed an interaction between ether and matter that resulted in a considerable modification in the behavior of the ether but none whatsoever in the motion of matter. This was only one of the difficulties encountered by the ether theory; in order to maintain the theory, scientists were forced to assign properties to the ether that were in direct conflict with all their usual notions of what was reasonable and possible. A decisive blow to the ether theory was struck by the "ether-drift" experiment of Michelson and his American chemist and physicist colleague, E. W. Morley (1838–1923).

1.2.1 *The experiment and its result†*

The motivation for the Michelson–Morley experiment was to test for the motion or drift of the ether past the earth during the earth's orbital motion around the sun. The phenomenon of aberration was regarded as proof that the earth, in its orbital motion, moves through the medium that carries light waves, as is evident from our analysis of that phenomenon in the last section.

If light consists of oscillations in the ether and therefore has a constant speed, say c, relative to the reference frame in which the ether is at rest, then the speed of light with respect to the earth c_E can be obtained from the galilean transformation law:

$$\mathbf{c}_E = \mathbf{c} - \mathbf{v}, \tag{1.17}$$

where $\mathbf{c}_E$ is the velocity of a light wave relative to the earth, $\mathbf{c}$ is its velocity relative to the ether, and $-\mathbf{v}$ is the velocity of the ether relative to the earth. Thus, $\mathbf{v}$ is the velocity of the earth with respect to the ether (Figure 1.22). The

---

* Other natural phenomena appeared to require the existence of other artificial substances if these phenomena were to be understood in terms of mechanical models. See, for example, A. Einstein and L. Infeld, *The Evolution of Physics*, Simon and Schuster, New York, 1938.
† Feynman, Leighton, and Sands (vol. 1), Addison-Wesley, Sec. 15–3, p. 15–3.
Kacser, Prentice-Hall, Sec. 2.6., p. 19.
Kittel, Knight, and Ruderman, McGraw-Hill, pp. 328–336.
Resnick and Halliday (Part 2), John Wiley, Sec. 43–7, p. 1090; Sec. 43–8, p. 1092.
Resnick, John Wiley, Sec. 1.5, p. 18.

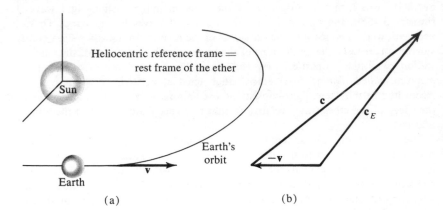

FIGURE 1.22 $c_E \neq c$, in general.

speed of light with respect to the earth c_E therefore varies, according to (1.17), between the limits $c - v$ and $c + v$, its value depending on the relative directions of the earth's velocity **v** and the propagation vector of the light wave; c is about 3×10^8 m/sec, and v, about 0.0003×10^8 m/sec.

At the time of the Michelson–Morley experiment, techniques were not sufficiently refined to measure the speed of light over a one-way path to better than one part in 10,000, the accuracy required to distinguish c_E from c directly (Problem 1.23). (For that matter, such an experiment is not feasible today. See Section 1.4.1.) Therefore, Michelson and Morley had to use a less direct method to detect the difference between c_E and c. Their method involved the measurement of the difference in the speeds of light in different directions relative to **v**. The difference itself could be measured even though the value of c_E in one direction could not.

The Michelson–Morley experiment involved the use of an interferometer that Michelson had invented previously. In a Michelson interferometer (Figure 1.23), the amplitude of the light from an extended source is divided into two beams by a lightly silvered mirror. The two beams are reflected from plane mirrors, one fixed and one movable, and recombined by the semitransparent mirror. The observer sees two images of the same source. If the two mirrors are not exactly perpendicular, the two images are separated by a wedge-shaped space, and the path difference between the light that strikes the two mirrors varies linearly across the wedge. The result is an interference pattern having the appearance of line fringes as shown in Figure 1.24. The darkest parts of the fringes result when the light from one path interferes destructively with the light from the other. If the movable mirror is displaced with the calibrated screw along the direction of the light beam, the fringes move across the field. A new fringe crosses a given point on the field when the difference in the path lengths of the two beams changes by one wavelength. Since the path length of the light is twice that of the displacement of the mirror, a new fringe crosses a given point on the field whenever the mirror is displaced a distance $\lambda/2$, where λ is the wavelength of the light. Thus, the distance d that the mirror is displaced when n fringes pass a given point is equal to

$$d = \frac{n\lambda}{2}. \tag{1.18}$$

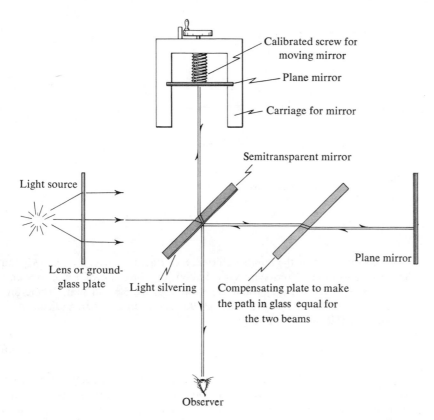

Calibrated screw for
moving mirror

Plane mirror

Carriage for mirror

Semitransparent mirror

Light source

Plane mirror

Lens or ground-
glass plate   Light silvering   Compensating plate to make
the path in glass equal for
the two beams

Observer

FIGURE 1.23 The Michelson interferometer.

The interferometer can be used to measure wavelengths or, alternatively, to determine small distances accurately in terms of known wavelengths.

The above analysis of the Michelson interferometer is based on the assumptions that the speed of light is the same in each beam and that this remains constant throughout the course of the experiment. However, if the interferometer is moving through the ether, as shown in Figure 1.25, the speed of the light going from S to M_1 is $c + v$, that from M_1 to S is $c - v$, and that from S to M_2 or from M_2 to S is $\sqrt{c^2 - v^2}$. As a result of these different speeds, there is a shift in the fringes from that that would occur if $\mathbf{v}$ were zero. The number of fringes shifted due to the motion alone is calculated in Example 1.2 and is given by

$$n = \frac{SM_1}{\lambda}\frac{v^2}{c^2}.$$ (1.19)

This shift and the shift due to the difference in path lengths, $|SM_1 - SM_2|$, are superposed, and it is not possible to measure the path lengths sufficiently precisely to disentangle these shifts. The interferometer cannot be brought to rest with respect to the ether, moreover, and thus the shift cannot be observed while the interferometer preserves its orientation relative to $\mathbf{v}$. However, a shift can be observed if the interferometer is rotated through 90° about an axis perpendicular to $\mathbf{v}$; in this circumstance, the number of fringes that should shift

FIGURE 1.24 Appearance of the line fringes observed in a Michelson interferometer.

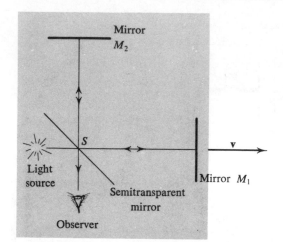

FIGURE 1.25 Schematic diagram of an interferometer traveling with the velocity **v** through the ether.

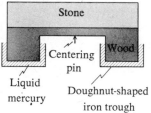

FIGURE 1.26 Cross-sectional view of the Michelson–Morley support for the interferometer.

is equal to the sum of the number shifted, n of Equation (1.19), when SM_1 and SM_2 rotate to orientations symmetrical about the direction of **v** and the same number n shifted when they rotate the same amount beyond that. Therefore, a rotation of the interferometer through 90° about an axis perpendicular to **v** introduces a shift of N fringes, with

$$N = 2n = 2 \frac{SM_1}{\lambda} \frac{v^2}{c^2}. \tag{1.20}$$

In the Michelson–Morley apparatus, a Michelson interferometer was mounted on a large stone slab that rested on a wooden float that in turn was supported by liquid mercury (Figure 1.26). The whole apparatus could be easily rotated slowly and continuously for hours at a time and an observer could walk around with the interferometer, thus all the while measuring any shift in the positions of the fringes. The effective distance SM_1 was increased to 11 m through the use of multiple reflections (Figure 1.27). Michelson and Morley used a sodium lamp with its characteristic yellow light of wavelength $\lambda = 5.89 \times 10^{-7}$ m. Therefore, the shift in the number of fringes expected for the sodium light was

$$N = \frac{2 \times 11 \text{ m}}{5.9 \times 10^{-7} \text{ m}} \left(\frac{3.0 \times 10^4}{3.0 \times 10^8}\right)^2 = 0.4. \tag{1.21}$$

The measurements of Michelson and Morley showed that *there was no shift* in the fringes that exceeded their small error limits, 0.02 at the most [2].

The null result of the Michelson–Morley experiment is incomprehensible on the basis of an ether theory of light transmission. However, it is not the ether theory primarily that is at fault; our analysis rests on the galilean transformation law, Equation (1.17), with the concept of the ether introduced only to define the reference frame in which the velocity of light **c** has the same magnitude in all directions. Thus, the Michelson–Morley experiment indicates that the galilean transformation law is not always valid. Because of the simplicity of the derivation of this law (see Problem 1.6), we therefore must reexamine the notions of space and time upon which this law is based.

SUMMARY   The Michelson–Morley experiment was designed to measure the speed with which the earth moves through the medium of transmission of light.

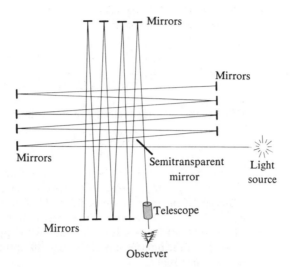

FIGURE 1.27 Light paths in the Michelson–Morley apparatus.

The null result of this experiment, combined with the results of other experiments, indicates the necessity of a major modification in the newtonian notions of space and time.

Example 1.2

Q. Calculate the fringe shift that results only from the motion through the ether of the interferometer (Figure 1.25).

A. Since the speeds of the light are different on the different paths, we calculate the times T_1 and T_2 required for the light to travel the paths S–M_1–S and S–M_2–S, respectively; then the number of fringes shifted as a result of the difference in path length and of the motion is given by (Figure 1.28)

$$n = \frac{c|T_2 - T_1|}{\lambda},\tag{1.22}$$

where λ is the wavelength of the light used. Along the path SM_1, the speed of light relative to the interferometer is $c - v$, so the time taken for light to go from S to M_1 is $SM_1/(c - v)$. Similarly, along M_1S the speed is $c + v$ and the transit time is $SM_1/(c + v)$. Therefore, the time required for light to travel the path S–M_1–S is

$$T_1 = \frac{SM_1}{c - v} + \frac{SM_1}{c + v} = 2SM_1 \frac{c}{c^2 - v^2}$$

$$\approx \frac{2SM_1}{c}\left(1 + \frac{v^2}{c^2}\right).\tag{1.23}$$

Along the paths SM_2 and M_2S, the speed of light is $\sqrt{c^2 - v^2}$ (Figure 1.29); hence

$$T_2 = 2SM_2 \frac{1}{\sqrt{c^2 - v^2}}$$

$$\approx \frac{2SM_2}{c}\left(1 + \frac{1}{2}\frac{v^2}{c^2}\right).\tag{1.24}$$

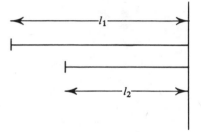

FIGURE 1.28 If light travels for a time $T_1 = l_1/c$ on one path relative to the ether and for a time $T_2 = l_2/c$ on another, the difference in path length relative to the ether is $|l_1 - l_2| = c|T_1 - T_2| = c|T_2 - T_1|$.

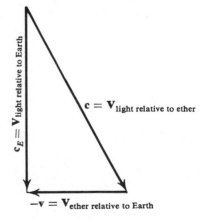

FIGURE 1.29 If c_E is perpendicular to **v**, then $c_E^2 = c^2 - v^2$.

The number of fringes shifted as a result of the difference in the path lengths and the motion is

$$n = \frac{2}{\lambda}\left| SM_1\left(1 + \frac{v^2}{c^2}\right) - SM_2\left(1 + \frac{v^2}{2c^2}\right)\right|$$

$$= \frac{2}{\lambda}\left|(SM_1 - SM_2) + \frac{2SM_1 - SM_2}{2}\frac{v^2}{c^2}\right|.$$

(1.25)

The number of fringes shifted owing to the motion alone is

$$\frac{2}{\lambda}\frac{2SM_1 - SM_2}{2}\cdot\frac{v^2}{c^2} \approx \frac{SM_1}{\lambda}\frac{v^2}{c^2},$$

(1.26)

if $SM_1 \approx SM_2$. This shift appears as a correction to the shift, of number $(2/\lambda)|SM_1 - SM_2|$, owing to the path-length difference, and cannot be observed if the orientation of the interferometer is kept fixed relative to the velocity $\mathbf{v}$. However, if the interferometer is rotated through $90°$, the shift (1.26) appears as a correction with opposite sign in (1.25); hence, rotation of the interferometer through $90°$ introduces a shift of the fringes in number equal to

$$N = \frac{2SM_1}{\lambda}\frac{v^2}{c^2}.$$

(1.27)

Problem 1.23

(a) Calculate c_E [of Equation (1.17)] as a function of the angle α between $\mathbf{c}_E$ and $\mathbf{v}$.

(b) Express the maximum possible value of $(c_E - c)/c$ as a percentage.

Problem 1.24

How many lines of sodium light would be shifted if the mirror M_2 were moved through 1.00 mm?

Problem 1.25

How far would the mirror M_1 of the Michelson interferometer have to be moved to give the expected shift, Equation (1.21), in the number of fringes?

Problem 1.26

Let θ be the angle between the direction of $\mathbf{v}$ and the line bisecting the angle M_1SM_2 (Figure 1.25). Assume $SM_1 = SM_2$.

(a) Calculate, as a function of θ, the number of fringes that would be expected to pass a given point relative to the view when $\theta = 0$.

(b) Plot the result of (a) on a graph and draw lines corresponding to $n = \pm 0.02$ within which the measurements of Michelson and Morley lie.

Problem 1.27

Calculate the shift that would have been expected in the Michelson–Morley experiment if that experiment had been performed at the position of perihelion of each of the planets listed.

Planet	Radius of orbit, km	Orbital period, days
Mercury	5.79×10^7	88.0
Jupiter	7.78×10^8	4.33×10^3
Neptune	4.50×10^9	6.02×10^4

1.3 The Kinematic Postulate of the Special Theory of Relativity

The assumption of a medium, the ether, for the transmission of light appeared very natural to the physicists of the nineteenth century. Faced with the contradiction in their interpretations of the phenomenon of aberration and the null result of the Michelson–Morley experiment, these scientists modified their models of the ether. Models were based on the assumption that only part of the ether was being dragged along by the earth or that the speed of light relative to its source is constant and equal to c.* However, the results of these other hypotheses were inconsistent with other experiments and eventually all had to be abandoned. The difficulty did not lie in any particular model of the ether that was constructed but rather in the assumption that all physical phenomena, in particular the wave-like behavior of light, could be explained on the basis of newtonian kinematics.

1.3.1 The invariance of the speed of light†

Consideration of the phenomenon of aberration and the null result of the Michelson–Morley experiment show that light waves do not satisfy the relation

$$\mathbf{v} = \mathbf{v}' + \mathbf{V} \tag{1.28}$$

based on our common-sense ideas of space and time (Figure 1.30). However, we should not be surprised if these ideas are not valid for the description of energy transmitted as fast as that of light. Our everyday experiences upon which these ideas are based include few phenomena in which perceivable speeds greater than 100 mi/hr appear and none in which speeds of any appreciable fraction of $c = 186{,}000$ mi/sec are apparent (Problem 1.28). It is possible that extrapolations of our common-sense ideas are valid beyond the range of everyday experience upon which they are based, but they need not be. In the case under consideration, extrapolation of our common-sense ideas leads to false conclusions, conclusions that disagree with experiments or, in other words, with the real world as it is.

The attempts that were made to patch up the newtonian theory of space and

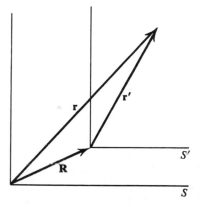

FIGURE 1.30 The equation $\mathbf{v} = \mathbf{v}' + \mathbf{V}$ follows immediately from the obvious relation $\mathbf{r} = \mathbf{r}' + \mathbf{R}$, according to our usual ideas on space and time.

* See the recent discussions of the evidence against this hypothesis in J. G. Fox, "Experimental Evidence for the Second Postulate of Special Relativity," *American Journal of Physics*, *30*: 297 (1962), and of an experiment to test this postulate in T. Alväger, F. J. M. Farley, J. Kjellman, and I. Wallin, "Test of the Second Postulate of Special Relativity in the GeV Region," *Physics Letters*, *12*: 260 (1964).
† Feynman, Leighton, and Sands (vol. 1), Addison-Wesley, Sec. 15–1, p. 15–1; Sec. 15–2, p. 15–2. Kacser, Prentice-Hall, Sec. 2.7, p. 23; Sec. 3.1, p. 25. Kittel, Knight, and Ruderman, McGraw-Hill, p. 336. Resnick, John Wiley, Sec. 1.9, p. 35; Sec. 1.10, p. 38.

time in order to obtain agreement with experiment failed. The evidence pointed out the necessity of a major modification in our ordinary ideas on the concepts of space and time. The accepted modification can be credited to the great Albert Einstein,* although some of his results had been anticipated independently by the Dutch theoretical physicist Hendrik Antoon Lorentz (1853–1928) and by the French mathematician and scientist Henri Poincaré (1854–1912) [3].†

Einstein realized very clearly that the difficulties discussed above pointed out a need for a major modification in the usual views of space and time. He concluded that the concept of the ether as a transmitter of light waves is artificial, since the effects of the ether could not be detected as shown, for example, by the Michelson–Morley experiment. There exists no privileged inertial reference system, and so the idea of an ether that defines such a preferred frame is superfluous. It is simply a property of space that light can propagate through it.

Waves in a material medium are transmitted as a result of the interactions between the particles in the material. A study of the behavior of these waves provides information on the properties of these interactions. On the other hand, light can be transmitted through space in which no matter is present. This suggests, by analogy, that a study of the behavior of light will provide us with information on properties of space. This is indeed the case; for example, the null result of the Michelson–Morley experiment indicates that the speed of light is the same in all inertial reference frames (since the existence of a preferred reference frame is inconsistent with the null result). This conclusion is consistent also with observations on aberration, for in that phenomenon, it is the direction and not the speed of the light that is observed; measurements of the speed give the result c. The postulate of the equality of the speed of light with respect to all inertial reference frames is not consistent, however, with the galilean transformation law, in particular with the equation for the galilean transformation of velocities, Equation (1.28). The contradiction is only apparent, though, since the range of experience upon which the galilean transformation law is based does not include perceptible speeds comparable to that of light. *The Michelson–Morley experiment extends our range of experience and shows that our usual concepts of space and time are not valid in that extended range.*

In 1905,‡ Einstein introduced modifications in the newtonian notions of space and time, and these modifications can be considered to be based on a total acceptance of the results of the Michelson–Morley experiment.§ The

* Einstein's only autobiography is presented in *Albert Einstein: Philosopher–Scientist*, P. A. Schilpp (Ed.), Tudor Publishing, New York, 1949. See also the biography by his colleague, L. Infeld, *Albert Einstein*, Scribner, New York, 1950. An interesting discussion of some of Einstein's views is given in M. J. Klein, "Einstein and some civilized discontents," *Physics Today, 18*; 38, January 1965.
† A discussion of these contributions is given by Max Born in "Physics and Relativity," *Jubilee of Relativity Theory, Helvetica Physica Acta, Supplement IV*: 244 (1956). See also G. Holton, "On the Origins of the Special Theory of Relativity," *American Journal of Physics, 28*: 627 (1960), reprinted in *Special Relativity Theory, Selected Reprints*, American Institute of Physics, New York, 1963.
‡ Einstein's 1905 paper on relativity was entitled "On the Electrodynamics of Moving Bodies" and appeared in *Annalen der Physik, 17*: 891 (1905). An English translation of this paper is given in *The Principle of Relativity* by A. Einstein, H. A. Lorentz, H. Minkowski, and H. Weyl, Dover, New York, 1924. It is reproduced in part on pp. 375–379 of Kittel, Knight, and Ruderman, McGraw-Hill, in Appendix B, p. 127 of Kacser, Prentice-Hall, and also in part, with helpful comments, in Appendix 3 of *Great Experiments in Physics*, M. H. Shamos (Ed.), Holt, Rinehart & Winston, New York, 1959.
§ It is not known what influence the null result of the Michelson–Morley experiment made directly to Einstein's thinking in 1905. See R. S. Shankland, "Conversations with Albert Einstein," *American Journal of Physics, 31*: 47 (1963), and "Michelson–Morley Experiment," *American Journal of Physics, 32*: 16 (1964).

concept of space and time introduced by Einstein is based on the *kinematic postulate of the special theory of relativity* [4]*:

▶ *The speed of light in a vacuum has the same value relative to all inertial reference frames and is independent of the relative velocity of the light source and the observer.*

Einstein's kinematic postulate and the predictions that follow from it are consistent with all experimental results to date, and in fact, the postulate has achieved the status of a law of physics. Therefore, our common-sense ideas of space and time are not universally valid; we must give up many of these, and in the next section, we shall show how drastic this change is.

SUMMARY Space and time have the properties that light can propagate through space and that the speed of light *in vacuo* has the same value relative to all inertial observers.

Problem 1.28

Estimate the highest speed with which you actually have seen something move. Calculate the percentage ratio of this speed to that of light.

Problem 1.29

Show that Einstein's postulate is not consistent with Equation (1.28).

Problem 1.30

Explain why it is not necessary to consider a property of invariance for the speed of sound.

1.4 The Relative Character of Simultaneity

A fundamental and important error in our customary ideas of space and time concerns simultaneity. To see this, we first introduce two pertinent definitions. An *event* is defined as an occurrence at a particular point in space at a particular time. Two events that take place at the same instant of time, though not necessarily at the same point in space, are said to occur *simultaneously*.

Common sense tells us that if two events occur simultaneously according to one person, these events must occur simultaneously from the point of view of all observers. However, common sense is based on our everyday experiences† and does not include perception of relative speeds near the speed of light (Problem 1.28). Therefore, our common-sense notions of simultaneity must be scrutinized in light of the evidence that our everyday notions of space and time are incorrect.

---

* Einstein also introduced a dynamic postulate (Chapter 2) that yields laws of motion valid in the modified framework of space and time. The theory of relativity based on these postulates is called special or restricted, to distinguish it from Einstein's theory of gravitation, which is called the general theory of relativity (Chapter 3).

The kinematic postulate is also called "the second postulate of special relativity," since Einstein listed this after the dynamic postulate in his 1905 relativity paper in *Annalen der Physik, 17*: 891 (1905).

† Einstein stated that "Common sense is that layer of prejudices laid down in the mind prior to the age of eighteen."

1.4.1   An analysis of the concept of simultaneity*

Our analysis will depend for its important features on the properties of light, since, as we argued before, a study of these properties can provide us with information on the properties of space and time.

We shall discuss two events that, according to one observer, occur at the same instant of time, and we shall consider whether or not these two events occur simultaneously from the point of view of another observer. In order to have a picture we can visualize, we shall consider a fictitious manned rocket ship that is traveling with a constant velocity, of magnitude comparable to the speed of light, relative to observer S (Figure 1.31). The frames of reference of both the ship and S are inertial. Observer S', traveling with the rocket, observes S from a window at a point midway between each end of the rocket.

The two events we shall consider occur at the ends of the rocket at that instant, according to S, at which S and S' are closest to one another. We shall suppose that each event is marked by a visible spark that results in a very short pulse of light that can be detected by both S and S'. Whether or not the sparks occur simultaneously according to S' will be left an open question for the moment.

How does S know that the sparks occurred at both ends of the rocket simultaneously? He certainly could not see the light from the sparks at the same instant the sparks occur, for light travels with a finite speed and therefore requires a finite time to reach S from each end of the rocket. However, if S has some method by which he can measure the time at which a distant event occurs, he could determine whether or not two distant events were simultaneous by comparing the times at which they took place. Therefore, the problem of determining whether or not two distant events are simultaneous reduces to our finding a means of measuring the time at which a distant event occurs.

The most obvious way to establish comparable time scales at two separated points A and B is to synchronize two identical clocks at one point, say A, and move one of these clocks to B. However, this procedure is valid only if the displacement of the clock has no effect on the timekeeping properties of the clock. Suppose, for example, that the clock is a pendulum. Then, in order that we could justify this means of comparing times at separated points, we would have to argue, using laws of mechanics, that the displacement of the oscillating pendulum has no effect on the timekeeping properties of that pendulum. However, we cannot argue in this way until we see what, if any, changes in our laws of mechanics are necessary as a result of the principle of the invariance of the speed of light. In the analogous circumstance in which two such pendulums are used to measure the speed of light in each of two inertial frames in relative motion, Newton's laws state that the measured speeds differ in general. Since these laws result in an incorrect description in that case, we cannot justify their use in predicting the effects of a displacement on the pendulum. Therefore, we must devise another method of comparing times at separated points, a method we can justify. Since our problem arises because of a property of light, we investigate the use of light pulses for determining times at distant points.

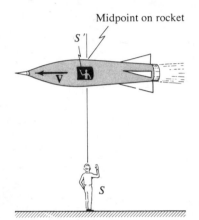

FIGURE 1.31   The observer S sees S' and the rocket moving past with speed V.

---

* Feynman, Leighton, and Sands (vol. 1), Addison-Wesley, Sec. 15–6, p. 15–7.
Kacser, Prentice-Hall, Sec. 2.5, p. 16; Sec. 3.1, p. 25; Sec. 3.2, p. 29.
Kittel, Knight, and Ruderman, McGraw-Hill, pp. 375–379.
Resnick, John Wiley, Sec. 2.1, p. 50.

With the use of light signals, S can establish the time t_1 of any event that occurs at a point in space different from his own position in the following way (Figure 1.32): Let d be the distance from S to the point at which the event occurs.

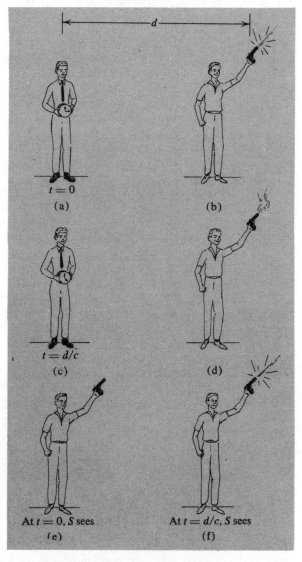

FIGURE 1.32 S sees the gun flash at $t = d/c$ and calculates that the gun was fired at $t = 0$. (a) S at $t = 0$. (b) Marksman at $t = 0$. (c) S at $t = d/c$. (d) Marksman at $t = d/c$. (e) The marksman as seen by S at $t = 0$. (f) The marksman as seen by S at $t = d/c$.

Let t_2 be the time at which light emitted at the event reaches S. The light traveled for the time interval d/c between its emission at the event and its reception at the position of S. Thus, the time interval $t_2 - t_1$ is equal to d/c or

$$t_1 = t_2 - \frac{d}{c}. \qquad (1.29)$$

Actually, this is a definition of the time at which a distant event occurs, and it is a valid definition only if it does not lead to any inconsistencies. Since this definition gives a comparison of times at two different points, for example the time of an event at B as measured by an observer at A, the definition is consistent if the relation between times as measured at the two points is reciprocal; this is true if the speed of light c_{AB} along AB is equal to that c_{BA} along BA.

How can we verify that these two speeds are equal? Consider a measurement of, say, c_{AB}, the speed of light transmitted along the line from A to B. This can be determined, from a measurement of the time t_A of emission of a light pulse at A and of the time t_B of its reception at B, to be $AB/(t_B - t_A)$. However, for this purpose, we need a method to compare the times at the two separated points A and B, and it is this method we are trying to establish. Thus, we can establish a time scale at distant points if we can show that $c_{AB} = c_{BA}$, which we can do if we have established a time scale at distant points. We break out of this circle by adopting the convention that the speed of light has the same value, namely c, in all directions relative to one inertial frame. With this convention, we define the time of a distant event as the time established by the method given above and embodied in Equation (1.29).

Since the speed of light c is the same relative to all inertial systems, we have obtained a consistent scheme for establishing a time scale at all points of any inertial reference frame. However, a time scale constructed in this way is not defined in an absolute manner, but one such time scale is defined relative to each inertial reference system.

We can conclude that S determines that the sparks occurred at both ends of the rocket simultaneously by considering the motions of the rocket and the light pulses (Figure 1.33) at various instants of time t relative to S. For $t < 0$, S', who is situated at the midpoint of the rocket, is approaching S [Figure 1.33(a)]. At the time $t = 0$ relative to S, S and S' are nearest each other [Figure 1.33(b)]; also at that instant, according to S, the sparks occur at both ends of the rocket, although S cannot be aware of that fact at that time. For $t > 0$, S' moves away from S and the light flashes move away from the ends. At $t = d/2c$, the light pulses are both at a distance $d/2$ from S [Figure 1.33(c)], and at $t = d/c$, the pulses meet at the position of S [Figure 1.33(d)]. Since each of these flashes traveled a distance d, S determines from Equation (1.29) the time at which each was emitted as $t = d/c - d/c = 0$ and concludes that the flashes were emitted simultaneously relative to his time scale.

Let us consider now the sequence of events *from the point of view of S'* (Figure 1.34). For this purpose, we must start with a relation between the point of view of S of some of the events and the point of view of S' of these same events. One relation is provided if we adopt the convention that the clock of S' reads $t' = 0$ at the event in which S and S' are nearest each other [Figure 1.34(c)]. Another relation results from the concept of coincidence of events. Consider the following events: the position of S at the instant at which the light flash from the front meets S and the position of S at the instant at which the light flash from the back meets S. According to S, these two events are coincident. These must also be coincident according to S', or we would be forced to give up our ideas of the identity of events, a very drastic step indeed and a step that is not necessary in order to obtain agreement with experiment.

Since S and S' agree on the coincidence of events, both S and S' accept that the light flashes meet S at a point on the rocket behind S'. According to S', the light pulses he sees originate from sparks at the ends of the rocket, and he sees the light pulse from the front of the rocket [Figure 1.34(d)] before he sees

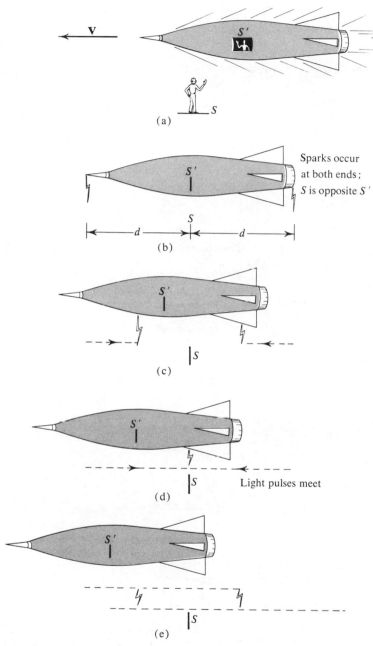

FIGURE 1.33 The motions of the rocket and the light pulses as viewed by S. (The length of the rocket is $2d$ according to S.) (a) $t < 0$. (b) $t = 0$. (c) $t = (d/2)/c$. (d) $t = d/c$. (e) $t = (3d/2)/c$.

the light pulse from the back end. Furthermore, the speed of light is the same in his reference frame as it is in that of S, so it takes the same amount of time for the light to travel from the front end of the rocket to S' at the midpoint as it does from the back end to S'. Hence, S' concludes that the spark struck at the front end of the rocket [Figure 1.34(b)] before it struck at the back end [Figure 1.34(d)].

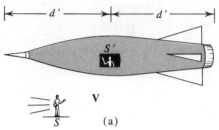

FIGURE 1.34 The motions of the
observer S and the light pulses as
viewed by S'. (The length of the rocket
is $2d'$ according to S'.) (a) $t' <
-(d'/2)/c$. (b) $t' = -(d'/2)/c$. (c) $t' = 0$.
(d) $t' = (d'/2)/c$. (e) $t' = d'/c$.

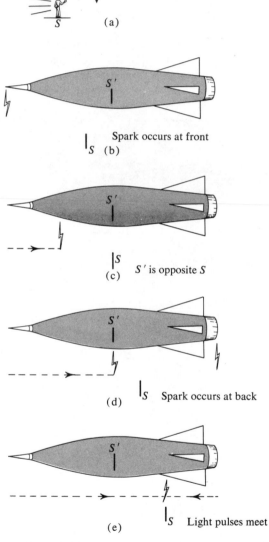

The positions relative to S' of S and the light flashes at the various instants
of time shown in Figure 1.34 can be deduced from the assumption, which will be
justified later, that the speed of S' relative to S is equal to the speed of S relative
to S'. In the case shown in Figure 1.34, this speed v is $\frac{1}{2}c$. We have adopted the
convention that S is nearest S' at time $t' = 0$ [Figure 1.34(c)]. At some later

time t_1', an event occurs in which S and the light flashes meet at a point on the rocket, and we must now determine where this point is. For this purpose, consider Figure 1.33. From 1.33(d), we see that the event in which the light flashes meet S occurs at the time $t = d/c$ according to S. In the time interval t, S' has moved a distance $vt = \frac{1}{2}c \cdot d/c = \frac{1}{2}d$ past S; thus, relative to S, the event occurs at a point P on the rocket halfway between S' and the end of the rocket. Therefore,* $vt_1' = d'/2$ or $t_1' = d'/c$ [Figure 1.34(e)]. Also, the light flash from the back of the rocket has traveled the distance $d'/2$ with speed c to arrive at P at the time $t_1' = d'/c$, so the spark at the back occurred at a time $t' = d'/2c$ [Figure 1.34(d)], a time interval $d'/2c$ before t_1'. Similarly, the light from the front has traveled a distance $3d'/2$ at the speed c to reach P at t_1', so the spark at the front occurred at $t' = d'/c - 3d'/2c = -d'/2c$ [Figure 1.34(b)].

We assumed that sparks occurred at both ends at the same instant according to S and concluded that the spark at the front end occurred first according to S'. Is our argument valid? A careful examination of the argument will not reveal any errors. Who is correct then, S or S'? Did the sparks occur at both ends at the same instant or did the spark at the front end occur first? The answer to the first question is that both are right. The answer to the second question is that *the sparks occurred simultaneously according to the point of view of S and that the spark at the front end occurred first according to the point of view of S'.*

We must conclude, therefore, that Einstein's principle of the invariance of the speed of light results in simultaneity having a relative character. *Simultaneity cannot be defined in an absolute sense, but rather can be defined only relative to an inertial frame of reference.* Two events that are simultaneous according to one observer may not be simultaneous relative to another. The name "theory of relativity" may be considered to be a reflection of this fact; some concepts, such as that of simultaneity, can be defined in fact only relative to an observer, and not in an absolute sense as suggested by newtonian kinematics. This does not mean, however, that Einstein's theory of relativity states that every concept must be defined relative to an observer or, as it is phrased frequently, that everything is relative. It is true that some concepts that, previous to Einstein's studies, were thought (incorrectly) to be absolute can be defined only relative to an observer, as we have shown is true for simultaneity. However, as we shall see shortly, Einstein's theory of relativity states that certain other entities can be defined in an absolute sense, independent of the observer; entities of this sort play a central role in the laws of physics.

It may appear at a first glance as if we have given up a certain amount of symmetry that was present in newtonian mechanics, since, for example, the two events considered above occur simultaneously according to S, whereas one of these events occurs before the other according to S'. This apparent asymmetry can be removed, however, if we consider the case in which the sparks occur at both ends of the rockets simultaneously according to S'. In this case, the light flashes meet at S' after he has passed S and therefore after the light pulse from the back has traveled past S. Hence, according to S, the spark

---

* It is conceivable that, relative to S', this point P on the rocket is not halfway between S' and the end of the rocket. For the time being, to keep the argument simple, we shall suppose that it is. It will be shown later, on the basis of the homogeneity of space and time and without further assumptions than used above, that this supposition is indeed valid (see Figure 1.57).

occurs at the back end of the rocket first (Problem 1.34). The two events, simultaneous according to S', are not simultaneous relative to S. Therefore, no asymmetry between the two reference frames is introduced by the theory of relativity: Two events can occur simultaneously in either frame of reference. However, events that occur simultaneously with respect to one reference frame may not occur simultaneously relative to the other.

SUMMARY   The times relative to an inertial observer of two spatially separated events are related by the kinematic postulate of special relativity, together with the convention that the speed of light has the same value in all directions relative to that observer. The simultaneity of two events relative to an inertial observer is defined by the equality of the times, relative to that observer, of the events. The times, relative to another inertial observer, of these two events may not be equal; thus, simultaneity has a relative character.

Problem 1.31

Explain why the above analysis does not apply to sound waves.

Problem 1.32

With the visible light emitted at the event, S sees an event at 4.00 P.M. that took place 5 mi from S. Calculate the time at which the event occurred according to S.

Problem 1.33

A second and identical rocket, carrying an observer S'', travels with velocity $-\mathbf{V}$ relative to S. Sparks occur at both ends of this second rocket when S is opposite S'', at time $t = 0$ (Figure 1.33).

(a) Draw a diagram showing the positions of the observer S'' and the light pulses at the times $t = -d/2c$, 0, $d/2c$, and d/c according to S. Assume the speed V is $\frac{1}{2}c$.

(b) Draw a diagram showing the positions of the observer S and the light pulses at the times $t'' = -d''/2c$, 0, $d''/2c$, and d''/c according to S'' for the case of (a); $2d''$ is the length of the rocket according to S''. Assume that the speed of S relative to S'' is $\frac{1}{2}c$.

Problem 1.34

(Use the results of Problem 1.33.)

(a) Draw a diagram showing the positions of the observer S and the light pulses at the times $t' = -d'/2c$, 0, $d'/2c$, and d'/c according to S' for the case in which sparks occur at both ends of the rocket simultaneously according to S'; $2d'$ is the length of the rocket according to S'. Assume that the speed of S relative to S' is $\frac{1}{2}c$.

(b) Draw a diagram showing the positions of the observer S' and the light pulses at the times $t = -d/2c$, 0, $d/2c$, and d/c according to S for the case of (a); $2d$ is the length of the rocket according to S. Assume that the speed of S' relative to S is $\frac{1}{2}c$.

(c) At what times do the sparks strike the front and the back of the rocket according to S?

1.4.2   *The relative character of time and length measurements**

Our analysis of the relative character of simultaneity also yields other consequences of Einstein's principle. First we shall consider measurements of time intervals by S and S'. Let us suppose that S has a number of clocks at rest in the S reference frame that show the same time simultaneously according to S and that S' also has a corresponding number of clocks synchronized according to S' and at rest in the S' reference frame. The clocks in one reference frame, say that of S', can be synchronized by the use of light signals in the manner described previously; S' sends out light pulses at a given instant, say $t' = 0$, and a clock at a distance d' from S' is set at the time d'/c at that instant at which the light signal arrives (Figure 1.35). The clocks in the two systems can be related by setting both t' of the clock at the position of S' and t of the clock at the position of S equal to zero at that instant at which S and S' are closest.

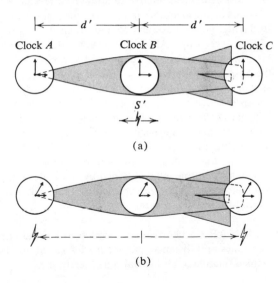

(a)

(b)

FIGURE 1.35 Synchronous clocks in S' reference frame. (a) Clocks A, B, and C. (b) Light signals sent out from clock B at $t' = 0$ reach clocks A and C at time $t' = d'/c'$.

A comparison of the two sets of clocks can be made with the use of Figures 1.33 and 1.34. Consider the time $t = 0$ according to S. This is the time according to S of the event at the front of the rocket in which a spark occurs. This event occurs at the time $t' = -d'/2c$ according to S'; $t = 0$ is also the time according to S of the event at which S is nearest S', and this event occurs at the time $t' = 0$ according to S'. Also, $t = 0$ is the time according to S of the event at the back of the rocket in which a spark occurs, and this event occurs at the time $t' = d'/2c$ according to S'. Figure 1.36 shows the clocks of S and S' that give the times of these events, and it can be seen from this that, according to the synchronized clocks of S, the clocks of S' are not synchronized. Similarly, according to the synchronized clocks of S', the clocks of S are not synchronized

---

* Feynman, Leighton, and Sands (vol. 1), Addison-Wesley, Sec. 15–4, p. 15–5.

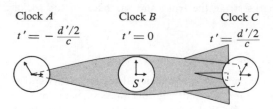

Clock A Clock B Clock C

$t' = -\dfrac{d'/2}{c}$ $t' = 0$ $t' = \dfrac{d'/2}{c}$

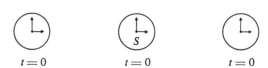

$t = 0$ $t = 0$ $t = 0$

FIGURE 1.36 A comparison of the clocks of S' and those of S at the time $t = 0$ in the frame of reference of S.

(Problem 1.35). We conclude that *time cannot be defined in an absolute manner*, but must be defined relative to an inertial reference frame.

We next consider measurements of the length of the rocket ship by S and by S'. By laying meter sticks end to end along its length, S' can measure the length of the rocket ship. Although S cannot do this since the rocket is moving past him, he can measure the length of the rocket in the following way: He marks the positions, in his reference frame, of both ends of the rocket simultaneously as it moves past him and then he lays meter sticks end to end to measure the distance between these marks. He must mark both ends simultaneously, for if he marks the front end first, his marks will give too small a value for the length, and if he marks the back end first, he will obtain too large a value for the length (Figure 1.37). However, if S marks both ends simultaneously in the S reference frame, then S' observes that S marked the front end first and hence concludes that the value obtained by S for the length of the rocket is too small. We conclude that *lengths are not absolute*, but must be defined relative to an inertial frame of reference.

SUMMARY The relative character of simultaneity is reflected in the facts that times and distances cannot be defined in an absolute manner but must be defined relative to an inertial reference frame.

Problem 1.35

Use the diagrams of Problem 1.34 to show a comparison of the clocks of S' at the midpoint and ends of the rocket at time $t' = 0$ with the corresponding clocks of S.

Problem 1.36

Use the diagrams of Problem 1.34 to show that S claims that the values obtained by S' for distances are too small.

Problem 1.37

Explain why it does not follow from $\mathbf{r} = \mathbf{r}' + \mathbf{R}$ (Figure 1.30) that $\mathbf{v} = \mathbf{v}' + \mathbf{V}$.

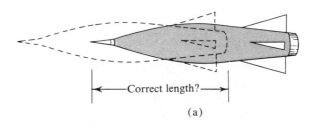

|←——Correct length?——→|

(a)

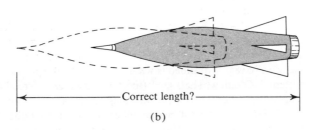

|←————————Correct length?————————→|

(b)

FIGURE 1.37 Two wrong ways to measure the length of a moving object. (a) S marks front end first. (b) S marks back end first.

Problem 1.38

Observer S determines the times t_1 and t_2 at which the front and back ends, respectively, of the rocket pass him and states that the length of the rocket is $V(t_2 - t_1)$. Does S' agree? Explain.

1.5   The Lorentz Transformation

The qualitative arguments given in the last section indicate the necessity of finding the transformation equations relating the position vector $\mathbf{r}$ and the time t of an event in one inertial reference frame S to those, $\mathbf{r}'$ and t', of the same event in another inertial reference frame S'. The transformation law that results from the kinematic postulate of special relativity is called *the Lorentz transformation*, after the theoretical physicist H. A. Lorentz.* Lorentz tried to explain all the macroscopic phenomena of optics and electrodynamics in terms of the microscopic behavior of electrons and atoms. In particular, Lorentz attempted to reconcile experiments such as that of Michelson and Morley with newtonian kinematics, and for this purpose he developed, on the basis of the laws of electromagnetism,† transformation laws having the form of those that now bear his name. Einstein, however, recognized that it was necessary to reexamine the newtonian concepts of space and time. The results of his examination yielded a transformation law that was identical in form to that given earlier by Lorentz, but it had a different interpretation and was derived on a sound

---

* According to Einstein, Lorentz was "the leading spirit" of theoretical physics at the turn of the century. Reminiscences of the man and his investigations, including his contribution to the enclosure of the Zuiderzee, are contained in *H. A. Lorentz, Impressions of His Life and Work*, G. L. de Haas-Lorentz (Ed.), North-Holland, Amsterdam, 1957.
† Short introductions to Lorentz's studies as they apply to the Lorentz transformation law are given in D. Bohm, *The Special Theory of Relativity*, W. A. Benjamin, New York, 1965, and in M. Born, *Einstein's Theory of Relativity*, (Rev. ed.), Dover, New York, 1962.

basis. In this section, we shall derive the Lorentz transformation law on the basis of Einstein's principle of the constancy of the speed of light.

1.5.1 Inadequacy of the galilean transformation law*

An *inertial reference system* is defined to include an inertial reference frame and a system of synchronous clocks at rest in that frame. For our purposes, it is convenient to think of the "scaffolding" of the reference frame as representing a set of coordinate lines. The reference frame can be imagined as an interlocking set of rigid bars that mark off the points with integral values of two of the coordinates (Figure 1.38). We can visualize a set of identical clocks attached at pertinent points to the frame and synchronized by the procedure outlined in Section 1.4.2. Furthermore, we can imagine that there are spectators at every point of the reference frame who can determine the position (x, y, z) and time t of each event coincident with its occurrence. The imaginary arrangement we have just outlined is what we mean by the phrase "an inertial reference system." Often, we shall say "(inertial) observer" instead of "inertial reference system" in order that the mental image given above be brought more vividly to mind. However, it must be remembered that such an observer is always present at every event; he is not seated at one point in space watching distant events when their light reaches his eye.†

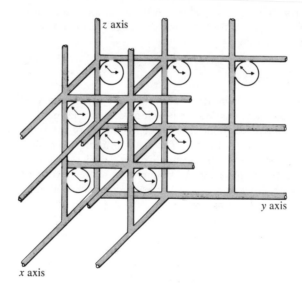

FIGURE 1.38 Visualization of an inertial reference system or observer: the reference frame scaffolding and a set of synchronous clocks.

We consider two inertial reference systems S and S', with S' moving at the constant speed V, relative to S, along the direction of the positive x axis. We assume that the two sets of clocks, one stationary and synchronous in S and the

* Feynman, Leighton, and Sands (vol. 1), Addison-Wesley, Sec. p. 15–2.
† Some of the following developments can be obtained by a consideration of single spectators in each reference frame, as in Section 1.4, each spectator being equipped with a clock and a radar transmitter and receiver. See H. Bondi, *Relativity and Common Sense*, Doubleday, New York, 1964 and Chapter 26 of D. Bohm, *The Special Theory of Relativity*, W. A. Benjamin, New York, 1965.

other stationary and synchronous in S', are adjusted so that, at the S time $t = 0$, the origins of S and S' coincide and the S' clock at the origin reads zero (Figure 1.39). The x and x' axes coincide (Figure 1.40). These restrictions are introduced at this point only to simplify the discussion; it does not require much effort to generalize our results to apply to inertial reference frames that are not so restricted.

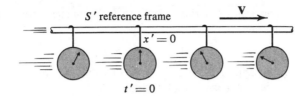

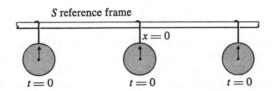

FIGURE 1.39 The adjustment of the sets of clocks: one set is synchronous in S' and the other synchronous in S. The clocks are seen as observed by S.

The transformation equations connecting the position vector and time of an event relative to the S inertial frame to those of the same event relative to the S' inertial frame must reduce, for values of V that we encounter in everyday life ($|V| \ll c$), to the galilean transformation law of newtonian mechanics:

$$\mathbf{r} = \mathbf{r}' + \mathbf{V}t, \qquad \mathbf{v} = \mathbf{v}' + \mathbf{V}, \qquad \mathbf{a} = \mathbf{a}', \qquad (|V| \ll c). \qquad (1.30)$$

We should add to this the equation

$$t = t', \qquad (|V| \ll c), \qquad (1.31)$$

since, for the relative speeds encountered in everyday experience, there exists a universal time that is defined independent of the inertial reference frame.

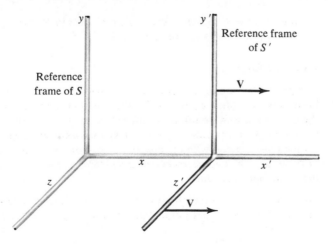

FIGURE 1.40 Reference frames S and S' as seen by S.

The galilean transformation equations can be rewritten in terms of the components of the position vectors $\mathbf{r}$ and $\mathbf{r}'$. Because of our assumptions, the vector $\mathbf{R}$ has zero y and z components and an x component equal to Vt or Vt'. We therefore obtain the following form for the galilean transformation equations for the particular case under consideration:

$$x = x' + Vt', \qquad y = y', \qquad z = z', \qquad t = t', \qquad (|V| \ll c). \qquad (1.32)$$

The transformation equations for velocities that follow from the galilean equations,

$$v_x = v_{x'} + V, \qquad v_y = v_{y'}, \qquad v_z = v_{z'}, \qquad (|V| \ll c), \qquad (1.33)$$

do not appear to agree with the fact that the speed of light is the same for all inertial observers in all directions, for if

$$v_x = c \qquad \text{and} \qquad v_y = v_z = 0, \qquad (1.34)$$

then

$$v_{x'} = c - V \neq c, \qquad v_{y'} = v_{z'} = 0. \qquad (1.35)$$

However, for V's having magnitudes with which we are familiar in everyday life, no experiment has been performed to date that can distinguish between $c - V$ and c. Thus, the galilean transformation law is valid for a description of our usual experiences of everyday life, but it is not valid for a description of those phenomena that involve relative speeds of magnitude comparable to the speed of light.

SUMMARY The galilean transformation law adequately describes the relation between the coordinates and times of an event relative to two different inertial systems *only* in those circumstances involving relative speeds much smaller than c.

Problem 1.39

A pulse of light, emitted at the origin at time $t = 0$ relative to S, lies on the surface of the sphere,

$$x^2 + y^2 + z^2 = c^2 t^2$$

at time t relative to S. On what surface relative to S' would the light lie if the galilean transformation were applicable?

Problem 1.40

A train 1 mi long travels at 60 mph past a station. An observer S' on the train uses a coordinate system with the origin at the rear of the train and with the positive x' axis pointing toward the front of the train. An observer S on the station platform uses a coordinate system with the origin at one end of the station platform and with the positive x axis pointing in the direction of motion of the train. The two observers set their clocks at $t = 0$ at that instant at which the two origins coincide.

(a) Find t and t' for that instant when the front of the train passes the origin of S.
(b) Find the coordinates x and x' of the rear of the train at the instant calculated in (a).

(c) Find the coordinates x and x' of the front end of the train at $t = t' = 0$.

(d) Find the coordinates x and x' of the front of the train and also those of the back at time $t = 1$ min.

Problem 1.41

Two events have coordinates (x_1, y_1, z_1) and t_1 and (x_2, y_2, z_2) and t_2 relative to an inertial observer S and coordinates (x'_1, y'_1, z'_1) and t'_1 and (x'_2, y'_2, z'_2) and t'_2 relative to S'. Find the relation between the distance between these events relative to S and that relative to S' on the basis that the galilean transformation equations are applicable. Explain why these distances are not necessarily equal.

1.5.2   Time dilatation and Lorentz contraction*

The Lorentz transformation law, which is valid for all possible speeds V, is derived on the basis of the kinematic postulate that a light signal traveling with speed c relative to S also travels with speed c relative to S'. There exists a variety of ways to deduce the Lorentz transformation law; we shall describe one of these in the text and introduce others in Problems 1.70 and 1.71. The derivation given in the text proceeds in the following way: Using properties of light pulses, we determine a comparison between lengths and also between time intervals, as measured by two inertial observers in relative motion. Then in Section 1.5.3, these results are collected together and applied to give the Lorentz transformation equations relating the position vector $\mathbf{r}$ and the time t of an event relative to S to those, $\mathbf{r}'$ and t', of the same event relative to S'.

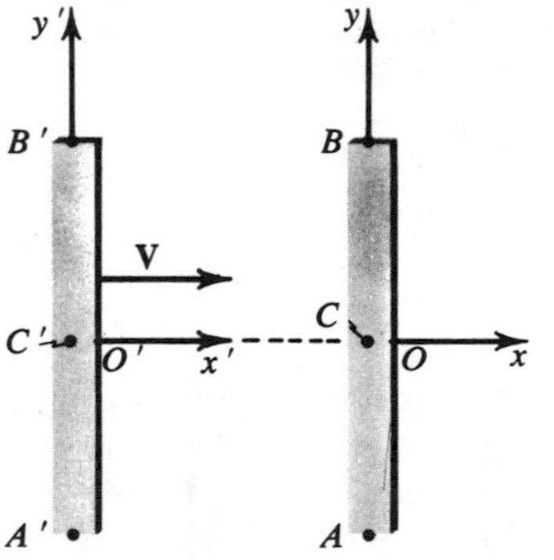

We begin by comparing the length of a meter stick $A'B'$ lying at rest along the y' axis of S' with that of a meter stick AB lying at rest along the y axis of S. [The meter is defined, relative to a frame of reference, as 1,650,763.73 wavelengths of the orange line in the spectrum of a sample of krypton-86 at rest relative to that frame, and this definition of the meter applies in either inertial system.] Let the midpoints of the meter sticks, C' and C, lie at O' and O, respectively (Figure 1.41). The points C' and C coincide at time $t' = 0 = t$.

Observer S determines the length of $A'B'$ by measuring the positions of A' and B' simultaneously relative to S. Let A'' be the position of A' relative to S when A' crosses the y axis, and let B'' be the corresponding position of B' (Figure 1.42). Since $C'B' = C'A'$ and since the direction of motion is perpendicular to the S' rod, then $A''C$ must equal $B''C$, or else one direction in space could be distinguished by physical means from another; no distinction has ever been observed among the different directions in space, and thus there is no experimental basis for violating the principle of the isotropy of space. Consider a light signal emitted at C at such a time that the signal reaches B' when B' coincides with B''. Since C moves along the x' axis, $CB' = CA'$, and the light signal reaches A' at the same time relative to S. Also, since $CB'' = CA''$, the light signal reaches A'' at the same time relative to S. Therefore, B' coincides with B'', and A' coincides with A'' simultaneously relative to S. Hence S measures the length of the S' stick $A'B'$ as equal to $A''B''$.

* Feynman, Leighton, and Sands (vol. 1), Addison-Wesley, Sec. 15–4, p. 15–5; Sec. 15–5, p. 15–7.
Kacser, Prentice-Hall, Sec. 4.1, p. 69.
Kittel, Knight, and Ruderman, McGraw-Hill, pp. 353–354 and 359–361.

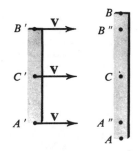

FIGURE 1.41   Meter sticks of S' and S lying along the y' and y axes, respectively.

FIGURE 1.42 A'' is the position of A' when it crosses the y axis, along AB. The figure was drawn on the (incorrect) assumption that $A''B'' < AB$.

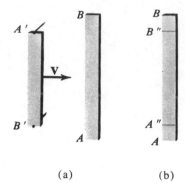

(a)   (b)

FIGURE 1.43 A method by which S can measure the length of $A'B'$. The figure was drawn on the (incorrect) assumption that $A''B'' < AB$. (a) The nails on the ends of the meter stick of S' move toward the meter stick of S. (b) The slashes left by the nails determine the length of $A'B'$ as measured by S.

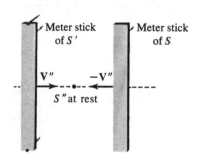

FIGURE 1.44 Relative to S'', the meter sticks are moving with the same speed, V''; thus, the nails at the ends of the meter stick of S' slash the ends of the meter stick of S.

One procedure by which S can measure the distance $A''B''$ is shown in Figure 1.43. Nails at A' and B' slash the meter stick of S at A'' and B'' as the meter stick of S' passes by, and S determines the length of $A'B'$ by a measurement of the distance between the slashes.

Observer S' could measure the length of the meter stick of S in an identical manner, and we might expect that the two results would be the same as a result of the fact, stated before, that the speed of S' relative to S is equal to the speed of S relative to S'. However, we have not proven this yet and, indeed, we do not need to introduce this fact to show that the results of the two measurements are the same. This can be shown through the use of a third inertial observer S''.

The measurement by S of the length of the S' stick is shown in Figure 1.44 from the point of view of the inertial observer S'' relative to whom both S and S' are moving with the same speed V'' along the x'' axis. Arguments similar to those given above show that A' and A'' coincide at the same time, relative to S'', that B' and B'' coincide. However, if A'' and B'' differ from A and B, say $A''B'' < AB$, then S'' can distinguish by an experiment the direction of $\mathbf{V}''$ from that of $-\mathbf{V}''$, contrary to the principle of the isotropy of space—that is, S'' observes that the slashes, physical marks observable to all, from the nails on the ends of the meter stick of S' occur at the ends of the meter stick of S. Hence $A''B'' = AB$, and therefore S determines that the length of the meter stick of S' is 1 m.

This result may be generalized. The length of each of the sticks may be chosen arbitrarily as long as the number of meters in length of one stick relative to the inertial system in which it is at rest is equal to the length of the other stick relative to its rest inertial system. Furthermore, the directions of the y and y' axes are arbitrary as long as these two directions are parallel and each is orthogonal to the direction of the relative motion. Therefore, we conclude that *a measuring stick has the same length relative to all inertial systems moving along directions perpendicular to the stick*.

Let us now investigate, from the point of view of S, the behavior of a clock that is at rest in the S' reference system. For this purpose, we construct a clock that uses light signals, thus depending on the speed of light for the ticking off of equal time intervals. This clock consists of a light pulse emitter and receiver situated at one end of a rod and a mirror at the other (Figure 1.45). If the distance between the mirror and the source and receiver is L, it takes a time $\Delta t = 2L/c$ for a pulse to travel from the source to the mirror and back to the receiver. By properly connecting the source and receiver so that the source emits a light pulse instantaneously when the receiver absorbs one, we obtain a clock that ticks off equal time intervals of magnitude $2L/c$. Let us suppose that S and S' possess a number of such clocks, each oriented so that the light pulses travel parallel to the respective y axes. Our previous conclusion shows that the distance in any one of these clocks between the mirror and the emitter-receiver is L according to both observers. Then according to S', for example, one of his clocks, say that shown in Figure 1.45, works as described above and regularly ticks off time intervals of duration $\Delta t' = 2L/c$. However, this device is in motion relative to S, so S observes that each light pulse travels a distance greater than L (Figure 1.46). Indeed, he observes that the time interval between the ticks of the S' clock is given by

$$\Delta t = \frac{2L}{\sqrt{c^2 - V^2}}.$$ (1.36)

Hence, S observes that the clock of S' is running slow and that *the time interval $\Delta t'$ as measured on the S' clock is related to the corresponding time Δt that S measures by the equation*

$$\Delta t = \frac{\Delta t'}{\sqrt{1 - (V^2/c^2)}}. \qquad (1.37)$$

Thus, for example, when the S' clock ticks off 1 sec,

$$\Delta t' = 1 \text{ sec}, \qquad (1.38)$$

then S observes that a greater time,

$$\Delta t = \frac{1}{\sqrt{1 - (V^2/c^2)}} \text{ sec} > 1 \text{ sec}, \qquad (1.39)$$

has elapsed. Therefore, the S' clock appears to S to be running slow. This phenomenon is known as *time dilatation* (or, sometimes, time dilation).

It is important to note that the clock of S' is timed by S with a number of clocks fixed in the S system (Figure 1.47).

We now compare length measurements performed by the two inertial observers S and S' on rods lying along the direction of relative motion of S and S'. Consider a rod $A'B'$ lying at rest relative to S' on the x' axis and moving past S with the speed V in the positive x direction (Figure 1.48). Let the length of the rod be l' relative to S' and l relative to S.

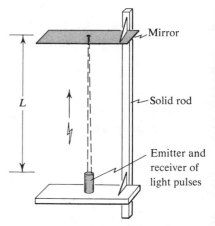

FIGURE 1.45 A light-pulse clock that ticks off time intervals of $\Delta t = 2L/c$.

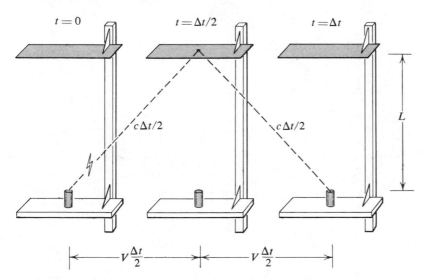

FIGURE 1.46 The dashed line is the path of the light pulse of the S' clock as observed by S. The time interval Δt between clocks is given by the relation $(c\,\Delta t/2)^2 = [V(\Delta t/2)]^2 + L^2$ as being $\Delta t = 2L/\sqrt{c^2 - V^2}$.

We can use the formula derived for time dilatations if we devise a method by which S' measures the length of the rod with light pulses and a single clock and then consider this method from the point of view of S. One such method is the following: A light-pulse emitter and receiver is placed at one end of the rod A' and a mirror at the other B' (Figure 1.49). With a clock at A', S' measures the time $\Delta t' = t_2' - t_1'$ for a light pulse, emitted at A' at time t_1', to pass from A' to B' and back to A' at time t_2'. The length of the rod l' relative to S' is given by

$$l' = \tfrac{1}{2}c\,\Delta t'. \qquad (1.40)$$

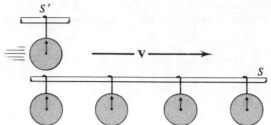

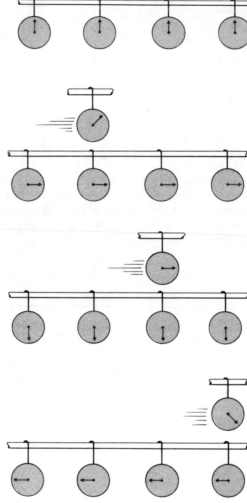

FIGURE 1.47 Time dilatation: S observes that each clock of S', such as the S' clock shown, is running slow.

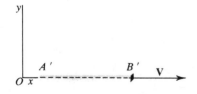

FIGURE 1.48 The rod $A'B'$ moves past S with velocity $\mathbf{V}$.

Now consider the sequence of events involved in this measurement from the point of view of S. The light flash is emitted at time t_1 relative to S when A' coincides with the point X [Figure 1.50(a)]. The flash strikes the mirror at time t relative to S when B' coincides with Y [Figure 1.50(b)] and returns to A' at time t_2 relative to S when A' coincides with Z [Figure 1.50(c)]. The time $\Delta t = t_2 - t_1$ relative to S for the light flash to go from A' to B' and back to A' consists of two parts, $t - t_1$ and $t_2 - t$, which we shall calculate separately.

The time required for the flash to travel from X to Y, XY/c, is $t - t_1$, and XY is equal to the length of the rod $RY = l$ plus the distance $XR = V(t - t_1)$

that the rod moves during that time interval. Hence,

$$t - t_1 = \frac{l}{c} + \frac{V}{c}(t - t_1) \tag{1.41}$$

or

$$t - t_1 = \frac{l/c}{1 - (V/c)}. \tag{1.42}$$

Similarly, $t_2 - t$ is the time required for light to travel from Y to Z, YZ/c, and YZ is equal to the length of the rod $ZS = l$ less the distance $YS = V(t_2 - t)$ that the rod moves during that interval of time. Therefore,

$$t_2 - t = \frac{l}{c} - \frac{V}{c}(t_2 - t) \tag{1.43}$$

or

$$t_2 - t = \frac{l/c}{1 + (V/c)}. \tag{1.44}$$

Emitter and receiver of light pulses

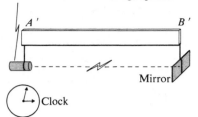

FIGURE 1.49 S' measures the length l' of the rod by measuring the time $\Delta t'$ required for light to travel to and fro; $l' = \frac{1}{2}c\,\Delta t'$ is obtained.

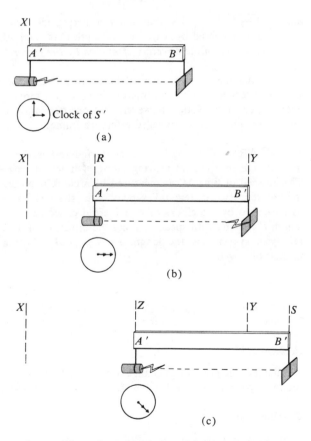

FIGURE 1.50   Sequence of events, from the point of view of S, involved in the measurement of the length $A'B'$ by S'. (a) The light flash is emitted at time t_1. (b) The light flash reaches the mirror at time t. (c) The light flash is received at A' at time t_2.

The total time interval Δt is given by

$$\Delta t = t_2 - t_1 = (t_2 - t) + (t - t_1)$$

$$= \frac{l/c}{1 - (V/c)} + \frac{l/c}{1 + (V/c)} \tag{1.45}$$

$$= \frac{2l/c}{1 - (V^2/c^2)}.$$

Observer S notices that his clocks read t_1 when one of the clocks of S' reads t_1' and that the S clocks read t_2 when the same S' clock reads t_2'. Hence,

$$\Delta t = \frac{\Delta t'}{\sqrt{1 - (V^2/c^2)}}, \tag{1.46}$$

and, from Equations (1.45) and (1.46),

$$l = \frac{1}{2}\Delta t c\left(1 - \frac{V^2}{c^2}\right) = \frac{1}{2}\Delta t' c\sqrt{1 - (V^2/c^2)}$$

$$= l'\sqrt{1 - (V^2/c^2)}. \tag{1.47}$$

Since $\sqrt{1 - (V^2/c^2)}$ is less than unity, a meter stick moving with S' (for which $l' = 1$ m) is observed by S to have a length that is less than 1 m. *The moving rod is contracted in the direction of motion by the factor* $\sqrt{1 - (V^2/c^2)}$ (Figure 1.51).

This phenomenon is known as the *Lorentz contraction*, since the theory of Lorentz predicted such a contraction in an object moving relative to the ether with a speed V.* Notice, however, that our derivation of this effect does not involve the ether (or a privileged reference frame).

SUMMARY The length of a measuring rod is the same relative to every inertial reference system moving along a direction perpendicular to the rod. The interval of time $\Delta t'$ as measured by a clock moving with speed V past an inertial reference system takes place over the time $\Delta t = \Delta t'[1 - (V^2/c^2)]^{-\frac{1}{2}}$ as measured by the clocks of that inertial observer. A measuring rod of rest length l' moving with speed V along the direction of its length past an inertial reference system has the length $l = l'[1 - (V^2/c^2)]^{\frac{1}{2}}$ as measured by that inertial observer.

Problem 1.42

A clock moves past you with the speed found in your answer to Problem 1.28. Calculate how long it takes for this clock to lose 1 sec relative to you.

Problem 1.43

Calculate the percentage change in the length of a rod moving with the speed found in your answer to Problem 1.28 in a direction parallel to its length.

Problem 1.44

An airplane travels at a speed of 1,500 mi/hr.

---

* A contraction of this sort had been postulated earlier by the Irish physicist G. F. FitzGerald (1851–1901); thus, sometimes the phenomenon is called the Lorentz–FitzGerald contraction.

$V = 0$

$V = \frac{1}{4}c$

$V = \frac{1}{2}c$

$V = \frac{3}{4}c$

$V = 0.9c$

$V = 0.99c$

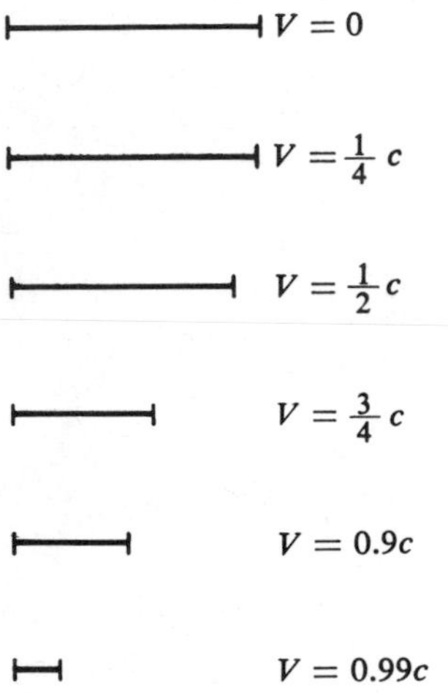

FIGURE 1.51   Lorentz contractions for various speeds V.

(a) Calculate the percentage change in its length due to Lorentz contraction relative to the earth.
(b) How far does the plane travel before the pilot's watch is slowed down by 1 sec relative to the earth?

Problem 1.45

An S' clock moves past the observer S, who measures that the S' clock loses 1 sec every hour.

(a) Calculate the speed of S' relative to S.
(b) The clocks of S are situated 1 m apart along the direction of motion of S'. How many clocks of S does the S' clock pass when it ticks off 1 sec?

Problem 1.46

When observer S measures a meter stick of S' that lies along the direction of motion of S', he notes a contraction of 1 mm. Find the speed of S' relative to S.

Problem 1.47

A cube with sides of length L moves with S'.

(a) Find the volume of the cube as measured by S.
(b) How fast would the cube have to be moving in order that its volume be halved according to S?

Problem 1.48

A meter stick, at rest in S', is inclined at an angle of 45° relative to S with respect to the direction of motion of the inertial frame S' relative to the frame S. The frame S' moves with a speed of $0.99c$ past the frame S. What is the length of the meter stick as measured by the observer S, and what is the angle between the meter stick and the x axis as observed by S?

Problem 1.49

A meter stick, at rest in S', is inclined at an angle θ with respect to the direction of motion of S' relative to S; S' moves past S with the speed $0.9c$. Calculate the length $l(\theta)$ of the meter stick as measured by the observer S, and draw a graph that shows the value of $l(\theta)$ for each value of θ.

Problem 1.50

The formula for time dilatation can be derived solely on the basis of the Lorentz contraction, as shown in this problem. Observer S' uses a light-pulse clock (Figure 1.45) with the light-pulse path, of length L', lying along the x' axis.

(a) Show that, according to S', the clock ticks off time intervals of length $\Delta t' = 2L'/c$.
(b) Let Δt_1 be the time as measured by S for the light pulse to travel from the emitter to the mirror. Show that, if proper account is taken of the Lorentz contraction,

$$\Delta t_1 = \frac{L'\sqrt{1 - (V^2/c^2)}}{c} + \frac{V\,\Delta t_1}{c};$$

obtain

$$\Delta t_1 = \frac{L'}{c}\sqrt{\frac{1 + (V/c)}{1 - (V/c)}}.$$

(c) Let Δt_2 be the time as measured by S for the light pulse to travel from the mirror to the receiver. Show that

$$\Delta t_2 = \Delta t_1 - \frac{V}{c}(\Delta t_1 + \Delta t_2),$$

and hence obtain

$$\Delta t_2 = \frac{L'}{c}\sqrt{\frac{1 - (V/c)}{1 + (V/c)}}.$$

(d) Show that S measures the time intervals between pulses as

$$\Delta t = \frac{2L'}{c\sqrt{1 - (V^2/c^2)}};$$

from this, derive the formula for the time dilatation.

Problem 1.51

The formula for the Lorentz contraction can be derived solely on the basis of time dilatation, as shown in this problem. A rod of length l' lying along the x' axis moves with S' past S, who observes the times t_1 and t_2 at which the ends of the rod pass the point x and determines the length l_1 from the formula $l_1 = V\,\Delta t$. Find l_1 in terms of l'. (For this problem, you will need to assume the fact, which will be proven shortly, that S moves with the velocity $-\mathbf{V}$ relative to S' if S' moves with the velocity $\mathbf{V}$ relative to S.)

1.5.3 A special set of Lorentz transformation equations*

An inertial observer S describes an event by the position vector $\mathbf{r}$, or the coordinates (x, y, z) of the point where the event took place, and the time t at which the event occurred. Another inertial observer S' describes the same event by a position vector $\mathbf{r}'$, or the coordinates (x', y', z'), and a time t' relative to the S' inertial system. There exists some relation between the inertial systems S and S', so, since the coordinates (x, y, z) and time t relative to S describe the *same* event as do the coordinates (x', y', z') and time t' relative to S', there must exist transformation equations that determine x', y', z', and t' in terms of x, y, z, and t:

$$\begin{aligned} x' &= x'(x, y, z, t), & y' &= y'(x, y, z, t), \\ z' &= z'(x, y, z, t), & t' &= t'(x, y, z, t). \end{aligned} \tag{1.48}$$

We shall derive now a formula for these transformation equations for the

* Feynman, Leighton, and Sands (vol. 1), Addison-Wesley, Sec. 15–2, p. 15–3; Sec. 15–5, p. 15–7; Sec. 15–6, p. 15–7.
Kacser, Prentice-Hall, Sec. 3.3, p. 37.
Kittel, Knight, and Ruderman, McGraw-Hill, pp. 346–350.
Resnick, John Wiley, Sec. 2.4, p. 65.

circumstance in which S and S' are related in a simple manner. This derivation of the equations for a Lorentz transformation depends on the consequences, such as the time dilatation and the Lorentz contraction, that we have deduced from the kinematic postulate of special relativity.

The form of these transformation equations is restricted by the kinematic postulate of special relativity; the explicit formula for any case depends also on the relation between the inertial systems S and S'. Let us consider the case involving the simplest nontrivial relation between the inertial systems S and S'. In this case, S and S' are in relative motion, or the problem reduces, in a trivial fashion, to a rotation of the axes or a translation of the origins. In order to avoid complications, we assume that the coordinate axes are aligned and the origins of the inertial systems are coincident at an initial time. Thus, we shall consider the special case in which the system S' moves with velocity $\mathbf{V} = V\hat{x}$ relative to S, the origins O and O' coincide at $t = t' = 0$, and the y' and z' axes coincide at $t = t' = 0$ with the y and z axes, respectively (Figure 1.52).

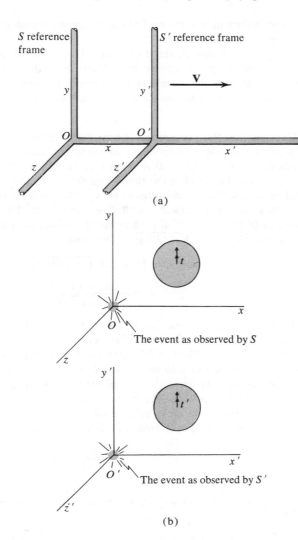

FIGURE 1.52 The relation between the inertial systems S and S' described in the text. (a) The motion of S' as observed by S. (b) The event that occurs at O at time $t = 0$ relative to S occurs at O' at time $t' = 0$ relative to S'.

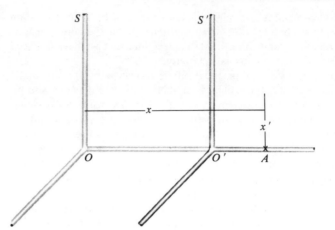

FIGURE 1.53 $x = Vt + x'\sqrt{1 - (V^2/c^2)}$.

We now deduce in turn the explicit set of Lorentz transformation equations for this special case.

Consider the y' coordinate of an event. This represents a distance relative to S', measured along a direction perpendicular to the direction of relative motion, and is equal to y, the corresponding distance for the event that S measures. The z' and z coordinates are similarly related, and hence

$$y' = y, \qquad z' = z. \tag{1.49}$$

Consider now the x coordinate at time t relative to S of the event that occurs at a point with the corresponding coordinate x' relative to S'. The value x is equal* to $OA = OO' + O'A$ (Figure 1.53), where OO' is the distance that O' has traveled from O in the time t and $O'A$ is the distance along the line of relative motion that S' measures to be x'. The point O' is moving with speed V relative to O and coincides with O at time $t = 0$. Hence, $OO' = Vt$. The length of $O'A$ relative to S' is x'; because of the Lorentz contraction, S measures this distance to be $x'\sqrt{1 - (V^2/c^2)}$. Therefore,

$$x = Vt + x'\sqrt{1 - (V^2/c^2)} \tag{1.50}$$

or

$$x' = \frac{x - Vt}{\sqrt{1 - (V^2/c^2)}}. \tag{1.51}$$

Finally, we must derive the transformation equation that determines the S' time t' of the event in terms of the coordinates (x, y, z) and time t of the event relative to S. The formula for time dilatation gives the rate at which a clock of S' appears to S to be slowing down, but it does not give directly the relation between the times t' and t of every event. It gives the relation only between the times for those events that occur at the origin O'; since the S and S' clocks at the origin are synchronized at $t = t' = 0$, then, according to S, the S' clock at the origin O' reads t_0' when the S time t_0 is given by

$$t_0 = \frac{t_0'}{\sqrt{1 - (V^2/c^2)}}. \tag{1.52}$$

---

* This is based on the assumption that the transformation equations are linear; this will be shown in Section 1.5.5.

However, we know how every clock in S is synchronized with that at O and how every clock in S' is synchronized with that at O'. This knowledge, together with the relation between the clocks at O and O', allows us to deduce an equation relating the times of any event.

Moreover, we need consider only the relation between those two sets of clocks that lie on the x and the x' axes. Any clock off one of these axes can be synchronized through the use of light signals with a corresponding clock on the axis of the relevant reference frame with the same x or x' coordinate. Since distances l and l' perpendicular to the axes are the same relative to both observers and since the process of synchronization involves only the time-interval changes l/c or l'/c, the observers agree that each clock off the axis is synchronized with the corresponding clock on the axis.

Suppose that when the S' clock at the origin reads t'_0 a light pulse is sent out from O' ($x' = 0$) along the x' axis, and that this light flash arrives at the point x' at time t'. According to S', the time t' is later than t'_0 by the time interval in which light travels the distance x'. Hence,

$$t' = t'_0 + \frac{x'}{c}. \tag{1.53}$$

On the other hand, according to S, the light signal was emitted at time t_0 and traveled the distance $x'\sqrt{1 - (V^2/c^2)}$ plus the distance $V(t - t_0)$ that the point labeled x' moved during the transit time of the light pulse. Therefore,

$$t = \frac{t'_0}{\sqrt{1 - (V^2/c^2)}} + \frac{x'\sqrt{1 - (V^2/c^2)} + V(t - t_0)}{c}; \tag{1.54}$$

thus

$$t[1 - (V/c)] = \frac{t'_0[1 - (V/c)]}{\sqrt{1 - (V^2/c^2)}} + \frac{x'\sqrt{1 - (V^2/c^2)}}{c}$$

$$= \frac{[t' - (x'/c)][1 - (V/c)] + (x'/c)[1 - (V^2/c^2)]}{\sqrt{1 - (V^2/c^2)}} \tag{1.55}$$

or

$$t = \frac{t' + (V/c^2)x'}{\sqrt{1 - (V^2/c^2)}}. \tag{1.56}$$

Substitution for x' from Equation (1.51) gives the result

$$t' = \frac{t - (V/c^2)x}{\sqrt{1 - (V^2/c^2)}}. \tag{1.57}$$

We conclude that the transformation equations, which relate to coordinates of an event relative to two reference systems whose origins coincide at time $t = t' = 0$, whose axes are parallel, and whose relative velocity is along the x or x' axes, are

$$x' = \frac{x - Vt}{\sqrt{1 - (V^2/c^2)}}, \qquad y' = y, \qquad z' = z, \qquad t' = \frac{t - (V/c^2)x}{\sqrt{1 - (V^2/c^2)}}. \tag{1.58}$$

These are a special case of the Lorentz transformation equations.

These equations can be solved for x, y, z, and t in terms of x', y', z', and t'.

The resulting inverse relations are

$$x = \frac{x' + Vt'}{\sqrt{1 - (V^2/c^2)}}, \qquad y = y', \qquad z = z', \qquad t = \frac{t' + (V/c^2)x'}{\sqrt{1 - (V^2/c^2)}}. \quad (1.59)$$

Alternatively, this inverse transformation can be obtained from Equation (1.58) by the replacement of $\mathbf{V}$, the velocity of S' relative to S, by $\mathbf{V}'$, the velocity of S relative to S':

$$x = \frac{x' - V't'}{\sqrt{1 - (V'^2/c^2)}}, \qquad y = y', \qquad z = z', \qquad t = \frac{t' - (V'/c^2)x'}{\sqrt{1 - (V'^2/c^2)}}. \quad (1.60)$$

A comparison of these sets of equations shows that the *S frame moves with the velocity* $-\mathbf{V} = -V\hat{x}'$ *relative to* S'. This equality of the speeds V and V' is not a trivial conclusion, since neither the unit of length nor the unit of time is directly comparable in S and S'.

The fact that the galilean transformation is valid for speeds that we normally encounter can be derived from the special Lorentz transformation equations

$$x' = \frac{x - Vt}{\sqrt{1 - (V^2/c^2)}}, \qquad y' = y, \qquad z' = z, \qquad t' = \frac{t - (V/c^2)x}{\sqrt{1 - (V^2/c^2)}}. \quad (1.61)$$

If V is much smaller than c, we can neglect the term V^2/c^2 as compared to unity in the denominators, and also, for not too large values of x, we can neglect the factor V/c^2 in the last equation. Thus, the galilean transformation equations,

$$x' = x - Vt, \qquad y' = y, \qquad z' = z, \qquad t' = t, \quad (1.62)$$

result as a first approximation to the Lorentz transformation equations. This result is important, for if the Lorentz transformation had not reduced to the galilean transformation for small values of V, we would know that the Lorentz transformation was not correct, because of the abundant evidence from everyday life that the galilean transformation is valid for values of V much smaller than c.

SUMMARY   There exist transformation equations relating the coordinates (x, y, z) and t relative to an inertial observer S of an event and the coordinates (x', y', z') and t' relative to another inertial observer S' of the same event. The form of these transformation equations is restricted by the kinematic postulate of special relativity, and the explicit equations for the transformation depend on the relation between S and S'. For the simplest nontrivial case,

$$x' = \frac{x - Vt}{\sqrt{1 - (V^2/c^2)}}, \qquad y' = y, \qquad z' = z, \qquad t' = \frac{t - (V/c^2)x}{\sqrt{1 - (V^2/c^2)}}.$$

Example 1.3

   Q. Two events in a particle's life occur, relative to an inertial observer S, at time t_1 with $ct_1 = 2$ m and at the point $x_1 = 1$ m, $y_1 = z_1 = 0$, and at time t_2 with $ct_2 = 5$ m and at the point $x_2 = 3$ m, $y_2 = z_2 = 0$.

   (a) Find the average speed relative to S with which the particle moved between these two events.

   (b) An inertial observer S' moves with velocity $\mathbf{V} = (4c/5)\hat{x}$ relative to S.

The origins of the two systems coincided at time $t = 0$ and $t' = 0$. Find the coordinates of the two events relative to S'.

(c) Find the distance between the two events relative to S' and also the time between the two events relative to S'.

(d) Find the average speed relative to S' with which the particle moved between these two events.

A. (a) The distance relative to S that the particle traveled between the two events is $x_2 - x_1 = 3 - 1 = 2$ m. The time $(t_2 - t_1)$ between the two events is given by

$$t_2 - t_1 = \frac{ct_2 - ct_1}{c} = \frac{5 - 2}{3.00 \times 10^8} = 1 \times 10^{-8} \text{ sec.} \tag{1.63}$$

Therefore, the average speed of the particle is $v = 2 \text{ m}/(1 \times 10^{-8} \text{ sec}) = 2 \times 10^8$ m/sec.

(b) The position relative to S' of the particle at the first event is

$$x_1' = \frac{x_1 - Vt_1}{\sqrt{1 - (V^2/c^2)}} = \frac{x_1 - (V/c)ct_1}{\sqrt{1 - (V^2/c^2)}} = \frac{1 - (4/5)2}{3/5} = -1 \text{ m.} \tag{1.64}$$

The time relative to S' of this event is given by

$$t_1' = \frac{t_1 - (V/c^2)x_1}{\sqrt{1 - (V^2/c^2)}} = \frac{1}{c} \frac{ct_1 - (V/c)x_1}{\sqrt{1 - (V^2/c^2)}} = \frac{1}{c} \frac{2 - (4/5)1}{3/5} = \frac{2}{c} \text{ m}$$
$$= \frac{2}{3} \times 10^{-8} \text{ sec.} \tag{1.65}$$

Similarly, the position and time relative to S' of the second event are

$$x_2' = \frac{3 - (4/5)5}{3/5} = -\frac{5}{3} \text{ m,} \qquad t_2' = \frac{1}{c} \frac{5 - (4/5)3}{3/5} = \frac{13}{9} \times 10^{-8} \text{ sec.} \tag{1.66}$$

(c) The distance relative to S' between the two events is given by

$$x_2' - x_1' = -\frac{5}{3} - (-1) = -\frac{2}{3} \text{ m.} \tag{1.67}$$

The time between the two events is given by

$$t_2' - t_1' = \frac{13}{9} \times 10^{-8} - \frac{2}{3} \times 10^{-8} = \frac{7}{9} \times 10^{-8} \text{ sec.} \tag{1.68}$$

(d) The average speed relative to S' of the particle between the two events is given by

$$\bar{v}' = \frac{x_2' - x_1'}{t_2' - t_1'} = -\frac{6}{7} \times 10^8 \text{ m/sec.} \tag{1.69}$$

Notice that this is not equal to $\bar{v} - V$.

Problem 1.52

Show, for V equal to the speed found in your answer to Problem 1.28, that the differences between the galilean transformation equations and the Lorentz transformation equations can be neglected.

Problem 1.53

The reference system S' moves with velocity $(3c/5)\hat{x}$ relative to S. Find the x' and t' coordinates of each of the events whose x and t coordinates are listed below. (Use $c = 3.0 \times 10^8$ m/sec.)

Event	E_1	E_2	E_3	E_4
x, m	1.5	3.2	2.1	1.5
t, sec	0.5×10^{-8}	0.3×10^{-8}	1.0×10^{-8}	1.0×10^{-8}

Problem 1.54

Using the data of Problem 1.53, calculate the distances between the positions of the following pairs of events in S':

(a) E_1 and E_2,
(b) E_2 and E_3,
(c) E_1 and E_4.

Explain why these results are not given by the formula for the Lorentz contraction.

Problem 1.55

Calculate the time intervals between the following pairs of events in S', using the data of Problem 1.53:

(a) E_1 and E_2,
(b) E_2 and E_3,
(c) E_3 and E_4.

Explain why these results are not given by the formula for time dilatation.

Problem 1.56

(a) Repeat the calculations of Problem 1.40 using the Lorentz transformation Equations (1.58).
(b) Calculate the percentage differences in your answers to Problem 1.40 and (a).
(c) Repeat the calculations of Problem 1.40 for the case in which the train's speed is $12c/13$.

Problem 1.57

(a) Consider the case in which S' moves with a velocity $\mathbf{V}$, not necessarily in the x direction, relative to S. Show that, under other conditions similar to those stated before,

$$\mathbf{r} = \mathbf{r}' + \frac{\mathbf{V}}{V\sqrt{1 - (V^2/c^2)}} \left\{ [1 - \sqrt{1 - (V^2/c^2)}] \frac{\mathbf{V} \cdot \mathbf{r}'}{V} + Vt' \right\},$$

$$t = \frac{t' + (\mathbf{V} \cdot \mathbf{r}'/c^2)}{\sqrt{1 - (V^2/c^2)}}.$$

Hint: Write $\mathbf{r} = \mathbf{r}_\| + \mathbf{r}_\perp$ and $\mathbf{r}' = \mathbf{r}'_\| + \mathbf{r}'_\perp$, where the subscripts $\|$ and $\perp$ denote components parallel and perpendicular to $\mathbf{V}$, respectively. In particular, show that $\mathbf{r}_\parallel = (\mathbf{V}/V)(\mathbf{V}\cdot\mathbf{r}/V)$. Then use the facts that $\mathbf{r}_\parallel$ changes as x does for the special case $\mathbf{V} = V\hat{x}$ studied in the text and that $\mathbf{r}_\perp$ remains unchanged.

(b) Show that

$$\mathbf{r}' = \mathbf{r} - \frac{\mathbf{V}}{V\sqrt{1 - (V^2/c^2)}}\left\{[1 - \sqrt{1 - (V^2/c^2)}]\frac{-\mathbf{V}\cdot\mathbf{r}}{V} + Vt\right\},$$

$$t' = \frac{t - (\mathbf{V}\cdot\mathbf{r}/c^2)}{\sqrt{1 - (V^2/c^2)}}.$$

1.5.4 Derivation of time-dilatation and Lorentz contraction formulas from Lorentz transformation equations*

Before we proceed to consider the general form of the Lorentz transformation equations, let us look again at a comparison by the observers S and S' of their clocks and measuring rods. We must, of course, arrive at the time-dilatation and length-contraction formulas, since we derived the Lorentz transformations using them. However, the Lorentz transformations afford a clear view of the circumstances in which these effects are observed, and they can be derived without specifically including these effects initially (see Problems 1.70 and 1.71). First we investigate, from the point of view of S, the behavior of a clock that is at rest in the S' reference frame.

In a problem of this type, we must examine carefully what information is given and also what information is desired. We know that the clock is stationary relative to S' and hence is situated at all times at a point, say that labeled (x'_0, y'_0, z'_0), in the frame of reference of S'. We wish to compare the time interval that elapses according to S between the instants when the clock, moving past S, reads t'_1 and when it reads t'_2. When the clock under consideration reads t'_1, this is an event labeled in the S' reference frame by the space coordinates (x'_0, y'_0, z'_0) and the time coordinate t'_1 and in the S frame by the coordinates (x_1, y_1, z_1) and t_1. The event that occurs at the clock as it reads t'_2 has the coordinates (x'_0, y'_0, z'_0) and t'_2 in S' and $(x_2, y_2, z_2) = (x_2, y_1, z_1)$ and t_2 in S. *Note:* This event corresponds to a different S clock than that at (x_1, y_1, z_1); see Figure 1.54.

The time according to S at which these events take place can be determined in terms of the coordinates in S' of these events from the Lorentz transformation equation

$$t = \frac{t' + (V/c^2)x'}{\sqrt{1 - (V^2/c^2)}}. \tag{1.70}$$

The events occur, according to S, at times t_1 and t_2, given by

$$t_1 = \frac{t'_1 + (V/c^2)x'_0}{\sqrt{1 - (V^2/c^2)}}, \qquad t_2 = \frac{t'_2 + (V/c^2)x'_0}{\sqrt{1 - (V^2/c^2)}}. \tag{1.71}$$

Therefore, the time interval between these two events that S measures is equal to

---

* Kacser, Prentice-Hall, Sec. 3.5, p. 47, Sec. 3.6, p. 54.
Kittel, Knight, and Ruderman, McGraw-Hill, pp. 354–358.
Resnick, John Wiley, Sec. 2.3, p. 62; Sec. 3.5, p. 47.

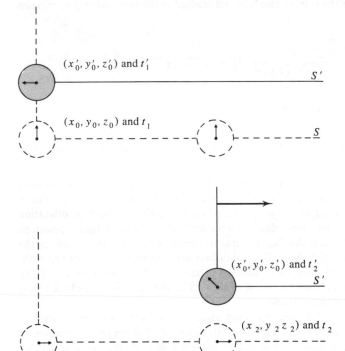

FIGURE 1.54   The measurement of the time interval $t_2 - t_1$ by S of the time required for the clock at (x_0', y_0', z_0') of S' to move its hands from t_1' to t_2'.

$$t_2 - t_1 = \frac{[t_2' + (V/c^2)x_0'] - [t_1' + (V/c^2)x_0']}{\sqrt{1 - (V^2/c^2)}}$$

$$= \frac{t_2' - t_1'}{\sqrt{1 - (V^2/c^2)}}.$$

$$(1.72)$$

This is the formula for time dilatation derived earlier through the use of the properties of a light-pulse clock.

Consider now the behavior of an S clock from the point of view of S'. The S clock is in motion with speed V relative to S', so the time interval $t_2' - t_1'$ measured by S' during which the reading on the S clock changes from t_1 to t_2 is given by

$$t_2' - t_1' = \frac{t_2 - t_1}{\sqrt{1 - (V^2/c^2)}}.$$

$$(1.73)$$

The Equations (1.72) and (1.73) appear to be inconsistent with each other, as does the resulting fact that each observer determines the other's clocks to be running slow. However, the equations are not inconsistent. Equation (1.72) relates the readings t_2' and t_1' on one clock of S' to the readings t_2 and t_1 on two clocks of S. The S clock that reads t_1 is coincident with the S' clock at the event at which it reads t_1', and the other S clock that reads t_2 is coincident with the same S' clock at the event at which it reads t_2'. Similarly, Equation (1.73) relates the readings t_2 and t_1 of one S clock with the readings of two different S' clocks. A comparison of these different circumstances is shown in Figure 1.55.

We now compare length measurements performed on the same rod by the two observers S and S'. The transformation equations for y and z are different than that for x, the coordinate in the direction of motion, so the result of a

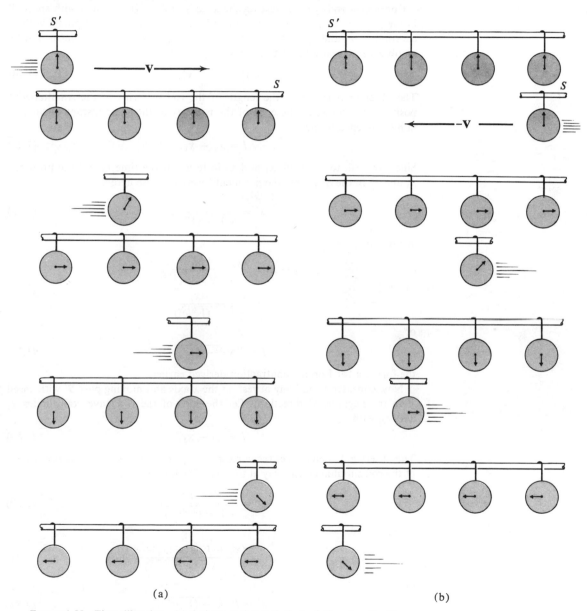

(a)   (b)

FIGURE 1.55   Time dilatation. (a) S observes that each clock of S', such as the S' clock shown, is running slow. (b) S' observes that each clock of S is running slow.

comparison of length measurements will depend on the angle of orientation of the rod with respect to the direction of **V**. The same result is obtained in a measurement of the length of a rod that is perpendicular to the direction of relative motion by observers S and S'. However, when the rod lies parallel to the direction of motion, the length of the rod as measured by an observer depends on the speed with which the rod moves relative to that observer.

Consider a rod lying at rest on the x' axis and moving past S with speed V in the positive x direction. The ends of the rod lie at the fixed points with x' coordinates x'_1 and x'_2 in the frame S', so the length of the rod, as measured by an observer at rest relative to the rod, is given by

$$l' = x'_2 - x'_1. \tag{1.74}$$

The observer S measures the length of the rod in terms of the simultaneous positions x_1 and x_2 of the ends of the rod, say at the same instant of time t_0, in his reference frame:

$$l = x_2 - x_1. \tag{1.75}$$

Since we wish to calculate x_1 and x_2 in terms of the time t_0 and the positions x'_1 and x'_2 in the S' frame, the pertinent formula to use here is

$$x' = \frac{x - Vt}{\sqrt{1 - (V^2/c^2)}}. \tag{1.76}$$

We obtain

$$x'_2 - x'_1 = \frac{(x_2 - Vt_0) - (x_1 - Vt_0)}{\sqrt{1 - (V^2/c^2)}}$$

$$= \frac{x_2 - x_1}{\sqrt{1 - (V^2/c^2)}} \tag{1.77}$$

or

$$l = l'\sqrt{1 - (V^2/c^2)}, \tag{1.78}$$

the formula for Lorentz contraction derived earlier.

Now consider a rod lying at rest on the x axis and moving past S' with speed V in the negative x' direction. Let the ends of the rod have coordinates x_1 and x_2, with

$$l = x_2 - x_1. \tag{1.79}$$

The observer S' determines the corresponding coordinates x'_1 and x'_2 of the ends of the rod simultaneously, say at time t'_0. Then, since

$$x = \frac{x' + Vt'}{\sqrt{1 - (V^2/c^2)}}, \tag{1.80}$$

we have

$$x_2 - x_1 = \frac{(x'_2 + Vt'_0) - (x'_1 + Vt'_0)}{\sqrt{1 - (V^2/c^2)}} \tag{1.81}$$

or

$$l' = l\sqrt{1 - (V^2/c^2)}. \tag{1.82}$$

The Equations (1.78) and (1.82) appear inconsistent, as does the resulting fact that each observer determines that the other's measuring sticks are shortened in the direction of motion. However, the equations are consistent, as the following analysis shows: The measurement of a moving rod by an observer involves two events, one in which the position of one end of the rod is marked, and the other in which the position of the other end is marked (Figure 1.56). An observer determines the length by marking the ends simultaneously relative to his own inertial system. However, these two events are not simultaneous according to the other observer. Therefore, the description of the two events, and in particular the differences in the coordinates $x_2 - x_1$ or $x_2' - x_1'$, is different for each of the observers. If S measures the length of a rod at rest relative to S', then S obtains a result that depends on his concept of simultaneity. According to S', S does not mark the ends of the rod simultaneously. Similarly, when S' measures the length of a rod at rest relative to S, the marks are not made simultaneously according to S.

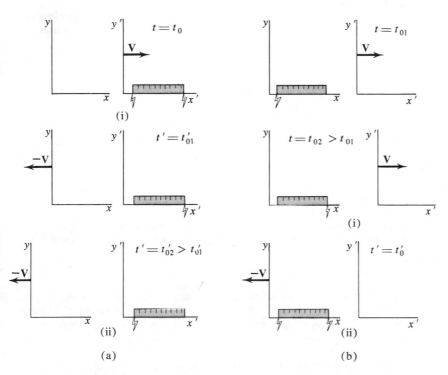

FIGURE 1.56 Each measurement of a length involves two events (designated by flashes) that are described differently by S and S'. (a) The measurement by S of the S' measuring rod. (i) Viewpoint of S of the measurement by S of the S' measuring rod. (ii) Viewpoint of S' of the measurement by S of the S' measuring rod. (b) The measurement by S' of the S measuring rod. (i) Viewpoint of S of the measurement by S' of the S measuring rod. (ii) Viewpoint of S' of the measurement by S' of the S measuring rod.

The circumstances are symmetrical with respect to the two observers: If the meter sticks lie along the x or x' directions, S determines the length of a meter stick of S' as being less than 1 m, and S' determines that the length of an S meter stick is less than 1 m. Furthermore, the Lorentz contraction factors are the same, namely, $\sqrt{1 - (V^2/c^2)}$, for both observers.

SUMMARY The formula for time dilatation can be derived from the Lorentz transformation equations by finding the times relative to an inertial observer of the two events at which a clock, moving relative to that observer, reads two different times. The formula for Lorentz contraction can be derived from the Lorentz transformation equations by finding the simultaneous positions relative to an inertial observer of the end points of a measuring rod moving relative to that observer. Furthermore, the Lorentz transformation equations show that there is no inconsistency in the fact that each of two inertial observers in relative motion observes the time dilatation and Lorentz contraction in the clocks and rods of the other.

Problem 1.58

A student is asked in an examination to calculate the time interval, as observed by S, in which a clock of S' ticks off the interval $t'_2 - t'_1$. The student uses the formula

$$t' = \frac{t - (V/c^2)x}{\sqrt{1 - (V^2/c^2)}}$$

instead of Equation (1.70) and obtains the result

$$(t'_2 - t'_1)\sqrt{1 - (V^2/c^2)} = t_2 - t_1$$

instead of Equation (1.72). Describe, as if you were correcting this student's paper, the error that he made and the physical situation to which his formulas are relevant.

Problem 1.59

An inertial observer measures the length of a meter stick, moving along the direction of its length, to be 57 cm. How fast is the meter stick moving?

Problem 1.60

The ends of a meter stick lie along the x' axis of the inertial frame of S' at the points $x'_1 = 2.00$ m and $x'_2 = 3.00$ m. Another inertial observer S measures the length of the meter stick at the time t_0 given by $ct_0 = 3.00$ m. The relations between the coordinates of S and S' are given by Equation (1.58) with $V = 4c/5$.

(a) Find the positions relative to S of the endpoints of the S' stick at the time of the measurement.
(b) Find ct'_1 and ct'_2, where t'_1 and t'_2 are the times relative to S' of the two events in the measurement by S.
(c) At time t'_0 given by $ct'_0 = 3.00$ m, S' measures the distance between the two points in S that are described in (a). Find the positions relative to S' of those points at the time t'_0.
(d) Find ct_1 and ct_2, where t_1 and t_2 are the times relative to S of the two events in the measurement by S' described in (c).

Problem 1.61

Observer S measures the length of an S' meter stick whose length lies along the direction of the relative motion of S and S'. The result is 92.8 cm.

(a) Observer S' measures the length of the meter stick of S. What result does he obtain?

(b) What time elapses, according to S', during which the hands of a clock of S move 1 hr ahead?

Problem 1.62

Observer S measures the length of an S' meter stick whose length lies along the direction of relative motion of S and S'. Using the procedure described above, S finds that the resulting marks on the S meter stick are 84.3 cm apart. Observer S' measures the distance between the marks.

(a) What result does S' obtain?

(b) Explain, with the aid of drawings of the events that occur, why S' does not obtain 1 m as the result.

Problem 1.63

Observer S uses a clock at O and compares the readings of this clock with the clocks of S' that are coincident with the S clock at various times t.

(a) How do these readings compare?

(b) Describe the corresponding circumstance for one clock of S'.

(c) Compare your answers to (a) and (b) with the formulas for the phenomena of time dilatation.

Problem 1.64

Observer S' measures the length of an S meter stick whose length lies along the direction of relative motion of S and S' and obtains 68.4 cm as a result. If S' marked the end of the rod simultaneously according to his clocks, what time interval occurred between these two events according to the clocks of S?

Problem 1.65

Observer S notices that 97 min elapse when the hands of an S' clock move ahead 1 hr.

(a) Find the speed of S' relative to S.

(b) How far, according to S, does the S' clock move in that interval?

1.5.5   The general form of the Lorentz transformation equations*

The special form of the Lorentz transformation Equations (1.58) applies only to the transformation between inertial systems whose origins coincide at $t = t' = 0$, whose axes are parallel, and whose relative velocity lies along the x or x' axis. These restrictions were introduced for convenience only, and we shall investigate now the general form of the Lorentz transformation equations.

---

* Feynman, Leighton, and Sands (vol. 1), Addison-Wesley, Sec. 15–7, p. 15–8; Sec. 17–1, p. 17–1; Sec. 17–2, p. 17–2.
Kacser, Prentice-Hall, Sec. 3.4, p. 45.
Kittel, Knight, and Ruderman, McGraw-Hill, pp. 346–348.
Resnick, John Wiley, Sec. 2.2, p. 56.
Taylor and Wheeler, W. H. Freeman, Sec. 8, p. 39.

Let the same event be labeled by the coordinates (x, y, z) and time t relative to S and the coordinates (x', y', z') and time t' relative to S'. We are concerned with the form of the transformation equations,

$$x' = x'(x, y, z, t), \qquad y' = y'(x, y, z, t),$$
$$z' = z'(x, y, z, t), \qquad t' = t'(x, y, z, t), \tag{1.83}$$

which is restricted by the kinematic postulate of special relativity and the facts that no one point in space can be distinguished from any other point: the principle of homogeneity of space; no one instant of time can be distinguished from any other instant: the principle of the homogeneity of time; and no one direction in space can be distinguished from any other direction: the principle of the isotropy of space.

The numbers x, y, and z are cartesian coordinates and designate lengths as determined by appropriate measuring rods. Similarly, the number t is a time interval as measured by an acceptable clock. Thus the transformation law (1.83) relates quantities that have the physical significance of lengths and time intervals, and upon which we can impose the conditions of the homogeneity and isotropy of space and the homogeneity of time. (These conditions would not be simple to impose if we considered transformations to spherical polar coordinates, for example.)

Assume, for the moment, that the functions, such as $x'(x, y, z, t)$, are not linear in x, y, z, and t. For example, x' might be quadratic in these variables as in the equation

$$x' = ax^2 + bt^2. \tag{1.84}$$

If this were the case, then the origin or some other point in S' could be distinguished, by an experiment, from other points, and we would be compelled to renounce the principle of homogeneity of space and time, an unwarranted step. An illustration of the effects of a nonlinear transformation is shown in Figure 1.57; in that case, the point O' is distinguishable from other points by the property that a length about O' relative to S' appears shorter relative to S than does the same length, relative to S', about any other point. This property of O' could be determined by an experiment and would distinguish O' from all other points, contrary to the principle of the homogeneity of space.

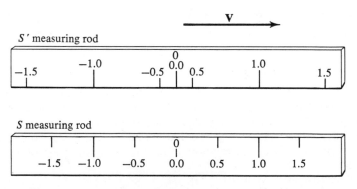

FIGURE 1.57   An example of a nonlinear relation between S and S' at time $t = 0$. If the relations between the coordinates of one event in S and the coordinates in S' of the same event were nonlinear as shown, one point in S' (the origin as shown in this figure) would appear different to S than other points.

We conclude that the transformation equations (1.83) are linear in x, y, z, and t, so we can write

$$x' = a_{x'x}x + a_{x'y}y + a_{x'z}z + a_{x't}t + x_0',$$
$$y' = a_{y'x}x + a_{y'y}y + a_{y'z}z + a_{y't}t + y_0',$$
$$z' = a_{z'x}x + a_{z'y}y + a_{z'z}z + a_{z't}t + z_0',$$
$$t' = a_{t'x}x + a_{t'y}y + a_{t'z}z + a_{t't}t + t_0', \tag{1.85}$$

where the a's, x_0', y_0', z_0', and t_0' are independent of x, y, z, and t, but they do depend on the relation between S and S'.

The event that, according to S, occurs at the origin O at time $t = 0$, takes place at the point with coordinates x_0', y_0', z_0' and at the time t_0' relative to S'. The origin O' of S' can be selected arbitrarily, so, for convenience, the co-ordinates x_0', y_0', z_0' often are chosen to be zero. Similarly, the time $t' = 0$ can be selected arbitrarily, and convenience often dictates the choice $t_0' = 0$.

We now consider the restriction that the kinematic postulate of special relativity places on the transformation equations (1.85). Consider a light signal emitted at the point $\mathbf{r}_1$ or (x_1, y_1, z_1) at time t_1 that reaches the point $\mathbf{r}_2$ or (x_2, y_2, z_2) at the time t_2 relative to S. Then,

$$|\mathbf{r}_2 - \mathbf{r}_1| = c|t_2 - t_1|. \tag{1.86}$$

Relative to the inertial observer S', the two events occur at the point $\mathbf{r}_1'$ or (x_1', y_1', z_1') at the time t_1' and at the point $\mathbf{r}_2'$ or (x_2', y_2', z_2') at the time t_2'. According to the kinematic postulate, the light signal travels with the speed c relative to both inertial systems; hence

$$|\mathbf{r}_2' - \mathbf{r}_1'| = c|t_2' - t_1'|. \tag{1.87}$$

The inverse relationship also is valid. If we define $\Delta \mathbf{r}$ and Δt by

$$\Delta \mathbf{r} = \mathbf{r}_2 - \mathbf{r}_1 \qquad \text{and} \qquad \Delta t = t_2 - t_1, \tag{1.88}$$

we have that

$$(\Delta s)^2 \equiv c^2(\Delta t)^2 - (\Delta \mathbf{r})^2 = 0 \tag{1.89}$$

if and only if

$$(\Delta s')^2 \equiv c^2(\Delta t')^2 - (\Delta \mathbf{r}')^2 = 0; \tag{1.90}$$

the symbols $(\Delta s)^2$ and $(\Delta s')^2$ are defined to be the quadratic expressions $c^2(\Delta t)^2 - (\Delta \mathbf{r})^2$ and $c^2(\Delta t')^2 - (\Delta \mathbf{r}')^2$, respectively.

The fact that the equations connecting the two sets of coordinates are linear allows us to determine a relation between

$$(\Delta s)^2 = c^2(\Delta t)^2 - (\Delta \mathbf{r})^2$$
$$= c^2(t_2 - t_1)^2 - (x_2 - x_1)^2 - (y_2 - y_1)^2 - (z_2 - z_1)^2 \tag{1.91}$$

and

$$(\Delta s')^2 = c^2(\Delta t')^2 - (\Delta \mathbf{r}')^2$$
$$= c^2(t_2' - t_1')^2 - (x_2' - x_1')^2 - (y_2' - y_1')^2 - (z_2' - z_1')^2 \tag{1.92}$$

for any two events and not just for those that can be connected by a light signal. The quantity $(\Delta s)^2$ is quadratic in $x_2 - x_1$, $y_2 - y_1$, $z_2 - z_1$, and $t_2 - t_1$ and, because of the linearity of the transformation equations, is thus quadratic in $x_2' - x_1'$, $y_2' - y_1'$, $z_2' - z_1'$, and $t_2' - t_1'$. Since $(\Delta s)^2$, a quadratic in $x_2' - x_1'$,

$y'_2 - y'_1$, $z'_2 - z'_1$, and $t'_2 - t'_1$, is zero if and only if $(\Delta s')^2$, also a quadratic in $x'_2 - x'_1$, $y'_2 - y'_1$, $z'_2 - z'_1$, and $t'_2 - t'_1$, is zero, we must have

$$(\Delta s)^2 = K(\mathbf{V})(\Delta s')^2, \tag{1.93}$$

where $K(\mathbf{V})$ is some function of the velocity $\mathbf{V}$ of S' relative to S. Since space is isotropic, $K(\mathbf{V})$ can depend only on the magnitude V of $\mathbf{V}$ and not on its direction:

$$K(\mathbf{V}) = \kappa(V). \tag{1.94}$$

A measurement of a length along a line perpendicular to the direction of relative motion gives the same results for both observers. The two events, the marking of each end, occur simultaneously for both observers according to our earlier arguments, and hence, for these two events, $\Delta t = \Delta t' = 0$. Since the two relative lengths are equal, $(\Delta \mathbf{r})^2 = (\Delta \mathbf{r}')^2$, and thus *for these two special events*

$$\kappa(V) = 1. \tag{1.95}$$

Moreover, since $\kappa(V)$ depends not on the events but only on the relative speed V of the two observers, this relation is valid in all cases. We conclude that

$$
\begin{aligned}
c^2(t_2 - t_1)^2 &- (x_2 - x_1)^2 - (y_2 - y_1)^2 - (z_2 - z_1)^2 \\
&= c^2(t'_2 - t'_1)^2 - (x'_2 - x'_1)^2 - (y'_2 - y'_1)^2 - (z'_2 - z'_1)^2.
\end{aligned}
\tag{1.96}
$$

This is the restriction on the transformation equations (1.85) that is imposed by the kinematic postulate of special relativity. *Any transformation of the form* (1.85) *that satisfies the restriction* (1.96) *is called a* Lorentz transformation.

The special Lorentz transformation (1.58) is derived from the restriction (1.96) in Problems (1.70) and (1.71).

SUMMARY   The coordinates of two events, (x_1, y_1, z_1) and t_1 and (x_2, y_2, z_2) and t_2, relative to one inertial reference system are related to the coordinates of the same two events, (x'_1, y'_1, z'_1) and t'_1 and (x'_2, y'_2, z'_2) and t'_2, relative to another inertial reference system by the condition

$$
\begin{aligned}
c^2(t_2 - t_1)^2 &- (x_2 - x_1)^2 - (y_2 - y_1)^2 - (z_2 - z_1)^2 \\
&= c^2(t'_2 - t'_1)^2 - (x'_2 - x'_1)^2 - (y'_2 - y'_1)^2 - (z'_2 - z'_1)^2
\end{aligned}
$$

according to the kinematic postulate of relativity. Any linear transformation between the coordinates of the two systems that satisfies this condition is called a Lorentz transformation.

Example 1.4

Q. Show that the transformation

$$x' = \frac{x - Vt}{\sqrt{1 - (V^2/c^2)}}, \qquad y' = y, \qquad z' = z, \qquad t' = \frac{t - (V/c^2)x}{\sqrt{1 - (V^2/c^2)}}$$

$$\tag{1.97}$$

is a Lorentz transformation.

A. Consider two events E_1 and E_2. The first, E_1, occurs at the point (x_1, y_1, z_1) at time t_1 relative to S and at the point

$$(x'_1, y'_1, z'_1) = \left(\frac{x_1 - Vt_1}{\sqrt{1 - (V^2/c^2)}}, y_1, z_1 \right) \tag{1.98}$$

at the time

$$t_1' = \frac{t_1 - (V/c^2)x_1}{\sqrt{1 - (V^2/c^2)}} \qquad (1.99)$$

relative to S'. Similar relations hold for the coordinates and time that describe E_2. Therefore,

$$c^2(t_2' - t_1')^2 - (x_2' - x_1')^2 - (y_2' - y_1')^2 - (z_2' - z_1')^2$$
$$= c^2\left(\frac{t_2 - (V/c^2)x_2 - t_1 + (V/c^2)x_1}{\sqrt{1 - (V^2/c^2)}}\right)^2 - \left(\frac{x_2 - Vt_2 - x_1 + Vt_1}{\sqrt{1 - (V^2/c^2)}}\right)^2$$
$$- (y_2 - y_1)^2 - (z_2 - z_1)^2$$
$$= c^2(t_2 - t_1)^2 - (x_2 - x_1)^2 - (y_2 - y_1)^2 - (z_2 - z_1)^2. \qquad (1.100)$$

Hence, the transformation (1.97) is a Lorentz transformation.

Example 1.5

Q. (a) Show that the transformation

$$t' = t, \quad x' = x\cos\theta + y\sin\theta, \quad y' = -x\sin\theta + y\cos\theta, \quad z' = z \qquad (1.101)$$

is a Lorentz transformation, and describe the relation between the S and the S' systems.

(b) Equation (1.101) is a linear transformation similar to the transformation

$$t' = \frac{t - (V/c^2)x}{\sqrt{1 - (V^2/c^2)}}, \quad x' = \frac{x - Vt}{\sqrt{1 - (V^2/c^2)}}, \qquad (1.102)$$
$$y' = y, \quad z' = z.$$

Discuss this similarity in more detail.

A. (a) That Equation (1.101) is a Lorentz transformation can be shown by direct substitution, as in the preceding example. Alternatively, we can show this by considering the relation between the S and S' systems. Since $t' = t$, we need consider only the spatial coordinates. It can be seen from Figure 1.58 that

$$x = x'\cos\theta - y'\sin\theta, \quad y = x'\sin\theta + y'\cos\theta \qquad (1.103)$$

or

$$x' = x\cos\theta + y\sin\theta, \quad y' = -x\sin\theta + y\cos\theta \qquad (1.104)$$

describe the relationship between the coordinates of a point P relative to two coordinate systems S and S' for which the S' axes are obtained by rotating the S axes about the z axis through the angle θ. Since distances between points do not change under a rotation of the axes,

$$(\Delta \mathbf{r})^2 = (\Delta \mathbf{r}')^2, \qquad (1.105)$$

and thus

$$c^2(\Delta t)^2 - (\Delta \mathbf{r})^2 = c^2(\Delta t)^2 - (\Delta \mathbf{r}')^2. \qquad (1.106)$$

Hence, the transformation (1.101) is a Lorentz transformation.

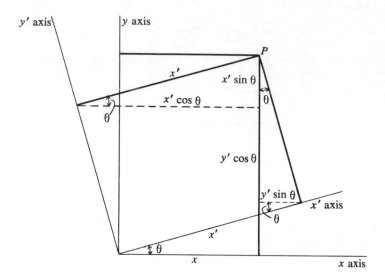

FIGURE 1.58 Relations between the two sets of coordinates of a point P, which are (x, y) in one coordinate system and (x', y') in another, obtained by a rotation through the angle θ.

(b) Consider Equation (1.102). Since the y and z coordinates do not change, we restrict our considerations to the t and x coordinates. Equation (1.102) describes a Lorentz transformation with

$$c^2(\Delta t)^2 - (\Delta x)^2 = c^2(\Delta t')^2 - (\Delta x')^2. \tag{1.107}$$

We can reduce this restriction to a familiar form by replacing t with λ/ic,

$$\lambda = ict \quad \text{and} \quad \lambda' = ict', \tag{1.108}$$

since, under this substitution, Equation (1.107) becomes

$$(\Delta\lambda)^2 + (\Delta x)^2 = (\Delta\lambda')^2 + (\Delta x')^2. \tag{1.109}$$

Since the sum of the squares of two numbers is equal to the square of the hypotenuse of a right-angle triangle with opposite sides having the length of those numbers, as shown in Figure 1.59, this relation is a reflection of the invariance of the distance between two points, specified by the rectangular coordinates (x_1, λ_1) and (x_2, λ_2) in one coordinate system, under a rotation of the coordinate axes through some angle α to another coordinate system. The transformation from one coordinate system to another is given by

$$x' = x \cos \alpha + \lambda \sin \alpha, \qquad \lambda' = \lambda \cos \alpha - x \sin \alpha. \tag{1.110}$$

If we substitute from Equation (1.108) for the λ's and compare the result with Equation (1.102), we find that

$$\tan \alpha = i\frac{V}{c}, \tag{1.111}$$

so that α, the "angle of rotation," is imaginary. Thus, the special Lorentz transformation (1.102) is sometimes said to be an imaginary rotation in the xt plane.

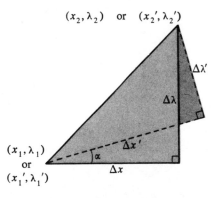

FIGURE 1.59 A circumstance for which $(\Delta x)^2 + (\Delta\lambda)^2 = (\Delta x')^2 + (\Delta\lambda')^2$.

Problem 1.66

The hyperbolic functions $\sinh \vartheta$ and $\cosh \vartheta$ are defined by the relations

$$\sinh \vartheta = \frac{e^{\vartheta} - e^{-\vartheta}}{2}, \qquad \cosh \vartheta = \frac{e^{\vartheta} + e^{-\vartheta}}{2},$$

which are similar to the relations

$$\sin \theta = \frac{e^{i\theta} - e^{-i\theta}}{2i}, \qquad \cos \theta = \frac{e^{i\theta} + e^{-i\theta}}{2}.$$

(a) Show that the special Lorentz transformation (1.102) can be written as

$$ct' = ct \cosh \vartheta + x \sinh \vartheta, \qquad x' = x \cosh \vartheta + ct \sinh \vartheta,$$

where

$$\tanh \vartheta = \frac{\sinh \vartheta}{\cosh \vartheta} = -\frac{V}{c}.$$

(b) Show that $\vartheta = i\alpha$, where α is the imaginary angle of rotation of the example above.
(c) Show that $\cosh^2 \vartheta - \sinh^2 \vartheta = 1$, and use this result to show that

$$1 - \frac{V^2}{c^2} = \frac{1}{\cosh^2 \vartheta}.$$

(d) Show, from the fact that $\tanh \vartheta = -V/c$, that

$$\cosh \vartheta = \frac{1}{\sqrt{1 - (V^2/c^2)}} \qquad \text{and} \qquad \sinh \vartheta = \frac{-V/c}{\sqrt{1 - (V^2/c^2)}}.$$

Problem 1.67

Show that, because the transformation equations (1.85) are linear, a motion with constant velocity relative to one inertial system appears as a motion with constant velocity relative to any other inertial system.

Problem 1.68

(a) Show that the transformation given by $x = x' + X$; $y = y' + Y$; $z = z' + Z$; $t = t' + T$, where X, Y, Z, and T are constants, is a Lorentz transformation.
(b) Describe the relation between the S and the S' coordinate systems.

Problem 1.69 *

(a) Show that each of the transformations given below is a Lorentz transformation:

(i) $x = -x'$, $\quad y = y'$, $\quad z = z'$, $\quad t = t'$,
(ii) $x = -x'$, $\quad y = -y'$, $\quad z = -z'$, $\quad t = t'$,
(iii) $x = x'$, $\quad y = y'$, $\quad z = z'$, $\quad t = -t'$,
(iv) $x = -x'$, $\quad y = -y'$, $\quad z = -z'$, $\quad t = -t'$.

(b) Describe the relation between the S and the S' coordinate systems connected by each of the transformations of (a).

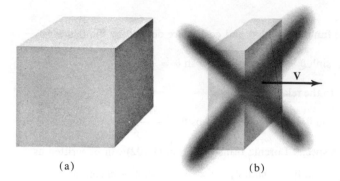

FIGURE 1.60 Prior to 1959, it was believed that the cube (a) would appear as in (b) when it moves past with velocity **V**. ($V = 0.9c$ for the case shown.) This belief is incorrect.

(a)   (b)

Problem 1.70

Derive the special Lorentz transformation (1.58) in the following way: Assume that x and t are related to x' and t' by the equation (see Equation 1.96)

$$c^2 t^2 - x^2 = c^2 t'^2 - x'^2$$

and that the transformation is linear:

$$x' = ax + bt,$$
$$t' = \alpha x + \beta t.$$

(a) Use the fact that the origin of S', $x' = 0$, corresponds to $x = Vt$ to obtain

$$x' = a(x - Vt),$$
$$t' = \alpha x + \beta t.$$

(b) Insert the transformation equations of (a) into the equation expressing the equality of the speed of light for all observers and solve for a, α, and β by equating the coefficients of the x^2, xt, and t^2 terms to each other.

Problem 1.71

Derive the special Lorentz transformation (1.58) as follows: Assume that x and t are related to x' and t' by the equation

$$c^2 t^2 - x^2 = c^2 t'^2 - x'^2.$$

This can be rewritten in the form

$$c^2 t'^2 + x^2 = c^2 t^2 + x'^2.*$$

(a) Show that, for some angle α,

$$x = x' \cos \alpha - ct \sin \alpha, \qquad ct' = ct \cos \alpha + x' \sin \alpha.$$

(b) Use the fact that the origin of S', $x' = 0$, corresponds to $x = Vt$ to obtain

$$\sin \alpha = -V/c \qquad \text{and} \qquad \cos \alpha = \sqrt{1 - (V^2/c^2)}.$$

---

* The use of this result in describing the kinematics of special relativity is given in detail in R. W. Brehme, "A Geometric Representation of Galilean and Lorentz Transformations," *American Journal of Physics*, *30*: 489 (1962). See also F. W. Sears, "Some Applications of the Brehme Diagram," *American Journal of Physics*, *31*: 269 (1963). This technique has also been used as the basis for a text by A. Shadowitz, *Special Relativity*, W. B. Saunders, Philadelpha, 1968.

(c) Show from (a) and (b) that

$$x' = \frac{x - Vt}{\sqrt{1 - (V^2/c^2)}}, \qquad t' = \frac{t - (V/c^2)x}{\sqrt{1 - (V^2/c^2)}}.$$

1.6   The Visual Appearance of Moving Objects

Until 1959, it was believed that the Lorentz contraction was the sole effect that needed to be considered when describing how you actually would see an object moving past you at a high speed. According to this belief, since an object, moving past with speed V, is shortened by the factor $\sqrt{1 - (V^2/c^2)}$ in the direction of motion, a viewer would see the object distorted by this compression. Figure 1.60 shows how it was believed that a cube would appear to a viewer watching the cube move past at a high speed.

This belief was maintained until 1959, when it was pointed out that the visual appearance is not determined by the simultaneous positions of all points on the object. Rather, the visual appearance is determined by the light from all points on the object that arrives simultaneously at the eye.

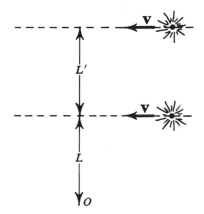

FIGURE 1.61   Instantaneous positions of two light sources moving past the viewer O.

1.6.1   *The difference between the instantaneous location of an object and its visual appearance*

The American physicist N. James Terrell (1923–) showed in 1959 that the Lorentz contraction is not the only effect that determines the visual appearance of a rapidly moving object.* The other fact that must be taken into account is that the eye records all the light that is received simultaneously, even though, if the object is extended, this light was emitted at different times from parts of the object at different distances from the observer. For example, suppose a moving object consists of two point sources of light, separated by the distance L', one traveling with speed V along a line at a distance L from a single viewer O and the other traveling with the same speed along a parallel line at a distance of $L + L'$ from O (Figure 1.61). The light from the closer source travels a shorter distance to O than the light from the distant source. Therefore, at any instant t, O sees the closer source as being ahead of the distant source, since the light arrives at the eyes of O simultaneously from a point P at which the close source occupied a time PO/c before t and from the point P', further back, at which the distant source occupied a time $P'O/c$ before t (Figure 1.62).

We can determine the visual appearance of a rapidly moving object by combining the effects of the Lorentz contraction in the direction of motion and the fact that the eye records the light that is received simultaneously even though this light was emitted at different times from different parts of the object. For example, consider a cube moving past O at a very high speed. We shall assume for simplicity that the cube is such a long distance away that the light from it that reaches O is essentially a parallel beam (Figure 1.63). Because of Lorentz contraction, the side of the cube nearest O and parallel to the direction of $\mathbf{V}$ is shortened by a factor of $\sqrt{1 - (V^2/c^2)}$. Also, as Terrell pointed out in

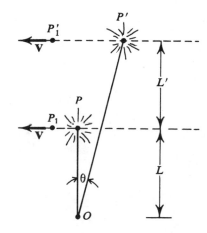

FIGURE 1.62 O sees the two sources as being at P and P' when the two sources are at P_1 and P_1'. ($V = 0.5c$ for the case illustrated.)

FIGURE 1.63   Cube moving past a viewer O.

---

* V. F. Weisskopf discusses Terrell's paper [5] in "The Visual Appearance of Rapidly Moving Objects," *Physics Today*, *13*: 24 September 1960, reproduced in *Special Relativity Theory, Selected Reprints*, American Institute of Physics, New York, 1963.

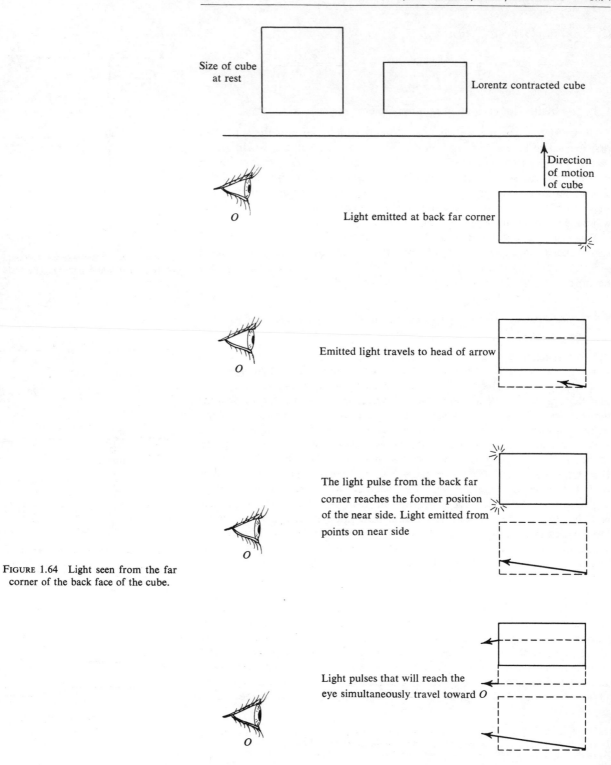

Size of cube
at rest

Lorentz contracted cube

Direction
of motion
of cube

Light emitted at back far corner

Emitted light travels to head of arrow

The light pulse from the back far
corner reaches the former position
of the near side. Light emitted from
points on near side

FIGURE 1.64   Light seen from the far
corner of the back face of the cube.

Light pulses that will reach the
eye simultaneously travel toward O

1959, O may see light from the back face of the cube, even if this back face is instantaneously in his line of sight. At first glance, it might not be expected that light from the far corner of the back face of the cube will reach O, because the whole back, all the way to that corner, is in his line of sight. However,

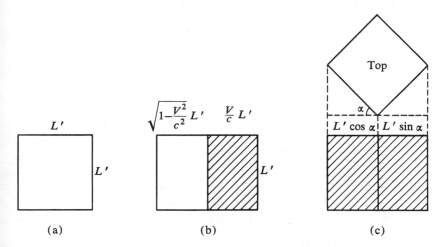

FIGURE 1.65   (a) Appearance of cube at rest. (b) Appearance of cube in motion. (c) Appearance of rotated cube at rest.

it must be kept in mind that the block is moving, so that as the light from that corner travels toward O, the block moves out of the way. We see from Figure 1.64 that O can see the far corner of the cube even though the block is in his line of sight at the instant the light was emitted.

Let us consider the appearance of a cube, with each side of length L', at the instant the cube appears to be directly opposite the viewer. Were the cube at rest, it would appear as in Figure 1.65(a). The motion of the cube results in the near side being shortened by the factor $\sqrt{1 - (V^2/c^2)}$ to the length $\sqrt{1 - (V^2/c^2)}\, L'$ and also in the viewer's seeing the far corner of the cube as if it were a distance $(V/c)L'$ behind the near rear corner (Figure 1.66). Thus the cube appears as shown in Figure 1.65(b). This appearance is identical to the appearance of the cube if it were at rest and rotated through the angle α given by $\sin \alpha = V/c$ [Figure 1.65(c)]. Hence, the cube does not appear distorted, but it does appear rotated. The cube does not appear to show any effects of the Lorentz transformation, although it is because of the Lorentz contraction of the near side that the appearance of the cube is not distorted.

SUMMARY   The instantaneous location of a rapidly moving extended object is determined by the simultaneous positions of all points on the object; it differs from that of a stationary object because of the effects of Lorentz contraction. The instantaneous visual appearance of a rapidly moving extended object is determined by the light that reaches the eye of the viewer simultaneously. This appearance differs from that of a stationary object because of the effects of the finite speed of light and the Lorentz contraction.

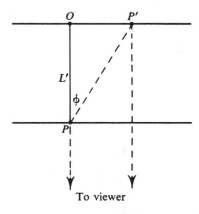

FIGURE 1.66 $OP' = L' \tan \phi = (V/c)L'$, since $\tan \phi = V/c$ (Problem 1.72).

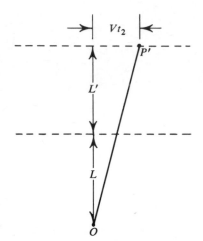

FIGURE 1.67
$OP' = \sqrt{V^2 t_2^2 + (L + L')^2}$.

Example 1.6

Q. Calculate the angle $\theta = \measuredangle POP'$ of Figure 1.62.

A. We choose the time $t = 0$ to be the instant at which the two sources are closest to O. Let t_1 be the time at which O sees the closer source at P. Then $t_1 = L/c$. Let $-t_2$, with $t_2 > 0$, be the time at which the distant source emits the light that O sees at time t_1. This light travels the distance $\sqrt{V^2 t_2^2 + (L + L')^2}$ in time $t_1 - (-t_2) = t_1 + t_2$ (Figure 1.67), so that

$$\sqrt{V^2 t_2^2 + (L + L')^2} = c(t_1 + t_2) = L + ct_2. \tag{1.112}$$

Hence,

$$V^2 t_2^2 + (L + L')^2 = L^2 + 2Lct_2 + c^2 t_2^2 \tag{1.113}$$

or

$$\left(1 - \frac{V^2}{c^2}\right) t_2^2 + \frac{2L}{c}\, t_2 - \frac{L'^2 + 2LL'}{c^2} = 0. \tag{1.114}$$

Therefore, the positive solution t_2 of this equation is

$$t_2 = \frac{-(2L/c) + \sqrt{(4L^2/c^2) + 4[1 - (V^2/c^2)][(L'^2 + 2LL')/c^2]}}{2[1 - (V^2/c^2)]}. \tag{1.115}$$

The angle θ is given by

$$\theta = \arctan \frac{Vt_2}{L + L'}$$

$$= \arctan \frac{(V/c)[\sqrt{L^2 + [1 - (V^2/c^2)](L'^2 + 2LL')} - L]}{[1 - (V^2/c^2)](L + L')}. \tag{1.116}$$

Problem 1.72

Consider the case for which $L \gg L'$ and L is sufficiently large that PO and PO' of Figure 1.62 are essentially parallel. Show that the angle ϕ (Figure 1.68) that PP' makes with the extension of OP is given by $\tan \phi = V/c$.

Problem 1.73

The distances L and L' of Figure 1.62 are equal. Draw a graph of θ versus V/c.

Problem 1.74

A cube with sides of 1.00 m travels past an observer with the speed $V = 4c/5$. The cube is sufficiently far from the observer that the light reaching the viewer's eye enters in a parallel beam. Find the time between the emissions at the far back corner and at the near side of light that reaches the viewer's eye simultaneously. How far does the cube travel in that time interval?

Problem 1.75

A box of length 1.62 m, height 0.54 m, and width 0.73 m moves with speed $0.7c$ in the direction of its length past an observer. The box is sufficiently far from the observer that the light reaching the viewer's eyes enters in a parallel

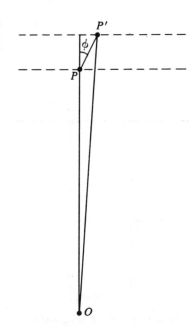

FIGURE 1.68 The angle ϕ through which the object appears to be rotated.

beam. Draw a scale diagram of the appearance of the box when it appears directly opposite the observer.

Problem 1.76

Determine the appearance of the box of Problem 1.75 when its rear corner is directly opposite an observer who is 1.38 m from the box at the closest point.

Problem 1.77

Describe the appearance of a "fast-moving" distant cube as "seen" by reflected sound waves. That is, determine the appearance, taking into account the finite speed of the radiation by which the cube is seen, but neglecting the effects of Lorentz contraction. Does the cube appear distorted?

1.7 Transformation Law for Velocities

Two inertial observers in relative motion do not agree on time-interval measurements nor on all length measurements, and so the relation between the velocities of an object that they measure is more complicated than that given by the galilean transformation law.

1.7.1 Derivation of the transformation law*

The Lorentz transformation law for velocities can be obtained from the Lorentz transformation equations for spatial coordinates and time. For simplicity, we shall consider the motion of an object along the x or x' axis (we can drop the arrow notation for vectors with the understanding that the labels, such as x, denote the components of the vectors in the positive x direction) that is moving with constant velocity v relative to S and with velocity v' relative to S'. The technique for obtaining the relationship between the velocities can be generalized in a straightforward manner to apply to velocities not along the direction of relative motion of the two inertial frames. This generalization is left as an exercise (Problem 1.78).

The average velocity of an object relative to a given reference frame is calculated from the coordinates of two events in the history of that object (Figure 1.69). For the case under consideration, we can omit consideration of the y and z coordinates. Let (x_1, t_1) and (x_2, t_2) be the coordinates relative to S of these two events. These are related to the coordinates relative to S', (x'_1, t'_1) and (x'_2, t'_2), respectively, of the same two events through the Lorentz transformation equations

$$x = \frac{x' + Vt'}{\sqrt{1 - (V^2/c^2)}}, \qquad t = \frac{t' + (V/c^2)x'}{\sqrt{1 - (V^2/c^2)}}. \qquad (1.117)$$

Hence, we obtain

---

* Feynman, Leighton, and Sands (vol. 1), Addison-Wesley, Sec. 16–3, p. 16–4.
Kacser, Prentice-Hall, Sec. 4.3, p. 69.
Kittel, Knight, and Ruderman, McGraw-Hill, pp. 350–353.
Resnick, John Wiley, Sec. 2.6, p. 79.
Taylor and Wheeler, W. H. Freeman, Sec. 9, p. 47.

$$v = \lim_{t_2 - t_1 \to 0} \frac{x_2 - x_1}{t_2 - t_1} = \lim_{t_2' - t_1' \to 0} \frac{(x_2' - x_1') + V(t_2' - t_1')}{(t_2' - t_1') + (V/c^2)(x_2' - x_1')}$$

(1.118)

$$= \lim_{t_2' - t_1' \to 0} \frac{[(x_2' - x_1')/(t_2' - t_1')] + V}{1 + (V/c^2)[(x_2' - x_1')/(t_2' - t_1')]}$$

since for the two events $t_2' - t_1' \to 0$ as $t_2 - t_1 \to 0$, or

$$v = \frac{v' + V}{1 + (Vv'/c^2)}.$$

(1.119)

The inverse transformation equation obtained if we solve (1.119) for v' is

$$v' = \frac{v - V}{1 - (Vv/c^2)}.$$

(1.120)

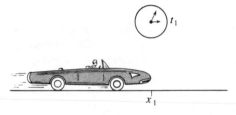

Event 1

FIGURE 1.69 The average velocity is calculated from the coordinates of two events in the history of an object.

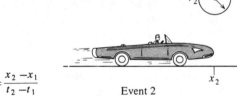

$$\bar{v} = \frac{x_2 - x_1}{t_2 - t_1}$$

Event 2

For speeds $|v|$ and $|V|$ much less than the speed of light c, the term Vv/c^2 in the denominator can be neglected, so

$$v' = v - V \qquad \text{or} \qquad v = v' + V \qquad \text{for } |v| \text{ and } |V| \ll c. \quad (1.121)$$

Thus, we obtain the familiar result of the galilean transformation law for the speeds with which we are familiar in everyday life. The deviations of the transformation law from expectations based on ordinary experience occur for speeds near that of light. In the limiting case in which an object is moving relative to S' with the velocity $v' = c$, the velocity relative to S is given by

$$v = \frac{c + V}{1 + (Vc/c^2)} = \frac{c + V}{1 + (V/c)} = c\frac{c + V}{c + V}$$

(1.122)

$$= c,$$

a result consistent with the hypothesis of the equality of the speed of light for all observers.

The case in which $|v'|$ is comparable to c can be studied if we rewrite Equation (1.119) in the form

$$v = v' + (c - v') \frac{Vc + Vv'}{c^2 + Vv'}$$

$$= v' + \alpha(c - v'), \qquad \alpha = \frac{Vc + Vv'}{c^2 + Vv'}. \tag{1.123}$$

We consider the case in which both v' and V are positive. Then, for $V < c$, the quantity designated by α is less than unity, and $\alpha(c - v')$ is not sufficient to increase v' up to the value c. Therefore, v is less than c for v' and V both positive and less than c.

SUMMARY   The transformation law for velocities can be calculated from the Lorentz transformation equations. The explicit form of this transformation law depends on the relation between the two pertinent inertial observers and the direction of the velocity relative to one of the reference systems.

Example 1.7

Q. (a) A pulse of light travels with the velocity $c\hat{y}'$ relative to an observer S'. If S' travels with the velocity $V\hat{x}$ relative to another observer S, determine the velocity of the pulse of light relative to S.

(b) Show explicitly that the speed of the light pulse relative to S is c.

A. (a) Let us assume that the light pulse starts from the origin of the S' reference frame at time $t' = 0$. Then, at time t', the light pulse is at the point with coordinates $x' = 0$, $y' = ct'$, $z' = 0$. The co-ordinates x, y, z of that point and the time t at which the light pulse is there, relative to S, are given by

$$t = \frac{t'}{\sqrt{1 - (V^2/c^2)}}, \qquad x = \frac{Vt'}{\sqrt{1 - (V^2/c^2)}}, \qquad y = ct', \qquad z = 0. \tag{1.124}$$

The velocity of the light pulse relative to S is given by the components

$$v_x = \frac{dx}{dt} = V,$$

$$v_y = \frac{dy}{dt} = \frac{dy}{dt'} \cdot \frac{dt'}{dt} = c\sqrt{1 - (V^2/c^2)} = \sqrt{c^2 - V^2},$$

$$v_z = \frac{dz}{dt} = 0. \tag{1.125}$$

(b) The speed v of the light pulse relative to S is given by

$$v^2 = v_x^2 + v_y^2 + v_z^2 = V^2 + (c^2 - V^2) = c^2; \tag{1.126}$$

thus, $v = c$.

Problem 1.78

The velocity of a particle is $\mathbf{v}$ relative to S and $\mathbf{v}'$ relative to S'. The velocity of S' relative to S is $\mathbf{V}$. Use the results of Problem 1.57 to determine $\mathbf{v}'$ in terms of $\mathbf{v}$ and $\mathbf{V}$:

$$\mathbf{v'} = \frac{\sqrt{1 - (V^2/c^2)}\,\mathbf{v} - (\mathbf{V}/V)\{[1 - \sqrt{1 - (V^2/c^2)}](-\mathbf{V}\cdot\mathbf{v}/V) + V\}}{1 - (\mathbf{V}\cdot\mathbf{v}/c^2)}. \qquad (1.127)$$

Problem 1.79

The velocity of S' relative to S is $\mathbf{V} = 0.99c\hat{x}$. An object moves with a speed of $0.99c$ in the negative x direction relative to S.

(a) What is the speed of separation of the origin of S' and the object as seen from S? Is your answer contradictory to the result deduced from Equation (1.123)? Explain.
(b) What is the speed of the object as seen by S'?

Problem 1.80

An observer S' moves with velocity $V\hat{x}$ relative to another observer S. Relative to a third observer S'', both S and S' move with the same speed V''; S moves with velocity $-V''\hat{x}''$ relative to S'' and S' with velocity $V''\hat{x}''$ relative to S''. Find V''.

Problem 1.81

Describe some science fiction effects that would occur if the speed of light were 60 mi/hr.

Problem 1.82

The velocity of an inertial observer S_2 is $V_{12}\hat{x}_1$ relative to another observer S_1. The velocity of a third inertial observer S_3 is $V_{23}\hat{x}_2$ relative to S_2.

(a) Find the velocity of S_3 relative to S_1.
(b) Let (x_1, t_1), (x_2, t_2), and (x_3, t_3) be the coordinates of an event relative to S_1, S_2, and S_3, respectively. Obtain the transformation equations relating (x_1, t_1) and (x_3, t_3) in the following two ways and show that they are equivalent:
  (i) First obtain the transformation equations relating (x_1, t_1) and (x_2, t_2) and those relating (x_2, t_2) and (x_3, t_3), and use these to obtain the equations relating (x_1, t_1) and (x_3, t_3).
  (ii) Use the result of (a) to obtain directly the equations relating (x_1, t_1) and (x_3, t_3).

Problem 1.83

Observer S' moves with velocity $V\hat{x}$ relative to S. Derive the transformation equation for the Lorentz contraction factor,

$$\sqrt{1 - (v'^2/c^2)} = \frac{\sqrt{1 - (v^2/c^2)}\,\sqrt{1 - (V^2/c^2)}}{1 - (v_x V/c^2)}, \qquad (1.128)$$

where

$$\mathbf{v} = v\hat{x}, \quad \mathbf{v'} = v'\hat{x}.$$

1.8   Events and Space-Time

The relativistic notions of space and time are formulated most clearly in terms of the concepts of events and of space-time. These can be introduced conveniently by analogy with the customary development of the corresponding ideas of position and space. After this introduction, however, it is necessary to use relativity theory to analyze the properties of space-time and the views of different inertial observers of space-time.

1.8.1 The terminology*

It is desirable to review briefly our ideas of points in space, vectors, coordinate systems, etc., before introducing the analogous but novel ideas of relativity theory.

Earlier in this chapter, we considered space as the arena within which objects undergo their motions. An object occupies a region of space at one instant, and that region of space exists even if there is no object there. The difficulties associated with finding simple rules for describing the regularities in the motions of real objects are bypassed in newtonian mechanics by the extrapolation to the concept of a point particle, an object of zero spatial dimensions; the totality of all possible positions of a point particle forms the continuum we call *space*.

Consider a number of inertial observers (who we shall assume are at rest relative to each other in order to avoid any difficulties with differences in their definitions of simultaneity). Although these observers may represent the position of a (point) particle or a point in space by different coordinates, nevertheless they all agree on the coincidence of the particle and that point in space. The position displacement† between the simultaneous positions of two particles is the same for all these observers; therefore, this position displacement can be represented by a vector that is defined independent of the reference frame of the observer. However, different observers may assign different components to the same (position displacement) vector. This is illustrated in Figure 1.70,

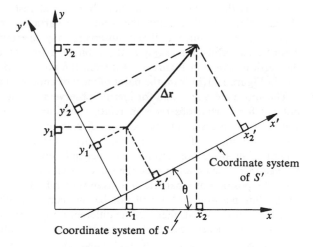

Coordinate system of S'

Coordinate system of S

FIGURE 1.70 The components of $\Delta\mathbf{r}$ are $(x_2 - x_1, y_2 - y_1)$ relative to the coordinate system of S and $(x_2' - x_1', y_2' - y_1')$ relative to the coordinate system of S'.

* Feynman, Leighton, and Sands (vol. 1), Addison-Wesley, Sec. 17–1, p. 17–1.
Kittel, Knight, and Ruderman, McGraw-Hill, pp. 366–368.
Taylor and Wheeler, W. H. Freeman, Sec. 1, p. 1; Sec. 6, p. 26.
† We introduce here the term "position displacement" in place of the equivalent and simpler term "displacement" that we have used up to now. This is done in order to reduce any confusion that may arise between the two types of displacements, the position displacement and the event displacement defined in this section.

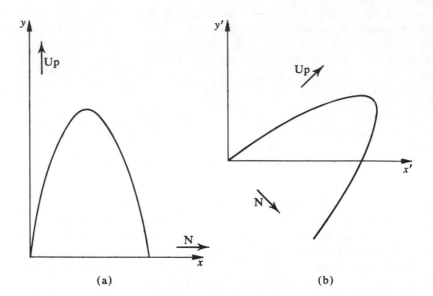

FIGURE 1.71   An orbit viewed by two different observers. (a) The orbit of a projectile near the surface of the earth as viewed by observer S. (b) The orbit of a projectile near the surface of the earth as viewed by another observer S'.

(a)   (b)

which shows the components of a vector $\Delta\mathbf{r}$ relative to the coordinate systems of two observers S and S'. The pairs of components $(\Delta x, \Delta y)$ and $(\Delta x', \Delta y')$ are related through the formulas [see Equations (1.103) and (1.104)]

$$\Delta x = \Delta x' \cos\theta - \Delta y' \sin\theta, \quad \Delta y = \Delta x' \sin\theta + \Delta y' \cos\theta. \quad (1.129)$$

The *orbit* of a particle is defined as the path in space of the particle's motion. The position vector of each point on the orbit, relative to a preassigned point in space, is the same for all inertial observers at rest relative to each other. Hence, relative to that point, the orbit of a particle is the same for all such observers, even though different observers may use different coordinates to describe the orbit—for example, the orbit of a projectile near the surface of the earth is shown in Figure 1.71(a) from the point of view of one observer S, and in Figure 1.71(b) from the point of view of another, namely, S'.

The above ideas are valid for inertial observers at rest relative to each other, but not for inertial observers in relative motion. For example, consider the view of two observers on the position-displacement vector between the simultaneous positions of two particles. Since whether or not two occurrences separated in space are simultaneous depends on the observer, that position-displacement vector will not be the same for all observers. If this vector represents the position of a particle at a given time relative to a preassigned point in space, this position vector will not be the same for all inertial observers in relative motion. What, then, do the different observers agree on? All observers see the same occurrences, each an incident such as a collision or an explosion that takes place in some region of space during some interval of time. All occurrences take place in a nonzero region of space and take some time to complete, but we can generalize from experience with occurrences that take place in a small region and are completed in a short time to the concept of a point occurrence or an event. A (*point*) *event* is an occurrence that takes place at a specific point in space and at a specific instant of time.

All inertial observers agree on the coincidence of events—that is, if two events take place at one point in space at the same time relative to one observer, they also occur simultaneously at some point in space relative to any other inertial observer. For example, if all the coordinates t_1 and (x_1, y_1, z_1) and t_2 and (x_2, y_2, z_2) of two events are equal according to S, then it follows that the coordinates of the two events in S', moving with velocity $V\hat{x}$ relative to S,

$$x_1' = \frac{x_1 - Vt_1}{\sqrt{1 - (V^2/c^2)}}, \quad y_1' = y_1, \quad z_1' = z_1, \quad t_1' = \frac{t_1 - (V/c^2)x_1}{\sqrt{1 - (V^2/c^2)}} \quad (1.130)$$

and

$$x_2' = \frac{x_2 - Vt_2}{\sqrt{1 - (V^2/c^2)}}, \quad y_2' = y_2, \quad z_2' = z_2, \quad t_2' = \frac{t_2 - (V/c^2)x_2}{\sqrt{1 - (V^2/c^2)}}, \quad (1.131)$$

are equal. Therefore, although the space and time coordinates of an event may differ for different inertial observers, each event is an entity that is independent of any particular set of coordinates that are used to describe it. In particular, the relation of one event to another, corresponding to a position displacement or the position vector of one point in space relative to another, is something that is the same for all observers even though different observers may assign different space and time components to describe the occurrence of one event relative to another. Figure 1.72 shows the components of the description of the event E_2 relative to the event E_1 with respect to the coordinate system of observer S. *These components describe the event displacement of E_2 relative to E_1.* The pairs of coordinates $(\Delta x, \Delta t)$ and $(\Delta x', \Delta t')$ of E_2 relative to E_1 with respect to S and S' are related through the formulas

$$\Delta x = \frac{\Delta x' + V\Delta t'}{\sqrt{1 - (V^2/c^2)}}, \quad \Delta t = \frac{\Delta t' + (V/c^2)\Delta x'}{\sqrt{1 - (V^2/c^2)}}. \quad (1.132)$$

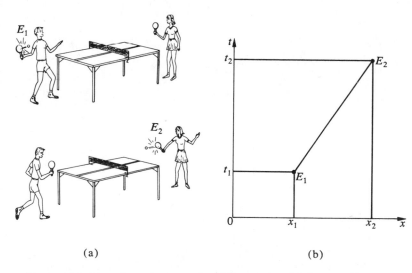

(a)   (b)

FIGURE 1.72 Two observers may assign different space and time coordinates to the two events, but the event displacement between the events, the relation of the event E_2 to E_1, is independent of the observers. (a) Events E_1 and E_2. (b) The components $(x_2 - x_1, t_2 - t_1)$ of the event displacement between E_1 and E_2.

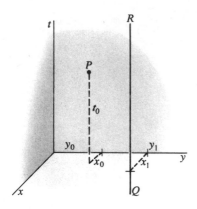

FIGURE 1.73 The event P occurs, relative to S, at the position (x_0, y_0, z_0) at the time t_0. Relative to S, QR is the point in space with coordinates (x_1, y_1, z_1).

The totality of all possible positions of a point particle forms what we call space. Since each point in space can be specified by three numbers, say the x, y, and z coordinates relative to one coordinate system, we say that space has three dimensions. Similarly, *the totality of all possible events forms space-time.* An event can be represented by an observer S as a point in space at an instant of time. Four numbers, the time and the three components of the spatial position vector, are needed to specify the event relative to the observer. Therefore, space-time has four dimensions.

It is impossible to construct in space four mutually perpendicular axes on which all coordinates of events can be plotted, just as it is impossible to construct three mutually perpendicular axes on a plane. However, on a plane, we can draw a perspective diagram of three mutually perpendicular axes; if we omit one spatial component, say the z component, we can draw the coordinate axes of space-time on a plane as shown in Figure 1.73. An event is represented by a point such as P on that graph. The point P, which represents a time and a position in space, is called a *world point* in space-time. A point in space is represented on such a graph by a line parallel to the t axis, such as the line QR shown. It must be kept in mind that QR is a fixed point in space relative to the observer S, but *not* a fixed point in space relative to all observers.

The history of a particle is a continuous sequence of events; it can be represented, relative to one observer S, as a curve in the four dimensions of space-time. This curve (Figure 1.74) is called the *world line* of the particle. The world line

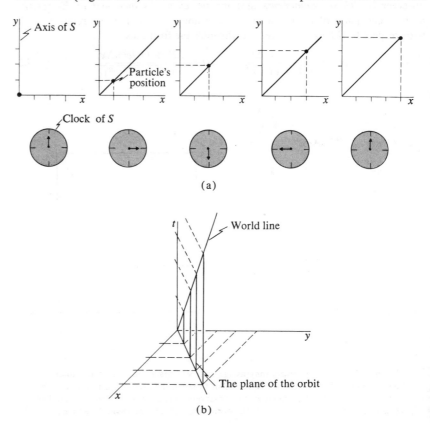

FIGURE 1.74 The world line of a particle moving with a constant velocity relative to an inertial observer S. (a) The position in space relative to S of a particle at various instants of time relative to S. (b) The world line of the particle of (a) as seen by an observer S. The orbit is the plane through the t axis and the projection of the world line onto the spatial region.

of a particle that moves with a variable velocity relative to S is curved (Figure 1.75). The world lines of all inertial observers relative to any one inertial observer are straight lines.

The world lines of light pulses play special roles in physics because of the equality of the speed of light for all inertial observers. A light pulse that starts at the origin of the reference frame of S at time $t = 0$ reaches the position (x, y, z) at time t where

$$t = \frac{\sqrt{x^2 + y^2 + z^2}}{c} = \frac{r}{c}. \qquad (1.133)$$

Because c is such a large speed compared to those that we perceive in everyday life, the world line of the light pulse can be approximated by a straight line with $t = r/\infty = 0$ for the description of familiar phenomena. This is valid also for the world line of any motion with a speed of the order of that of light. Therefore, to show the pertinent features of such fast motions on a space-time diagram, it is necessary to use relative scales for the time axis and the spatial axes that would appear preposterous for everyday use. The most convenient scale to use is that in which the time axis is marked off in the units of ct, that is, in units of distance. We introduce the variable τ, defined by

$$\tau = ct, \qquad (1.134)$$

which measures off time, for example, in meters or miles.

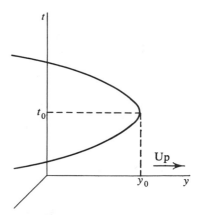

FIGURE 1.75 The world line, relative to an observer S, of a particle that rises to the height y_0 at time t_0 and is in free fall near the earth's surface.

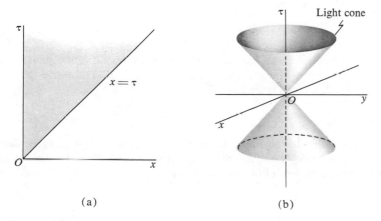

(a)   (b)

FIGURE 1.76   (a) The world line of a light pulse that goes through the event O, the origin in space at time $\tau = 0$, and moves in the positive x direction. (b) The world lines of all light pulses through O form the light or null cone relative to O, described by $\tau^2 - (x^2 + y^2 + z^2) = 0$.

In these units, the world line of a light pulse is a straight line that makes an angle of $45°$ with the τ axis (Figure 1.76). The world lines of all light pulses through an event O form the *light cone*, also called the null cone, relative to O. The world line of a particle traveling with the constant velocity $\mathbf{v}$ is a straight line that makes an angle θ, given by

$$\tan \theta = \frac{|\Delta \mathbf{r}|}{\Delta \tau} = \frac{v \, \Delta t}{c \, \Delta t} = \frac{v}{c} = \beta, \qquad (1.135)$$

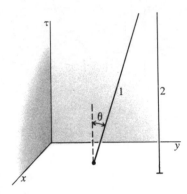

FIGURE 1.77 The world line of two particles traveling with constant velocity relative to S. The speed of particle 1 is of the same order as c, and that of particle 2 is of the order of the speeds perceived in everyday life.

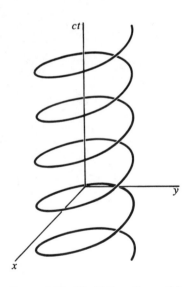

FIGURE 1.78 World line of a particle undergoing circular motion with constant speed.

with the τ axis. If v is negligible compared to c, then $\tan \theta \approx 0$ and $\theta = 0$; the world lines of the motions (with constant velocity) perceived in everyday life appear in this space-time diagram to be parallel to the τ axis (Figure 1.77).

SUMMARY The totality of all possible occurrences forms the four-dimensional continuum that we call space-time. The analogues, in space-time, of positions, position displacements, points, and orbits in space are, respectively, events, event displacements, world points, and world lines.

Example 1.8

Q. A particle undergoes circular motion in the xy plane with constant speed relative to an inertial observer S. Draw the world line of the particle on the space-time diagram of S.

A. Let us choose the spatial origin, $x = y = z = 0$, as the center of the motion so that the position vector of the particle is given by the components

$$x = R \cos \omega t, \qquad y = R \sin \omega t, \qquad z = 0. \qquad (1.136)$$

The speed of the particle is $R\omega$, where R is the radius of the circle of motion. We suppress the z axis in the space-time diagram. Since $x^2 + y^2 = R^2$, the particle travels on the surface of a right cylinder of radius R in space-time with its axis along the ct axis ($x = y = 0$). We set $ct = \tau$ so that

$$x = R \cos \frac{\omega}{c} \tau, \qquad y = R \sin \frac{\omega}{c} \tau, \qquad t = \frac{1}{c} \tau, \qquad (1.137)$$

and from this, we can see that the world points rotate about the ct axis on a circle whose plane moves up the axis. The world line is the helix shown in Figure 1.78.

Problem 1.84

Calculate the following times, in meters: Your age, the time since Julius Caesar, the time it takes light to pass an atom of radius $\sim 10^{-8}$ cm, the time it takes light to pass a nucleus of radius $\sim 10^{-12}$ cm.

Problem 1.85

(a) Calculate the parameter $\beta = v/c$ for the following speeds: The speed with which you entered this room, 60 mi/hr, the speed of an orbiting satellite (~ 5 mi/sec).

(b) Calculate the angle $\theta = \arctan \beta$ for the speeds of (a).

Problem 1.86

In an alternate formalism to that given above, distances are measured in units of time—for example, the distance L meters is given in seconds by $L/c = L/(2.998 \times 10^8)$ sec.

(a) What is the physical significance of L/c sec?

(b) Calculate the following distances in seconds: Your height, the distance around the earth, the distance to the sun.

Problem 1.87

Draw a space-time diagram showing world lines of two particles that undergo a collision and then separate.

Problem 1.88

The world line of particle 1 relative to an inertial observer S is described by the equations

$$x = \frac{3}{5} ct, \qquad y = z = 0.$$

The world line of particle 2 relative to an inertial observer S is described by the equations

$$x = -\frac{4}{5} ct, \qquad y = z = 0.$$

(a) Draw the world lines of the particles on the space-time diagram of S.
(b) What equations describe the world line of particle 1 relative to an inertial system fixed to particle 2?
(c) Draw the world lines of particle 1 and the point $x = y = z = 0$ in S on the space-time diagram of particle 2.

Problem 1.89

Draw the world line, on the space-time diagram of S, of a particle that starts from rest at time $t = 0$ and undergoes an acceleration along the positive x direction until it acquires a speed $v \approx c$.

Problem 1.90

A pulse of light is reflected back and forth between two mirrors at the end of a meter stick lying along the positive x' axis of an inertial observer S' (Figure 1.79). The light pulse is at the mirror at $x' = 0$ at $t' = 0$.

(a) Draw the world line of the light pulse on the space-time diagram of S'.
(b) Find the coordinates, relative to an inertial observer S, of the events at which the light pulse reverses direction; S' moves with velocity $3c\hat{x}/5$ relative to S.
(c) Draw the world line of the light pulse on the space-time diagram of S.

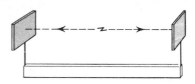

FIGURE 1.79   Reflected light pulse.

1.8.2   *The relation between the space-time diagrams of two inertial observers**

The relation between the space-time diagrams of two inertial observers, S and S', in relative motion with respect to each other, can be determined from

---

* Kacser, Prentice-Hall, Sec. A.1, p. 112.
Kittel, Knight, and Ruderman, McGraw-Hill, p. 370.
Resnick, John Wiley, Sec. A–1, p. 188.

the Lorentz transformation law. For simplicity, we consider the case in which S' is moving with speed V along the positive x direction and for which the event $x = y = z = \tau = 0$ (with $\tau = ct$) is described by S' as $x' = y' = z' = \tau' = 0$. In this case, we can omit consideration of the respective y and z axes and coordinates, since these are the same for both observers. The relevant parts of the space-time diagrams are the $x\tau$ and $x'\tau'$ planes, which coincide. The transformation law for the coordinates of an event,

$$x' = \frac{x - Vt}{\sqrt{1 - (V^2/c^2)}}, \qquad t' = \frac{t - (V/c^2)x}{\sqrt{1 - (V^2/c^2)}}, \qquad (1.138)$$

can be simplified through the introduction of the parameters

$$\beta = \frac{V}{c}, \qquad \gamma = \frac{1}{\sqrt{1 - (V^2/c^2)}}. \qquad (1.139)$$

These parameters are related by the equations

$$\gamma = (1 - \beta^2)^{-\frac{1}{2}}, \qquad \beta = \frac{\sqrt{\gamma^2 - 1}}{\gamma}. \qquad (1.140)$$

In terms of these parameters, the Lorentz transformation law (1.138) becomes

$$x' = \gamma(x - \beta\tau), \qquad \tau' = \gamma(\tau - \beta x). \qquad (1.141)$$

The τ' axis of S' is seen by S as the line $x' = 0$ or $x = \beta\tau$. This is the world line of the spatial origin of S'. The x' axis of S' is the line $\tau' = 0$ or $\tau = \beta x$. These two axes appear to S as lines that make equal angles, each given by $\theta = \arctan\beta$, with the τ and x axes, respectively (Figure 1.80). The relative scales on the axes can be determined from the Lorentz transformation equations or, as shown in Example 1.9, from the formulas for time dilatation and Lorentz contraction.

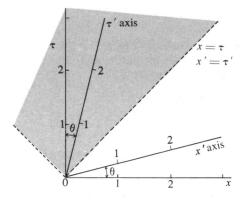

FIGURE 1.80   The S' axes on the space-time diagram of S. $\beta = \frac{1}{4}$ for the case shown.

The light cone and the coordinate lines of S' can also be drawn on the space-time diagram of S (Figure 1.81). The light cone is given by the equation $x = \pm\tau$ on the space-time diagram of S and by the equation $x' = \pm\tau'$ on the space-time diagram of S'. The coordinate line, $x' = K$, a constant, appears in the space-time diagram of S as the line

$$x - \beta\tau = \frac{1}{\gamma}K; \qquad (1.142)$$

this line is parallel to the τ' axis. The line $x' = $ a constant represents a point in space to S', since the spatial x' coordinate remains the same for all times τ'. This line does not represent a point in space to S but is the world line of a particle traveling with the constant speed V. The line $\tau' = $ a constant appears in the space-time diagram of S as a line parallel to the x' axis and does not represent to S a line of constant time τ. Example 1.9 illustrates one method for determining the scales and the coordinate lines of the space-time diagram of S' on that of S.

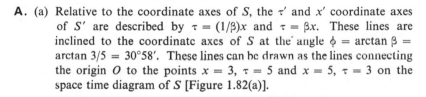

(a)

SUMMARY   The space-time diagram of one inertial observer appears to have oblique axes relative to another inertial observer in relative motion. The direction of the axes and the scales can be determined from the Lorentz transformation or its consequences.

Example 1.9

Q. An inertial observer S' travels with the velocity $3c\hat{x}/5$ relative to another inertial observer S.

(a) Draw the x' and τ' coordinate axes of S' on the space-time diagram of S.

(b) Use the formulas for time dilatation and Lorentz contraction to mark off the scales on these axes.

(c) Mark on this diagram the events for which $x' = 2$, $\tau' = 1$ and $x' = 1$, $\tau' = 3$.

A. (a) Relative to the coordinate axes of S, the τ' and x' coordinate axes of S' are described by $\tau = (1/\beta)x$ and $\tau = \beta x$. These lines are inclined to the coordinate axes of S at the angle $\phi = \arctan \beta = \arctan 3/5 = 30°58'$. These lines can be drawn as the lines connecting the origin O to the points $x = 3$, $\tau = 5$ and $x = 5$, $\tau = 3$ on the space time diagram of S [Figure 1.82(a)].

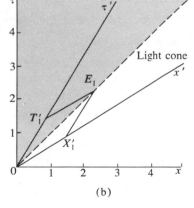

(b)

FIGURE 1.81   (a) Lines of constant x and lines of constant τ on the space-time diagram of S. (b) Lines of constant x' and lines of constant τ' on the space-time diagram of S.

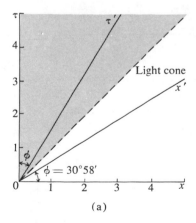

(a)

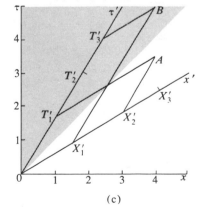

(c)

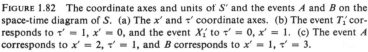

FIGURE 1.82   The coordinate axes and units of S' and the events A and B on the space-time diagram of S. (a) The x' and τ' coordinate axes. (b) The event T_1' corresponds to $\tau' = 1$, $x' = 0$, and the event X_1' to $\tau' = 0$, $x' = 1$. (c) The event A corresponds to $x' = 2$, $\tau' = 1$, and B corresponds to $x' = 1$, $\tau' = 3$.

(b) Let T_1' be the event on the τ' axis with the τ' coordinate unity. The τ coordinate of T_1' is given by the time-dilatation formula $\tau' = \tau\sqrt{1 - \beta^2}$ to be $\tau_1 = 1/\sqrt{1 - \beta^2} = \gamma$. Therefore, on the space-time diagram of S, $OT_1' = \tau_1/\cos\phi = \sqrt{(1 + \beta^2)/(1 - \beta^2)} = \sqrt{34}/4 = 1.458$. Let X_1' be the event on the x' axis with the x' coordinate unity. Then the following argument or the use of the formula for Lorentz contraction (Problem 1.92) shows that, on the space-time diagram of S, $OX_1' = OT_1'$. The point E_1 at the intersection of the line through T_1' parallel to OX_1' and the line through X_1' parallel to OT_1' lies on the light cone $x' = \tau'$. Because of this construction, the x' coordinate of X_1' equals that of E_1, and this in turn is equal to the τ' coordinate of E_1 and hence that of T_1'. Therefore, $OX_1' = OT_1'$ [Figure 1.82(b)].

(c) The events X_1', X_2', X_3', ... on the x' axis at distances from O of OT_1', $2 \times OT_1'$, $3 \times OT_1'$ are marked off. Similarly, the events T_1', T_2', T_3', ... on the y' axis at distances from O of OT_1', $2 \times OT_1'$, $3 \times OT_1'$ are shown. The event with coordinates $x' = 2$, $\tau' = 1$ is at the intersection of the line through X_2' parallel to OT_1' and the line through T_1' parallel to OX_1'. The event with coordinates $x' = 1$, $\tau' = 3$ is at the intersection of the line through X_1' parallel to OT_1' and the line through T_3' parallel to OX_1' [Figure 1.82(c)].

Problem 1.91

An inertial observer S' travels with the velocity $12c\hat{x}/13$ relative to another inertial observer S.

(a) Draw the x' and τ' coordinate axes of S' on the space-time diagram of S.
(b) Mark off the scales on these axes.
(c) Mark on this diagram the following events:

Event	E_1	E_2	E_3	E_4
x'	1.0	-0.5	0.5	-0.5
τ'	1.0	1.0	-1.0	-0.5

Problem 1.92

An inertial observer S' travels with the velocity $\beta c\hat{x}$ relative to another inertial observer S. Let X' be the event with coordinates $x' = 1$, $\tau' = 0$ and X be the event $x' = 1$, $\tau = 0$. *Note:* It is τ, not τ', that is zero at X.

(a) Show, from the Lorentz contraction formula, that $OX = \sqrt{1 - \beta^2}$ on the space-time diagram of S.
(b) Show that the acute angle that XX' makes with the x axis is $\pi/2 - \phi$ on the space-time diagram of S.
(c) Apply the trigonometric law of sines to obtain the length OX' on the space-time diagram of S.

Problem 1.93

The earth travels with speed 2.98×10^4 m/sec in its orbit around the sun. Calculate the angle between the τ' and τ axes for the case in which $\beta c = 2.98 \times 10^4$ m/sec.

Problem 1.94

An inertial observer S' travels with the velocity $-4c\hat{x}/5$ relative to another inertial observer S.

(a) Draw the x' and τ' coordinate axes of S' on the space-time diagram of S.
(b) Find the coordinates relative to S of the following events:

Event	E_1	E_2	E_3	E_4	E_5	E_6	E_7	E_8	E_9
x'	1	0	1	2	0	2	3	0	3
τ'	0	1	1	0	2	2	0	3	3

(c) Draw the scales on the axes and draw in the coordinate lines $x' = 1$, $x' = 2$, $x' = 3$ and $\tau' = 1$, $\tau' = 2$, $\tau' = 3$.

Problem 1.95

(a) Show that the quantity $\tau^2 - x^2$ has the same value relative to all inertial observers that are in relative motion in the x directions and that have origins coincident at time zero.
(b) Draw, on the space-time diagram of S, the curves $\tau^2 - x^2 = 1$ and $\tau^2 - x^2 = -1$. Explain why these curves are called calibration curves.

1.8.3   Space-time view of time dilatation and Lorentz contraction*

The space-time diagram of S' as seen by S (Figure 1.81) shows Lorentz contraction and time dilatation in the following way: Consider a rod of unit length, with its ends at $x' = 0$ and $x' = 1$, moving with S' past S. The world lines of the endpoints of the rod are shown on the space-time diagram of S in Figure 1.83. Three events, O, B, and C, in the history of the endpoints of the rod are marked; O and B are the positions of the endpoints at the time $\tau = 0$, and O and C are the positions of the endpoints at the time $\tau' = 0$. Therefore, according to S, the rod appears at OB at time $\tau = 0$, and thereafter moves parallel to OB between the lines OD and BC; according to S', the rod appears at OC at time $\tau' = 0$, and thereafter moves parallel to OC between the lines OD and BC. It can be seen from the figure that OB is less than unity, so the rod appears contracted to S. On the other hand, the event B corresponds to a value of τ' less than zero, and thus according to the clocks of S', the event B occurred before the event O. Also, since the time τ of the event C is greater than zero, the event C occurred after the events O and B, according to the clocks of S.

It can be seen from the space-time diagram (Figure 1.83) that, to both inertial observers S and S', the world lines of the ends of the rod are the straight lines OD and BC. However, the coordinates that one observer assigns one event on these world lines differ (except for the event O) from the coordinates assigned to that event by the other observer.

Time dilatation is illustrated in Figure 1.84. The τ' axis is the world line of a clock situated at the origin in the space of S'. The events T_1', T_2', and T_3' mark the successive tickings of that clock. These events have τ coordinates successively

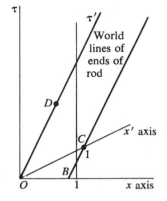

FIGURE 1.83   World lines of the ends of a rod of unit length moving with S'.

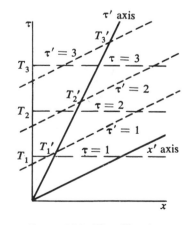

FIGURE 1.84   Time dilatation.

---

* Resnick, John Wiley, Sec. A–2, p. 193.

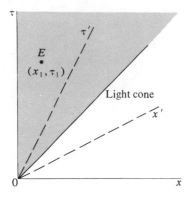

FIGURE 1.85 The event E occurs after $\tau = 0$ according to S and after. $\tau' = 0$ according to S'.

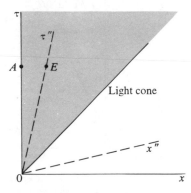

FIGURE 1.86 If $c \tan \measuredangle AOE = V''$, the event E occurs at $x'' = 0$ relative to an observer S''.

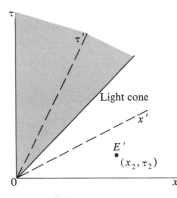

FIGURE 1.87 The event E' occurs at a point in space different from $x = 0$ according to S, and at a point in space different from $x' = 0$ according to S'.

greater than 1, 2, and 3, respectively, so, according to the clocks of S, that S' clock appears to be running slow. On the other hand, the events T_1, T_2, and T_3 mark the successive tickings of a clock at the origin in space of S, and these events have τ' coordinates successively greater than 1, 2, and 3, respectively. According to the clocks of S', that clock appears to be running slow.

SUMMARY Time dilatation and Lorentz contraction can be exhibited on a space-time diagram. These phenomena result from the obliquity of the axes of two inertial observers in relative motion and the relation between the scales on these axes.

Problem 1.96

(a) Draw the lines of constant x and constant τ on the space-time diagram of S' for the case in which S' moves with velocity $3c\hat{x}/5$ relative to S.

(b) Show the phenomena of Lorentz contraction and time dilatation on the diagram of (a). Show the Lorentz contraction of a rod at rest in S and of a rod at rest in S'.

Problem 1.97

(a) Draw the space-time diagram of S in terms of the coordinates x and $T = ct/3$ instead of x and τ. Show the x' and T' axes of S' if S' moves with the velocity $c\hat{x}/2$ relative to S.

(b) On your diagram of (a), show the Lorentz contraction of a rod of unit length moving with S'.

1.8.4 The partitioning of space-time by a light cone*

Let us now consider an event, say that marked E in Figure 1.85, that occurs within the light cone ($\tau_1/x_1 > 1$) at a time later than $\tau = 0$ according to S ($\tau_1 > 0$). This event also occurs at a time τ'_1 later than $\tau' = 0$ according to any observer S' moving with the velocity $V\hat{x}$, $|V| < c$, relative to S, since $\tau'_1 = \gamma\tau_1[1 - \beta(x_1/\tau_1)] > 0$. However, the point in space at which this event occurs depends on V. For example, for an observer S'' for which $V'' = c \tan \measuredangle AOE = cx_1/\tau_1$ (Figure 1.86), the event E occurs at the origin of the S'' space, $x''_1 = \gamma[x_1 - (x_1/\tau_1)\tau_1] = 0$. Thus, although the spatial coordinate of the event E relative to S', x', is less than zero, that relative to S, x, is greater than zero, and that relative to S'', x'', is equal to zero, the event E in each case occurs at a time greater than zero. Relative to the time $\tau = 0$, $\tau' = 0$ or $\tau'' = 0$, the event E occurs in the future. This is true for all events within that branch of the light cone, and hence that region of space-time can be called the *absolute future* relative to the event O. Similarly, all events within the other branch of the light cone occur in the *absolute past* with respect to O.

Now we consider an event, say that marked E' in Figure 1.87, that occurs outside the light cone ($|x_2/\tau_2| > 1$) at a point in space different from $x = 0$.

---

* Feynman, Leighton, and Sands (vol. 1), Addison-Wesley, Sec. 17–3, p. 17–4.
Kacser, Prentice-Hall, Sec. 4.5, p. 77.
Kittel, Knight, and Ruderman, McGraw-Hill, pp. 369–371.
Resnick, John Wiley, Sec. A–3, p. 196.
Taylor and Wheeler, W. H. Freeman, Sec. 7, p. 37.

Let S' be any other inertial observer moving with velocity $V\hat{x}$ relative to S and for which the event O has coordinates $x' = \tau' = 0$. It is still assumed that $|V| < c$. Since the world line $x' = 0$ lies within the light cone, the event E' occurs at a point x'_2 in space different from $x' = 0$ according to S': $x'_2 = \gamma x_2[1 - (\beta\tau_2/x_2)] \neq 0$. The time at which this event occurs, however, depends on the observer. For example, for an observer S''' for which $V''' = c \tan \measuredangle A'OE' = c\tau_2/x_2$ (Figure 1.88), the event occurs at time $\tau''' = \gamma[\tau_2 - (\tau_2/x_2)x_2] = 0$. Therefore, although the time at which the event E' occurs is before that of O according to S', later than that of O according to S, and simultaneous with O according to S''', the event E' occurred at a point in space different from O according to each observer. This is true for all events outside the light cone, and so the region of space-time outside the light cone can be called the *absolute elsewhere* relative to O. The division of space-time by the light cone into absolute past, absolute elsewhere, and absolute future is shown in Figure 1.89.

If units for the time and space axes are used that are appropriate for the description of the motions perceived in everyday life, the angular opening of the light cone becomes almost 180°; in the limiting case, the two branches of the

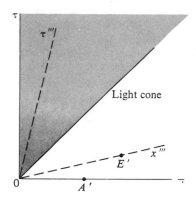

FIGURE 1.88 If $c \tan \measuredangle A'OE' = V'''$, the event E' occurs at the time $\tau''' = 0$.

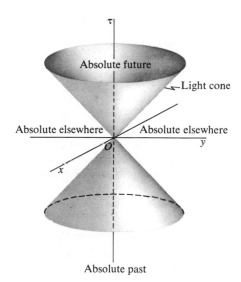

FIGURE 1.89 Division of space-time by the light cone.

cone degenerate into one plane (Figure 1.90) that separates the absolute future from the absolute past. All lines such as that corresponding to time $\tau' = 0$ lie squashed between the two degenerate cones or in the plane, and hence it appears that time is absolute, the same for every observer. This result is in agreement with the galilean transformation that is valid for all speeds $V \ll c$.

SUMMARY   Space-time is divided by the light cone with its vertex at an event O into three distinct regions: the absolute future relative to O, the absolute elsewhere relative to O, and the absolute past relative to O.

Example 1.10

Q. An event E occurs at the world point with coordinates $x = 5$, $\tau = 3$ relative to an inertial observer S (Figure 1.91).

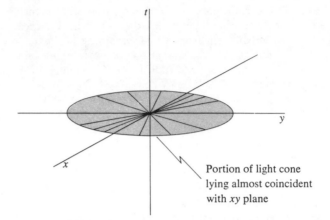

FIGURE 1.90 Division of space-time, with the axes marked in units appropriate for everyday experience, by the light cone.

Portion of light cone lying almost coincident with xy plane

(a) Find the speed of the inertial observer S' for which E and O occur at the same time.

(b) Find the spatial distance between E and O relative to S'.

A. (a) Let the event O occur at the time $t' = 0$ relative to S'. Then, since E and O occur at the same time, E occurs at the time $t' = 0$ and lies on the x' axis. The x' and τ' axes are symmetrically situated about the light cone, and since the event $x = 5$, $\tau = 3$ lies on the x' axis, the event E', $\tau = 5$, $x = 3$, lies on the τ' axis. Event E' represents one event in the life of the S' clock that is at O at time $t = 0$, so the speed V of that clock relative to S is given by $V = 3c/5$. Since the τ' axis lies in the region between the positive τ and x axes, the motion of S' is along the positive x direction.

(b) The position of E relative to S' is given by

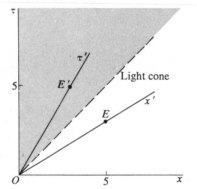

FIGURE 1.91 The event E and the coordinate axes of S'.

$$x' = \frac{x - \dot{V}t}{\sqrt{1 - (V^2/c^2)}} = \frac{x - (V/c)\tau}{\sqrt{1 - (V^2/c^2)}} = \frac{5 - (3/5)3}{4/5} = 4. \qquad (1.143)$$

Therefore, the spatial distance between E and O relative to S' is 4.

Problem 1.98

An event E occurs at the world point with coordinates $x = -13$, $\tau = 12$ relative to an inertial observer S.

(a) Find the speed of the inertial observer S' for which E and O occur at the same time.

(b) Find the spatial distance between E and O relative to S'.

(c) Draw the coordinate axes of S' on the space-time diagram of S and mark off the scales on these axes.

Problem 1.99

An event E occurs at the world point with coordinates $x = 4$, $\tau = 5$ relative to an inertial observer S.

(a) Find the speed of the inertial observer S' for which E and O occur at the same position in space.

(b) Find the time between the events E and O relative to S'.

Problem 1.100

An event E_1 can be said to be the cause of an event E_2, the effect, only if the event E_1 occurs (in time) before the event E_2. Suppose that the effect E_2 is triggered by a signal propagated from the cause E_1. Let v be the speed at which this signal is propagated. Show that $v \le c$. *Hint:* Consider the alternative, $v > c$. Show that, in some inertial reference systems, the event E_2 appears as the cause and not the effect.

Problem 1.101

Show that the world line of a particle, after an event E, must lie inside the light cone relative to E (Figure 1.92) if the particle's speed never exceeds c.*

Problem 1.102

(a) Two searchlights, a distance $2L$ apart on the ground, are directed toward each other. The two beams are swung at the angular rate of ω Hz upward in the vertical plane containing the two searchlights. Calculate the speed of the point of intersection of the two beams as a function of t, the time after they were horizontal. When is this speed equal to c? Is it ever greater than c? *Hint:* Let x equal the distance from the midpoint between the two searchlights to the point of intersection of the beams at time t. Show that $\tan \omega t' = x/L$, where $t' = t - \sqrt{x^2 + L^2}/c$. Then show that $v = dx/dt$ can be written as $c \cos \alpha(t')/\sin [\omega t' + \alpha(t')]$, where $\tan \alpha(t') = c \cos \omega t'/\omega L$. Note that $\omega t' \to \pi/2$ for $x \to \infty$ and that $v_{t'=0} = \omega L$.

(b) The two searchlights of (a) are directed upward at $t = 0$ and rotated toward each other in the vertical plane containing the two searchlights at the constant angular rate of ω Hz. Calculate the velocity of the point of intersection of the beams and discuss your answer.

(c) Is there any contradiction between the answer to (a) and that of Problem 1.100?†

1.8.5 Space-time intervals ‡

The region of space-time in which an event E lies relative to the event O can be determined from the coordinates in space, (x, y, z), and in time, τ, of the event E with respect to O of any observer S. If E lies on the light cone, then the quadratic form s^2, defined by

$$
\begin{aligned}
s^2 &= c^2 t^2 - (x^2 + y^2 + z^2) \\
&= \tau^2 - (x^2 + y^2 + z^2),
\end{aligned} \tag{1.144}
$$

is equal to zero. This quadratic expression has the same value when evaluated for the event E for all observers (Section 1.5),

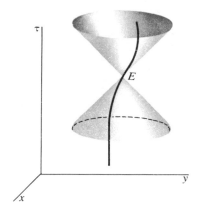

FIGURE 1.92 The world line through E lies inside the light cone with vertex at E.

---

* An experiment that demonstrates directly that the speed of electrons is limited by c, regardless of their energies, is shown in the film *The Ultimate Speed* by W. Bertozzi, produced by Educational Services Incorporated, Watertown, Mass. See also W. Bertozzi, "Speed and Kinetic Energy of Relativistic Electrons," *American Journal of Physics*, 32: 551 (1964).
† For other examples of speeds greater than c, see M. A. Rothman, "Things That Go Faster Than Light," *Scientific American*, 203: 142, July 1960.
‡ Feynman, Leighton, and Sands (vol. 1), Addison-Wesley, Sec. 16–2, p. 16–3; Sec. 17–2, p. 17–2. Kittel, Knight, and Ruderman, McGraw-Hill, p. 369. Taylor and Wheeler, W. H. Freeman, Sec. 6, p. 26.

$$\tau^2 - (x^2 + y^2 + z^2) = \tau'^2 - (x'^2 + y'^2 + z'^2); \qquad (1.145)$$

thus for all events on the light cone, s^2 is equal to zero for all observers.

Consider now an event E in the absolute future with respect to O. There exists an inertial reference system S'' (Figure 1.86) for which this event occurs at the same point in space as does O but later in time; in this reference frame, $x'' = y'' = z'' = 0$ and $\tau'' > 0$. Hence

$$s^2 = (\tau'')^2 > 0, \qquad \tau'' > 0. \qquad (1.146)$$

Since s^2 has the same value relative to any inertial observer, the coordinates of E relative to S satisfy the condition

$$\tau^2 - (x^2 + y^2 + z^2) > 0. \qquad (1.147)$$

Also, since E occurs in the absolute future, $\tau > 0$. Similarly, it can be shown that the coordinates relative to S of any event E'' in the absolute past also satisfy Equation (1.147) and the condition $\tau < 0$.

Now consider an event E' in the absolute elsewhere. There exists an inertial reference frame S''' (Figure 1.88) in which E' occurs at the same time as O; in that frame, $\tau''' = 0$ and $x'''^2 + y'''^2 + z'''^2 > 0$. Hence

$$s^2 = -(x'''^2 + y'''^2 + z'''^2) < 0. \qquad (1.148)$$

Since s^2 has the same value relative to any inertial frame, the coordinates of E' relative to S satisfy the condition

$$\tau^2 - (x^2 + y^2 + z^2) < 0. \qquad (1.149)$$

The coordinates (x, y, z) and τ specify the event displacement of the event E relative to the event O. An event displacement for which $s^2 > 0$ is called a *time-like* event displacement, one for which $s^2 = 0$ is called a *null* event displacement, and one for which $s^2 < 0$ is called a *space-like* event displacement* (Figure 1.93). The square root of the absolute value of s^2, $\sqrt{|s^2|}$, is called the *interval* between the two events O and E.†

If the event displacement between two events in space-time is space-like, there exists an inertial reference system in which the two events occur simultaneously. The distance between the two points in space relative to that reference frame is called the *proper length* separating those events. The proper length between the two space-like events is given by $\sqrt{-s^2}$.

If the event displacement between two events in space-time is time-like, there exists an inertial reference system in which the two events occur at the same position in space. One can imagine a clock fixed in this system, which therefore moves with constant velocity relative to S from the event O to the event E and which measures time, in units of length, as the interval $\sqrt{s^2}$ past O if $t > 0$ or as $-\sqrt{s^2}$ before O if $t < 0$. Consider now a particle that moves with constant velocity relative to S from the event O to the event E. The time τ_0 that would be read by a clock fixed to the particle,

$$\tau_0 = \sqrt{s^2} \qquad (t > 0) \qquad (1.150)$$

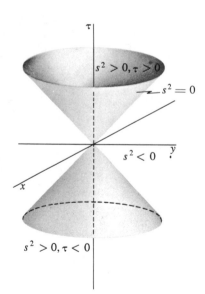

FIGURE 1.93   The values of $s^2 = \tau^2 - (x^2 + y^2 + z^2)$ in the regions of space-time.

* Some physicists define s^2 by the equation $s^2 = (x^2 + y^2 + z^2) - c^2 t^2$, whereupon their conditions for time-like and space-like event displacements are opposite to those given here.
† Some physicists call $\sqrt{|s^2|}$ the separation between the two events.

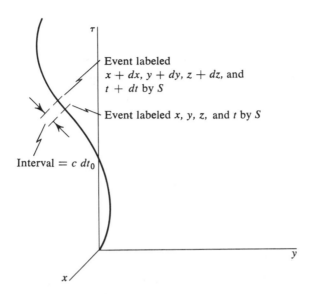

Event labeled
$x + dx$, $y + dy$, $z + dz$, and
$t + dt$ by S

Event labeled x, y, z, and t by S

Interval $= c\,dt_0$

FIGURE 1.94 The world line of a particle undergoing accelerated motion relative to S.

is called the *proper time* of the particle. This proper time interval is the smallest measure of the time between the two events O and E in the particle's life; the time between these two events relative to any inertial observer is equal to or greater than τ_0. In the usual units of time, the proper time of the particle when it coincides with the event with coordinates relative to S of (x, y, z) and t is given by

$$t_0 = \sqrt{t^2 - [(x^2 + y^2 + z^2)/c^2]} \qquad (t > 0). \qquad (1.151)$$

We can associate a proper time relative to O with a particle whose velocity is not constant with respect to S by introducing the small change in the proper time during each small segment of the motion in which the speed of the particle is appreciably constant with respect to S (Figure 1.94). If the particle's speed relative to S at the event labeled x, y, z and t is approximately constant during the motion to the event at $(x + dx, y + dy, z + dz)$, and $t + dt$, the change in the proper time dt_0 associated with the separation of those two events can be approximated by

$$\begin{aligned} dt_0 &= \sqrt{(dt)^2 - (1/c^2)[(dx)^2 + (dy)^2 + (dz)^2]} \\ &= \sqrt{1 - [v^2(t)/c^2]}\,dt, \end{aligned} \qquad (1.152)$$

where $\mathbf{v}(t)$, with components dx/dt, dy/dt, dz/dt, is the velocity of the particle with respect to S in its motion from the first event to the second (Figure 1.95). The total proper time throughout its motion from the event O to the event with time t according to the clocks of S is therefore the sum of the small changes dt_0 in the proper time throughout the motion; we write

$$t_0 = \int_0^t \sqrt{1 - [v^2(t)/c^2]}\,dt \qquad (1.153)$$

where, it must be noted, $v(t)$ is a function of t.

The proper time of any event on a particle's world line depends on the history of the particle; thus, there is, in general, no simple transformation equation

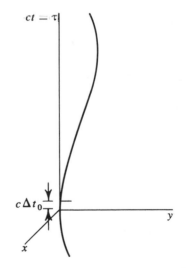

FIGURE 1.95 The proper time interval Δt_0 in the reference frame in which the particle is instantaneously at rest at $t_0 = 0$.

relating the proper time on a world line to the corresponding time as determined by an inertial observer. However, the world line of a particle is a curve in space-time, and *the proper time τ_0 along that world line is an appropriate parameter to use in describing the world line*. Thus, the world line can be represented relative to an observer S by the equations

$$x = x(\tau_0), \qquad y = y(\tau_0), \qquad z = z(\tau_0), \qquad \tau = \tau(\tau_0). \qquad (1.154)$$

SUMMARY   The quadratic form $s^2 = \tau^2 - (x^2 + y^2 + z^2)$ is positive, zero, or negative if the event displacement described by the coordinates (x, y, z), and τ is time-like, null, or space-like, respectively. If the event displacement is space-like, $\sqrt{-s^2}$ is the proper length, or the distance between the two events relative to that inertial system in which the events occur at the same time. The event displacements between two events along a particle's world line are time-like, and $\tau_0 = \sqrt{s^2}$ measures the proper time between the events, the time measured by an inertial clock that moves from one event to the other.

Example 1.11

Q. An object, at the event O at time $\tau = 0$, travels with the velocity $4c\hat{x}/5$ relative to S for a time of 5 m, and then abruptly reverses its velocity to $-4c\hat{x}/5$ and returns to the same point in the reference frame of S [Figure 1.96(a)]. Describe the proper time of the particle that elapses during this trip.

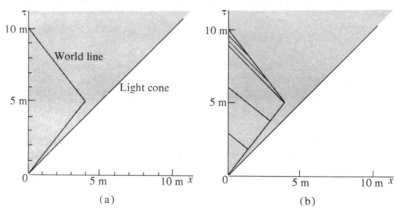

(a)   (b)

FIGURE 1.96   The twin paradox. (a) World line of a particle that, relative to S, travels for a time of 5 m with velocity $4c\hat{x}/5$, and then for the same length of time with velocity $-4c\hat{x}/5$. (b) The proper time of the particle during the motion of (a) is 6 m. At each meter of time, the particle emits a light signal that travels to S on one of the world lines shown.

A. The proper time interval dt_0 during the time interval dt as determined by S is

$$dt_0 = \sqrt{1 - (v^2/c^2)}\, dt, \qquad (1.155)$$

where $v = |4c\hat{x}/5| = 4c/5$ during the first 5 m of time and $v = |-4c\hat{x}/5| = 4c/5$ during the second 5 m of time. Therefore, throughout the motion

$$dt_0 = \sqrt{1 - (4/5)^2} \, dt = \frac{3}{5} \, dt; \qquad (1.156)$$

the total proper time T_0 is [Figure 1.96(b)]

$$T_0 = \int_{t=0}^{t=10\,\text{m}} dt_0 = \int_0^{10\,\text{m}} \frac{3}{5} \, dt = 6 \, \text{m}. \qquad (1.157)$$

Thus, the proper time is less than the time observed by an inertial observer.

This is the basis of the famous "twin paradox."[*] One twin goes on a space journey and returns to the other twin, who stayed at home. Time is measured in each case by the pulse rate and the aging process. When the traveler returns home, he has aged by the amount

$$T_0 = \int_0^T \sqrt{1 - [v^2(t)/c^2]} \, dt, \qquad (1.158)$$

while the twin that stayed home has aged by

$$T = \int_0^T dt > T_0. \qquad (1.159)$$

The paradox appears if we note that, to the traveler, the stay-at-home twin moves off and then returns. However, there is an observable difference in the motions; the traveler experiences accelerations that distinguish his motion from that of the other twin.

Problem 1.103

The earth travels with the speed 2.98×10^4 m/sec in its orbit around the sun. Assume that the sun is at rest in an inertial system. Calculate the age of a man who lives 70 years as determined by clocks in the sun's rest system.

Problem 1.104

The Pleiades nebula is 130 parsecs away. How fast would a space traveler have to move relative to the earth in order that he reach this nebula in 70 years? 1 parsec $= 3.09 \times 10^{16}$ m.

Problem 1.105

The x and τ coordinates, in meters, of three events relative to S are listed below. Their y and z coordinates are the same.

Event	E_1	E_2	E_3
x	1.5	2.1	2.8
τ	2.3	1.5	0.8

(a) Determine whether the event separation between each pair of events is time-like, space-like, or null.

---

[*] See A. Schild, "The Clock Paradox in Relativity Theory," *The American Mathematical Monthly*, 66:1, 1959. This paradox is also discussed in J. Bronowski, "The Clock Paradox," *Scientific American* 208: 134, February 1963 and the papers on the twin paradox contained in *Special Relativity Theory, Selected Reprints*, American Institute of Physics, New York, 1963.

(b) Find the proper length between each space-like separation.
(c) Find the proper time along each time-like separation.
(d) Find the velocity relative to S of the reference system in which each space-like separation is along the x' axis.
(e) Find the velocity relative to S of the reference system in which each time-like separation is along the τ' axis.

Problem 1.106

A space traveler wishes to travel to a star that is 15×10^{13} km away; he wishes to age only 10 years in the process. If he did this and returned to earth at the same speed, how long would he have been traveling according to a clock at rest on the earth?

1.8.6   Minkowski geometry*

Space-time is a four-dimensional continuum in which we can define points (events) and straight lines. However, the axioms that these points and lines satisfy are not those of Euclidean geometry, the geometry applicable, for example, to this sheet of paper. A Euclidean four-dimensional space is conceivable and, therefore, is one of the geometries of interest to mathematicians; in such a 4-space, the distance d between two points labeled (τ_1, x_1, y_1, z_1) and (τ_2, x_2, y_2, z_2) is given by

$$d^2 = (\tau_2 - \tau_1)^2 + (x_2 - x_1)^2 + (y_2 - y_1)^2 + (z_2 - z_1)^2, \quad (1.160)$$

and the distance between the points calculated from this formula is the same for all coordinate systems. On the other hand, in the geometry of space-time, the geometry that describes the real world, d^2 calculated for two points as given by Equation (1.160) does not have the same value in all coordinate systems; rather, the "distance" in space-time between two events (τ_1, x_1, y_1, z_1) and (τ_2, x_2, y_2, z_2), defined as the interval s,

$$s^2 = (\tau_2 - \tau_1)^2 - [(x_2 - x_1)^2 + (y_2 - y_1)^2 + (z_2 - z_1)^2], \quad (1.161)$$

is the same for all observers.

In a space with Euclidean geometry, two lines that are orthogonal in one coordinate system are orthogonal in all. However, two straight lines, such as the τ and x axes defined by $x = y = z = 0$ and $\tau = y = z = 0$, that subtend an angle of 90° in one coordinate system on a space-time diagram do not subtend an angle of 90° relative to the space-time diagram of every other inertial observer (Figure 1.97). Therefore, orthogonality in space-time cannot be defined in terms of 90° angles that retain their values relative to all coordinate systems. Orthogonality in space-time can be defined, however, in terms of 90° angles relative to one coordinate system, and this definition can be stated in a manner that does not depend in any way on that coordinate system—that is, we can define orthogonality in space-time such that, if it is a property of two lines in one coordinate system, it is a property of those two lines in all coordinate systems. This point is illustrated by the following examples.

---

* Feynman, Leighton, and Sands (vol. 1), Addison-Wesley, Sec. 17–1, p. 17–1.
Kittel, Knight, and Ruderman, McGraw-Hill, p. 366.

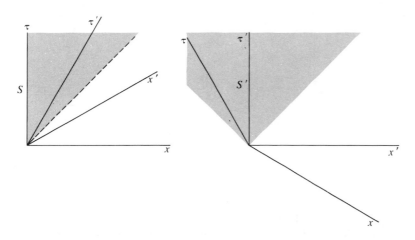

FIGURE 1.97   The two lines, the τ and x axes, that subtend an angle of 90° on the space-time diagram of S do not do so on the space-time diagram of S'.

Consider two perpendicular line segments $\overrightarrow{OP}$ and $\overrightarrow{OQ}$ at one instant of time in the space of an inertial observer S (Figure 1.98). The event displacement between the events O and P is space-like relative to all inertial observers, and furthermore, since the interval is invariant,

$$s_{OP}^2 = -\overrightarrow{OP}^2. \tag{1.162}$$

Similarly, the event displacement between the events O and Q is space-like, and the invariant interval is given by

$$s_{OQ}^2 = -\overrightarrow{OQ}^2. \tag{1.163}$$

The condition relative to S, that the spatial line segments $\overrightarrow{OP}$ and $\overrightarrow{OQ}$ subtend an angle of 90°, can be expressed by the pythagorean theorem:

$$\overrightarrow{OP}^2 + \overrightarrow{OQ}^2 = \overrightarrow{PQ}^2.$$

This can be rewritten in terms of the invariant intervals as

$$s_{OP}^2 + s_{OQ}^2 = s_{PQ}^2; \tag{1.164}$$

by definition, two space-like event displacements, those between events O and Q

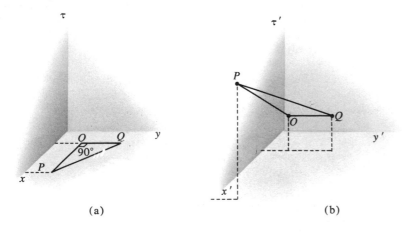

FIGURE 1.98   Two orthogonal space-like event displacements as seen by two different inertial observers. [The dashed lines in (b) are parallel to the axes.]

(a)   (b)

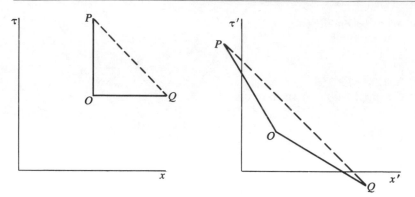

FIGURE 1.99   The event displacements between events O and P and events O and Q are orthogonal. The dashed line is a null event displacement.

and events O and P, are orthogonal if the event displacement between events P and Q is space-like and if the respective intervals satisfy Equation (1.164).

Consider now three events, O, P, and Q, for which O and P occur at the same point in space and O and Q occur at the same time relative to an inertial observer S (Figure 1.99). On the space-time diagram of S, the two event displacements subtend an angle of 90°. This condition can be expressed in an invariant way as follows: If the intervals $\sqrt{|s_{OP}^2|}$ and $\sqrt{|s_{OQ}^2|}$ are equal, the event displacement from the event P to the event Q is null. By definition, a time-like event displacement is orthogonal to a space-like event displacement if the endpoints of equal-interval event displacements along those directions are connected by a null event displacement. Thus, for example, the τ and x axes in one coordinate system are orthogonal by this definition even though, on a space-time diagram relative to another coordinate system, they meet at an angle different from 90°.

Two time-like event displacements are never orthogonal; since the vertex angle of the light cone is 90°, there exists no coordinate system in which they subtend an angle of 90°.

The above examples show the manner in which orthogonality can be defined in the geometry of space-time. These definitions can be combined into one definition through use of the notion of 4-vectors, considered in Section 1.9. This definition also gives the condition for the orthogonality of a null event displacement with another event displacement.

The geometry that is defined by these relations of distance and orthogonality is called Minkowski geometry,* after the Russian–German mathematician H. Minkowski (1864–1909), who formulated Einstein's special theory of relativity in terms of this geometry of space-time.

SUMMARY   The geometry of space-time is called Minkowski geometry. The distances of this geometry and the conditions for orthogonality can be specified in a manner that is independent of any coordinate system.

Example 1.12

Four event displacements, one time-like and three space-like—OT, OX, OY, and OZ—which are mutually orthogonal, can be used as coordinate axes for a

---

* See the illuminating discussion of this geometry in A. Schild, "The Clock Paradox in Relativity Theory," *The American Mathematical Monthly*, 66: 1, 1959.

reference system in space-time. If we choose OT as the time axis, the event T has the coordinates $(\tau, 0, 0, 0)$ where $\tau = \sqrt{s_{OT}^2}$. Since OT is orthogonal to OX, OY, and OZ, the events X, Y, and Z all have zero time coordinates [Problem 1.108(c)]. If we can choose OX as the x axis, the event X has the coordinates $(0, x, 0, 0)$, where $x = \sqrt{-s_{OX}^2}$; the events Y and Z have zero x coordinates [Problem 1.108(a)]. We therefore can choose Y as the y axis, and the event Y has the coordinates $(0, 0, y, 0)$, where $y = \sqrt{-s_{OY}^2}$; then Z must have coordinates $(0, 0, 0, z)$, with $z = \sqrt{-s_{OZ}^2}$, and it must lie on the z axis.

Problem 1.107

A sphere in a Euclidean 4-space is defined by the condition $d^2 = c^2$. Define a "sphere" in a Minkowski 4-space, and draw the intersection of one in the coordinate system S with the plane $y = z = 0$.

Problem 1.108

(a) Let OP and OQ be two space-like line segments that are orthogonal in a Minkowski 4-space. Let (τ_1, x_1, y_1, z_1) and (τ_2, x_2, y_2, z_2) be their coordinates relative to a coordinate system S (which you may choose on the basis of convenience). Show that

$$\tau_1 \tau_2 - (x_1 x_2 + y_1 y_2 + z_1 z_2) = 0.$$

(b) Show that if S' is moving with the velocity $\beta c \hat{x}$ relative to S, then

$$\tau_1' \tau_2' - (x_1' x_2' + y_1' y_2' + z_1' z_2') = 0$$

is a consequence of the equation of (a).

(c) Let OP be a time-like line segment that is orthogonal in a Minkowski 4-space to the space-like line segment OQ. Show that the coordinates of OP and OQ satisfy the equation of (a).

(d) Two line segments in a Minkowski 4-space are said to be orthogonal if their coordinates (τ_1, x_1, y_1, z_1) and (τ_2, x_2, y_2, z_2) relative to any coordinate system S satisfy

$$\tau_1 \tau_2 - (x_1 x_2 + y_1 y_2 + z_1 z_2) = 0.$$

Show that a null line segment is orthogonal to itself.

1.9   4-Vectors

The experiences of everyday life suggest that some physical entities—for example the time of an event—can be defined in an absolute way independent of the inertial observer. With the additional insight into the nature of time obtained from our studies, we have seen that time, and other entities, cannot be defined in an absolute manner but can be defined only relative to an observer. This does not mean, however, that everything can be defined only relative to an observer; in fact, our new insight into the nature of space and time provides us with a method of determining which entities are independent of the inertial observer. One such entity is the event displacement between two events, say the displacement from the event O to the event E that is represented by the coordinates (x, y, z) and τ relative to the inertial observer S; an event displacement is the same for all inertial observers, even though another observer may use different coordinates to represent that displacement. In this section, we shall

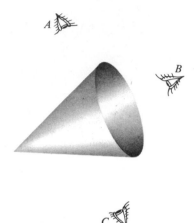

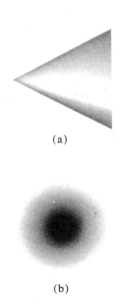

FIGURE 1.100   Different spectators A, B, and C view a translucent cone. The picture of the cone is the view of C.

(a)

(b)

FIGURE 1.101 A and B see the cone as (a) a triangle and (b) a circle, respectively.

see how to recognize other such entities and some ways in which such entities can be introduced into our study of mechanics.

1.9.1   Definition and properties of 4-vectors*

Entities that are independent of the observer are very important in the formulation of physical laws. A law of physics is a statement about behavior—for example about how an object moves—and thus should refer only to what that behavior is undergoing and not to the observer who witnesses it. Entities that can be defined only relative to an observer depend on the circumstances of the observer, as well as on what is observed; they should not play a central role in the statement of a law of physics. Thus, while it matters that our new insight into the nature of space and time shows us which entities can be defined in a relative manner only, it is also important that we be able to recognize which entities can be defined independent of the observer.

Let us consider an analogous circumstance from everyday life. An object, say a translucent plastic cone as shown in Figure 1.100, is viewed by different spectators. The way in which the object appears to each spectator depends on his point of view (Figure 1.101). Since the appearance of the object is different for different viewers, how can we determine that the object exists independent of the viewer?

If we know that the object exists independent of the viewer and if we know that it is a translucent cone, we can state how that object appears from each point of view. Now we reverse this procedure: If A, from his knowledge of what he sees and of the relative position of B, can state how the object appears to B,† and if B can do the same for C, etc., we can say that the object exists independent of the viewer, as the following argument suggests. The appearance of the object is different for different viewers. These differences in appearance are not a property of the object alone; they depend also on the viewpoint of the spectator. Since each spectator, from his own observations, can determine what every other spectator sees, there must be something there that does not depend on the point of view of the spectator. The features in the views that are common to all spectators are features of the object. These and the object that is the totality of all such features exist independent of the viewer.

Let us examine circumstances in the familiar case in which all inertial observers under consideration are at rest relative to each other. For such observers, there are some entities, called scalars, that are described by a number and a unit and that have the same value for all such observers. For example, the length of a rod or the time interval between two specified events are scalars with respect to inertial observers at rest relative to each other. Moreover, there exist more complicated entities than scalars that do not depend on the observer. These entities are described by each observer in terms of a set of components that are defined relative to the observer. For example, although different observers may represent a position-displacement vector by different components (Figure 1.102),

---

* Feynman, Leighton, and Sands (vol. 1), Addison-Wesley, Sec. 17–4, p. 17–5; Sec. 17–5, p. 17–7. Kacser, Prentice-Hall, Sec. A.2, p. 114; Sec. A.3, p. 116.
Kittel, Knight, and Ruderman, McGraw-Hill, pp. 371–374.
† The construction of the appearance of an object from another point of view on the basis of knowledge of the appearance of the object from one point of view can be carried out by a computer. See M. L. Minsky, "Artificial Intelligence," *Scientific American, 215*: 246, September 1966.

$$x = x' \cos\theta - y' \sin\theta, \qquad y = x' \sin\theta + y' \cos\theta, \qquad z = z', \quad (1.165)$$

the vector exists independent of the observer. For position-displacement vectors, this results from the fact that the observers can communicate to each other what features are characteristic of the positions of the endpoints; thus, they know that they are assigning their respective sets of coordinates, such as (x, y, z), to describe the given position P relative to the given position O.

The vector by itself determines one scalar, the length of the vector given by $\sqrt{x^2 + y^2 + z^2}$ relative to S. This is indeed a scalar, since it follows from the transformation law (1.165) that

$$x^2 + y^2 + z^2 = x'^2 + y'^2 + z'^2. \qquad (1.166)$$

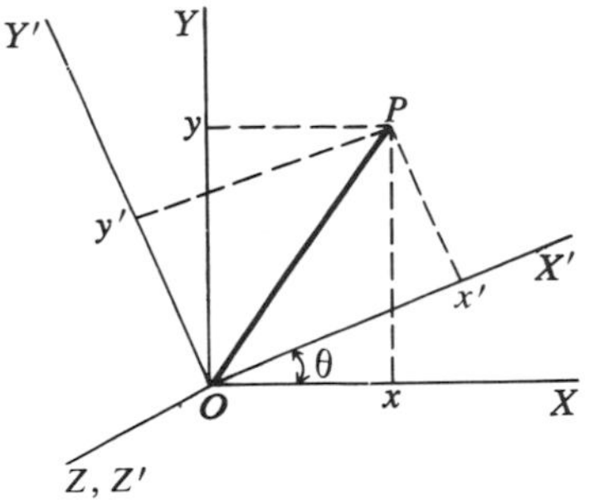

Thus the expression $x^2 + y^2 + z^2$ has the same value when evaluated in any coordinate system, and we say that the square of the length of a vector is *invariant in form under the transformation law* (1.165). Similarly, two vectors **a** and **b** determine a scalar **a**·**b**, the dot or scalar product given by

$$\mathbf{a} \cdot \mathbf{b} = (a_x \hat{x} + a_y \hat{y} + a_z \hat{z}) \cdot (b_x \hat{x} + b_y \hat{y} + b_z \hat{z})$$
$$= a_x b_x + a_y b_y + a_z b_z. \qquad (1.167)$$

The expression $a_x b_x + a_y b_y + a_z b_z$ is also invariant in form under the transformation law (1.165):

$$a_x b_x + a_y b_y + a_z b_z = a_{x'} b_{x'} + a_{y'} b_{y'} + a_{z'} b_{z'}. \qquad (1.168)$$

Consider now the event displacement from event O to event E. In principle, observers can communicate to each other what features are characteristic of each event; they can know that they are assigning their respective sets of co-ordinates, such as (x, y, z) and τ, to describe the given event E relative to the given event O. Furthermore, the Lorentz transformation provides a relation, analogous to Equation (1.165), between the coordinates (x, y, z) and τ relative to the inertial system S of the event E with respect to O and the coordinates (x', y', z') and τ' relative to the inertial system S' of the same event separation. For example,

$$x' = \gamma(x - \beta\tau), \qquad y' = y, \qquad z' = z, \qquad \tau' = \gamma(\tau - \beta x) \quad (1.169)$$

for the special case in which the relative velocity of S' with respect to S is $\beta c \hat{x}$. Hence, the inertial observer S can calculate the components relative to S' of the event displacement in terms of the S components and the measurable relation between the systems S and S'. Therefore, since one observer can tell how another describes the same event displacement, event displacements exist independent of the observer.*

Let the coordinates with respect to S of the event E_1 relative to O be (x_1, y_1, z_1) and τ_1, and those of the event E_2 relative to E_1 be (x_{12}, y_{12}, z_{12}) and τ_{12}; then the coordinates (x_2, y_2, z_2) and τ_2 with respect to S of the event E_2 relative to O are $(x_1 + x_{12}, y_1 + y_{12}, z_1 + z_{12})$ and $(\tau_1 + \tau_{12})$ (Figure 1.103). Since the Lorentz transformation law for event displacements is linear and homogeneous in the coordinates relative to S, then, if the coordinates with respect to S' of the respective event displacements are indicated by a prime, we have, for example,

* This point is discussed further in Section 1.9.4.

FIGURE 1.102 The vector $\overrightarrow{OP}$ is independent of the observers S and S', at rest relative to each other, although it is represented by the components (x, y) relative to S and the different components (x', y') relative to S'.

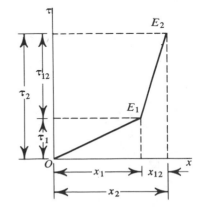

FIGURE 1.103 The coordinates x_2 and τ_2 of E_2 with respect to O are given by $x_2 = x_1 + x_{12}$ and $\tau_2 = \tau_1 + \tau_{12}$.

$$x_{2'} = \gamma(x_2 - \beta\tau_2) = \gamma[(x_1 + x_{12}) - \beta(\tau_1 + \tau_{12})]$$
$$= \gamma(x_1 - \beta\tau_1) + \gamma(x_{12} - \beta\tau_{12}) \qquad (1.170)$$
$$= x_{1'} + x_{12'}.$$

Similarly,

$$y_{2'} = y_{1'} + y_{12'}, \qquad z_{2'} = z_{1'} + z_{12'}, \qquad \tau_{2'} = \tau_{1'} + \tau_{12'}. \qquad (1.171)$$

Hence, for all inertial observers, event displacements combine in a linear manner, analogous to the way in which position-displacement vectors in space combine, relative to observers at rest with respect to each other. However, an event displacement is represented to an observer by four coordinates, whereas a position-displacement vector is represented by only three. Therefore, we say that an event displacement is a 4-vector.

There exist other entities that share the properties of event displacements that we have described. Thus, in general, *a 4-vector is defined as an entity that, relative to every inertial observer, is represented by four components that transform in the same way that the components of event displacements do.* All 4-vectors, like event displacements, are independent of the inertial observer, even though different observers may use different components to represent a given 4-vector.

The event-displacement 4-vector has four components τ, x, y, and z, which can be rewritten, for simplification of our later work, as

$$\tau = x_0, \qquad x = x_1, \qquad y = x_2, \qquad z = x_3. \qquad (1.172)$$

The components of this 4-vector are, therefore, (x_0, x_1, x_2, x_3); they can be referred to collectively by the symbol x_μ, where the Greek symbol μ is understood to range over 0, 1, 2, and 3.* Another 4-vector has, relative to an inertial observer S, one time-like component, say F_0 analogous to $\tau = x_0$, and three space-like components, F_1, F_2, and F_3 similar to $x = x_1$, $y = x_2$, and $z = x_3$. If another inertial observer is moving with the velocity $\beta c \hat{x}$ relative to S, the components $F_{0'}$, $F_{1'}$, $F_{2'}$, and $F_{3'}$ of this 4-vector relative to S' are given by

$$F_{0'} = \gamma(F_0 - \beta F_1), \quad F_{1'} = \gamma(F_1 - \beta F_0), \quad F_{2'} = F_2, \quad \text{and} \quad F_{3'} = F_3. \qquad (1.173)$$

The 4-vector can be represented by its components relative to the inertial observer S as (F_0, F_1, F_2, F_3) or as $(F_0, \mathbf{F})$, where $\mathbf{F}$ is equal to $F_1 \hat{x} + F_2 \hat{y} + F_3 \hat{z}$. Alternatively, that 4-vector can be designated † by $\underline{F}$ in handwritten material or by a script letter $\mathscr{F}$ in printed matter. For example, the displacement of the event E relative to O (Figure 1.104) can be designated by $\underline{OE}$, or by $(\tau, x, y, z) = (x_0, x_1, x_2, x_3)$, or even by x to indicate that it is the four-dimensional generalization of the coordinate x of a point on a line.

Two 4-vectors, say a and b, are equal relative to one inertial observer S if they have the same components relative to that observer. If this is true, then the 4-vector

$$c = a - b \qquad (1.174)$$

has all components relative to S equal to zero. Since the Lorentz transformation law is linear and homogeneous in the four components, the 4-vector c is equal

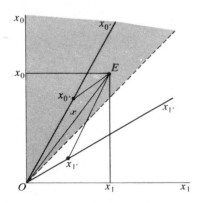

FIGURE 1.104 The coordinates of the 4-vector x are (x_0, x_1, x_2, x_3) relative to S and $(x_{0'}, x_{1'}, x_{2'}, x_{3'})$ relative to S'.

---

* In an alternative notation, the time component is written as $x_4 = ict$, where $i^2 = -1$. This alternative notation will be discussed shortly.
† The notation selected here is only one choice of a number of possibilities. We use script letters for 4-vectors to distinguish them from 3-vectors, in boldface type, and from other physical quantities, in italic type.

to zero in every inertial system. Therefore, if two 4-vectors are equal relative to one observer, they are equal relative to all observers, and we can write

$$a = \ell. \tag{1.175}$$

Since every 4-vector $\mathscr{F}$ transforms in a manner identical to x, we can associate with $\mathscr{F}$ a scalar, corresponding to $\tau^2 - (x^2 + y^2 + z^2)$,

$$\mathscr{F}^2 = F_0^2 - (F_1^2 + F_2^2 + F_3^2), \tag{1.176}$$

which we call *the norm of the vector*. The vector $\mathscr{F}$ is said to be time-like, space-like, or null if $\mathscr{F}^2$ is positive, negative, or zero, respectively.

Just as the length squared of a position-displacement vector in space can be generalized to the scalar product of two position-displacement vectors [see Equations (1.166) and (1.167)], so also the definition of the norm of a 4-vector can be generalized to yield a scalar that represents a product of two 4-vectors. Let a and ℓ be 4-vectors; we define the *scalar product of the two 4-vectors* by

$$a \cdot \ell = a_0 b_0 - (a_1 b_1 + a_2 b_2 + a_3 b_3). \tag{1.177}$$

This is equal to $a_{0'} b_{0'} - (a_{1'} b_{1'} + a_{2'} b_{2'} + a_{3'} b_{3'})$, since, for example,

$$
\begin{aligned}
a_{0'} b_{0'} &- (a_{1'} b_{1'} + a_{2'} b_{2'} + a_{3'} b_{3'}) \\
&= \gamma(a_0 - \beta a_1)\gamma(b_0 - \beta b_1) - [\gamma(a_1 - \beta a_0)\gamma(b_1 - \beta b_0) + a_2 b_2 + a_3 b_3] \\
&= \gamma^2(1 - \beta^2)a_0 b_0 - [\gamma^2(1 - \beta^2)a_1 b_1 + a_2 b_2 + a_3 b_3] \\
&= a_0 b_0 - (a_1 b_1 + a_2 b_2 + a_3 b_3).
\end{aligned}
\tag{1.178}
$$

Therefore, the scalar product of two 4-vectors, as defined by (1.177), is a scalar in space-time.

SUMMARY   Some dynamical variables, of which an event displacement is the prototype, that characterize the behavior or properties of a physical system are represented by a set of four components defined relative to an inertial observer. The set of components is called a 4-vector if the components defined relative to inertial observers are connected by Lorentz transformations in the same way that event displacements are. One 4-vector determines one number, the norm, which has the same value relative to all inertial observers. Two 4-vectors together determine another such invariant, the scalar product.

Problem 1.109

Find the norms of a and ℓ and the scalar product $a \cdot \ell$ for each set of a and ℓ whose components relative to an inertial observer S are listed below:

(a) $a = (4, 3, 3, 3)$ $\ell = (2, 6, 1, 7)$.
(b) $a = (5, 2, 1, 1)$ $\ell = (7, 6, 2, 2)$.
(c) $a = (6, 3, 2, 4)$ $\ell = (2, 4, 1, 2)$.

Problem 1.110

The components relative to an inertial observer S of two 4-vectors p and q are $(3, 3, 0, 0)$ and $(0, 0, 2, 0)$.

(a) Draw the 4-vectors on a space-time diagram of S.
(b) An inertial observer S' travels with the velocity $4c\hat{x}/5$ relative to S. Find the components of p and q relative to S'.

(c) Find p^2, q^2 and $p \cdot q$ from the information given initially, and also calculate them from your answer to (b).

(d) Would you say that p and q are orthogonal?

Problem 1.111

The spatial components of a 4-vector $\imath$ are equal to the spatial components of the event separation x in every inertial reference system. Show that $\imath = x$. *Hint:* Show that $\imath - x$ cannot have only time-like components in every inertial reference system.

Problem 1.112

The scalar product of two 4-vectors, a and ℓ, is zero: $a \cdot \ell = 0$.

(a) Assume that a is time-like. Show that ℓ is space-like and that the directions of a and ℓ on a space-time diagram in the plane of a and ℓ make equal angles with the light cone. *Hint:* Consider the components of the vectors in that inertial system in which a lies along the time axis.

(b) Assume that a and ℓ are both space-like. Show that the spatial components a and ℓ are perpendicular in that inertial reference system in which a has a zero time component.

(c) Draw the 4-vectors a and ℓ on a space-time diagram for the case of (a) and on another such diagram for the case of (b).

1.9.2   The 4-velocity *

As we stressed before, 4-vectors are important because they are entities that do not depend upon an inertial observer nor upon his circumstances. In order that we can formulate laws of physics that do not depend upon an inertial observer, we must find 4-vector generalizations of the 3-vectors in space, such as the velocity vector **v**, that are used in newtonian mechanics to describe motions. Sometimes, the fact that the last three components of a 4-vector form a 3-vector in space can be used to find the required generalization. (Unlike the spatial components, the time-component of a 4-vector, such as $\tau = ct$, is a scalar under the restricted transformations between inertial observers at rest relative to each other.) For example, the components of the position vector **r** form the last three components of the event 4-vector x, so x is the 4-vector generalization of **r**; indeed, there is no other 4-vector that has spatial components equal to the position vector **r** in every inertial system (Problem 1.111).

Let us use these ideas to find the relativistic generalization of the velocity vector relative to S,

$$\mathbf{v} = \frac{d\mathbf{r}}{dt}. \tag{1.179}$$

For this purpose, we examine the manner in which this 3-vector is calculated. The position vector $\mathbf{r}(t)$ is determined as a function of the time, a scalar under rotations of the (spatial) coordinate axes. The ratio of the change of position vector $\Delta\mathbf{r}$ to the change in the (scalar) time Δt is evaluated, and the usual limiting procedure applied. If we restrict our considerations to the transformations

---

* Kacser, Prentice-Hall, Sec. A.4, p. 118.

between observers at rest relative to each other, then **v** is a 3-vector, since $\Delta\mathbf{r}$ is a vector in space and Δt is a scalar.

This procedure can be generalized to yield the velocity 4-vector. The components of the position vector **r** are the last three components of the event vector x. We wish to express the event vector along the world line of the particle as a function of a scalar that reduces, in circumstances familiar from everyday life, to the time t. The time t is not a scalar, since it depends on the observer, but there is one scalar that is equal to t in circumstances that involve speeds much smaller than c. This is the proper time t_0 defined by the proper time interval

$$\Delta t_0 = \sqrt{1 - (v^2/c^2)}\,\Delta t, \qquad (1.180)$$

the time interval measured in that inertial frame in which the particle is instantaneously at rest, where v is the (instantaneous) speed of the particle relative to the inertial observer S. Therefore, we describe the world line of the particle by the equation

$$x = x(t_0), \qquad (1.181)$$

the relativistic generalization of the equation for the particle's position as a function of time

$$\mathbf{r} = \mathbf{r}(t). \qquad (1.182)$$

The 4-velocity is defined, therefore, as the rate of change of the 4-vector x with respect to the scalar t_0:

$$v = \frac{dx}{dt_0}. \qquad (1.183)$$

This 4-vector has components

$$\left(c\,\frac{dt}{dt_0}, \frac{dx}{dt_0}, \frac{dy}{dt_0}, \frac{dz}{dt_0} \right) = \frac{1}{\sqrt{1 - (v^2/c^2)}}\,(c, \mathbf{v}) \qquad (1.184)$$

and the norm

$$v^2 = v \cdot v = \frac{1}{1 - (v^2/c^2)}\,(c^2 - v^2) \qquad (1.185)$$

$$= c^2.$$

The two velocity vectors $\mathbf{v}(t)$ and $v(t_0)$ have a different interpretation. The first, $\mathbf{v}(t)$, is the rate of change of the particle's position in space with respect to the time of the observer S. Other observers will obtain vector functions of the time different from $\mathbf{v}(t)$ for the time rate of change of position (Figure 1.105), since different observers may not agree on simultaneity; also, the velocity 3-vector **v** corresponds, for each observer, to a displacement in space. On the other hand, the 4-vector v is the same for all observers. Since v corresponds to the direction in space-time of the separation Δx between two neighboring events on the world line of the particle (Figure 1.106), it is the tangent vector v to the world line, with norm c^2. Thus although v does not have the simple interpretation of the time rate of change of position, it has the advantage, for our purposes, of being the same to all observers.

SUMMARY   The 4-velocity v of a particle is defined as the derivative with respect to the proper time t_0 of the event-displacement vector $x(t_0)$ of the

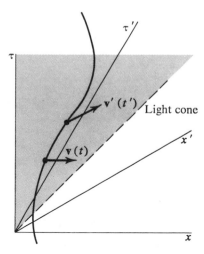

FIGURE 1.105   Relative to each observer, the velocity 3-vector corresponds to a displacement in space that is represented on the diagram by a line segment joining two simultaneous events relative to the respective observer.

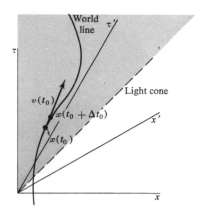

FIGURE 1.106   The 4-velocity $v(t_0)$ is the tangent vector to the world line at the event $x(t_0)$ and has norm c^2.

particle. The velocity $v(t_0)$ is a 4-vector with norm c^2 that is tangent to the world line of the particle at the event specified by the proper time t_0.

Example 1.13

Q. The inertial observer S' travels with the constant velocity $\beta c\hat{x}$ with respect to S. The 3-velocities of a particle relative to S and relative to S' are $v(t)\hat{x}$ and $v'(t')\hat{x}'$, respectively. Find the transformation law relating v and v' (Section 1.7) from the fact that v is a 4-vector.

A. The components of v relative to S and to S' are, respectively,

$$v = \frac{1}{\sqrt{1 - (v^2/c^2)}}\,(c, \mathbf{v}) \tag{1.186}$$

and

$$v = \frac{1}{\sqrt{1 - (v'^2/c^2)}}\,(c, \mathbf{v}'). \tag{1.187}$$

The components v_0, v_1 are related to the components $v_{0'}$, $v_{1'}$, through

$$v_{1'} = \gamma(v_1 - \beta v_0), \qquad v_{0'} = \gamma(v_0 - \beta v_1) \tag{1.188}$$

or, with $\beta = V/c$,

$$\begin{aligned}
\frac{v'}{\sqrt{1 - (v'^2/c^2)}} &= \frac{v - V}{\sqrt{1 - (v^2/c^2)}\,\sqrt{1 - (V^2/c^2)}}, \\
\frac{c}{\sqrt{1 - (v'^2/c^2)}} &= \frac{c - (V/c)v}{\sqrt{1 - (v^2/c^2)}\,\sqrt{1 - (V^2/c^2)}}.
\end{aligned} \tag{1.189}$$

We replace $1/\sqrt{1 - (v'^2/c^2)}$ on the right-hand side of the first equation by $1/c$ times the left-hand side of the second to obtain

$$v' = \frac{v - V}{1 - (Vv/c^2)}. \tag{1.190}$$

Note that we also obtain the transformation law for the Lorentz contraction and time dilatation factor $\sqrt{1 - (v'^2/c^2)}$:

$$\sqrt{1 - (v'^2/c^2)} = \frac{\sqrt{1 - (v^2/c^2)}\,\sqrt{1 - (V^2/c^2)}}{1 - (Vv/c^2)}. \tag{1.191}$$

Problem 1.113

(a) Draw, on a space-time diagram of S, the world line of a particle that moves with velocity $-3c\hat{x}/5$ until $t = 0$ relative to S, and thereafter with velocity $3c\hat{x}/5$.

(b) Find the proper time t_0 of the particle in terms of the time t as measured by S.

Problem 1.114

A particle travels, relative to an inertial observer S, on a circle of radius 2 m with the constant speed $v = 4c/5$.

(a) Find the proper time t_0 of the particle in terms of the time t as measured by S.

(b) Show that, relative to an appropriately chosen set of coordinate axes for S,

$$x(t) = 2 \cos\frac{2ct}{5}, \qquad y(t) = 2 \sin\frac{2ct}{5}.$$

(c) Find the components of $x(t_0)$ relative to S.

(d) Find $v(t_0)$.

(e) Show explicitly from the results of (d) that $v^2 = c^2$.

Problem 1.115

An inertial observer S measures the motion of an object at times $\tau_1 = 1$, $\tau_2 = 2$, and $\tau_3 = 3$. At these times, the particle is traveling in the positive x direction along the directions in space-time specified by the event displacements with coordinates (τ, x) given by $(1, 0)$, $(1, \frac{1}{2})$, and $(1, \frac{9}{10})$, respectively.

(a) Find the corresponding 4-velocity vectors. *Hint:* Use the fact that $v \cdot v = c^2$ or, equivalently, $(dx/d\tau_0) \cdot (dx/d\tau_0) = 1$.

(b) Draw these 4-velocity vectors on the space-time diagrams of S. Explain why they do not have the same length.

Problem 1.116

(a) Show that the acceleration 4-vector is given by $a = dv/dt_0$.

(b) Show that, relative to an inertial observer S, the components of a are

$$\frac{1}{(1 - v^2/c^2)^{1/2}} \left[\frac{(v/c)(dv/dt)}{[1 - (v^2/c^2)]^{3/2}}, \frac{(v/c^2)(dv/dt)\mathbf{v}}{[1 - (v^2/c^2)]^{3/2}} + \frac{1}{\sqrt{1 - (v^2/c^2)}} \frac{d\mathbf{v}}{dt} \right].$$

(c) Show that, if $v \ll c$, a has the components $(0, \mathbf{a})$, where $\mathbf{a} = d^2\mathbf{r}/dt^2$.

(d) Show that $v \cdot a = 0$. Is a time-like, null, or space-like?

Problem 1.117

Throughout its motion, a particle experiences a constant acceleration $\mathbf{a}$ of $32\hat{x}$ in feet per second per second relative to that reference system in which the particle is instantaneously at rest. Calculate the acceleration 4-vector a relative to an observer with respect to which, at the instant under consideration, the particle is traveling at the speed $c/2$ along the direction of the acceleration $\mathbf{a}$.

1.9.3 Conditions satisfied by the transformations of sets of components among different observers

We have seen that 4-vectors exist independent of the observer, even though different observers describe a 4-vector with different values of the components. There also exist other entities that do not depend on the observer and that, like 4-vectors, are described by each observer in terms of a number of components, the values of which depend upon the observer. How can we find out if a set of components relative to an observer describes an entity that is independent of that observer?

Consider an entity that exists in space-time and is independent of the observer. We can say that the entity exists independent of each observer if we know how to transform the components relative to one observer into the components relative to another. The manner in which these components transform is not arbitrary. Their transformation law must share some features of the Lorentz transformation law for the components of event displacements between the two observers. Let us examine the conditions that such transformation laws must satisfy.

For simplicity, we shall restrict our considerations for the moment to the transformations between inertial observers whose spatial axes coincide at some event O. Thus, for the time being, we omit consideration of the transformations between observers at rest relative to each other—for example, those differing only in the orientation of their spatial axes. As a result of this restriction, the relation between two inertial systems is defined by one vector, their relative velocity.

We are interested in an entity, which we denote by **R**, that exists independent of the observer. An inertial observer S_1 describes **R** in terms of a number of components, and we denote these components collectively by the symbol $\mathbf{R}_{S_1}$. Consider an observer S_2 moving with velocity $\mathbf{V}_{21}$ relative to S_1. If **R** exists independent of the observer, there is a transformation law that determines the components $\mathbf{R}_{S_2}$ relative to S_2 in terms of the components $\mathbf{R}_{S_1}$ and the velocity $\mathbf{V}_{21}$. There is an operation, which we denote by $\mathscr{L}_{\mathbf{V}_{21}}$, that transforms the set of components $\mathbf{R}_{S_1}$ into the set $\mathbf{R}_{S_2}$. We indicate the operation by the sign $\circ$ and write

$$\mathbf{R}_{S_2} = \mathscr{L}_{\mathbf{V}_{21}} \circ \mathbf{R}_{S_1}. \tag{1.192}$$

Note that the order of the two entities, $\mathscr{L}_{\mathbf{V}_{21}}$ and $\mathbf{R}_{S_1}$, is important.

This transformation law satisfies certain conditions as a result of the fact that the transformation relates components referred to different inertial systems in space-time. These conditions arise in the following manner: To each inertial reference system, we can associate in a unique manner a vector **V**, where **V** is the velocity of that system with respect to one (arbitrarily) chosen inertial reference frame S (Figure 1.107). If we consider each vector **V** to represent a point in space relative to a point P as the origin, the totality of these points fills a sphere of radius c centered at P up to but not including the surface (Figure 1.108). Each transformation function $\mathscr{L}_{\mathbf{V}_{21}}$ corresponds to the displacement from the point defined by $\mathbf{V}_1$ to the point defined by $\mathbf{V}_2$.* Note, however, that

FIGURE 1.107 Each inertial frame S_1 can be related to a point P inside a sphere of radius c about P. (Note that the point P is related to the frame S.) (a) The reference frame of S_1 moves with velocity $\mathbf{V}_1$ relative to the chosen reference frame S. (b) The vector $\mathbf{V}_1$ designates a point P_1 in space relative to P; P_1 lies inside a sphere of radius c centered at P since $V_1 < c$.

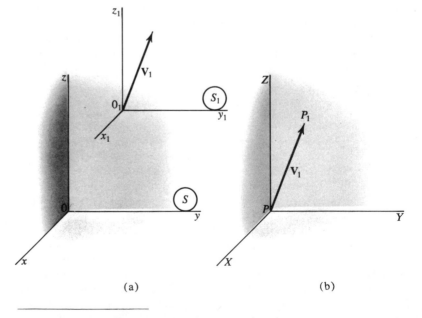

(a)   (b)

---

* See I. Grossman and W. Magnus, *Groups and Their Graphs*, New Mathematical Library, Vol. 14, Random House, New York, 1964.

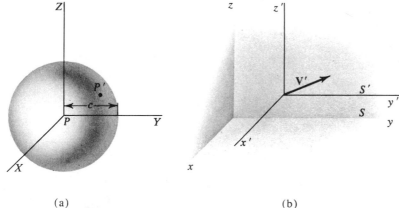

(a)   (b)

FIGURE 1.108 Every inertial frame S_1 can be related to a point P_1 inside the sphere (Figure 1.107) and every point P' inside the sphere can be related to an inertial reference frame S'. (a) To each point P' inside a sphere of radius c centered at P, there corresponds a vector $\mathbf{V}' = \overrightarrow{PP'}$. (b) To each vector $\mathbf{V}'$, there corresponds an inertial reference frame S' moving with velocity $\mathbf{V}'$ relative to S.

$\mathbf{V}_{21} \neq \mathbf{V}_2 - \mathbf{V}_1$, but rather $\mathbf{V}_{21}$ is given in terms of $\mathbf{V}_2$ and $\mathbf{V}_1$ by the Lorentz transformation equations for velocities, Equation (1.127). In order that this correspondence be consistent, the transformation functions must combine in a manner reflecting that by which the corresponding displacements combine (by vector addition). For example, the displacement from the point $\mathbf{V}_1$ to $\mathbf{V}_2$ followed by the displacement from $\mathbf{V}_2$ to $\mathbf{V}_3$ is equal to the displacement from $\mathbf{V}_1$ directly to $\mathbf{V}_3$. Therefore, the transformation of the components of $\mathbf{R}$ from S_1 to S_2 followed by the transformation from S_2 to S_3 must be equal to the transformation from S_1 directly to S_3 (Figure 1.109):

$$\mathscr{L}_{\mathbf{V}_{32}} \circ \mathscr{L}_{\mathbf{V}_{21}} \circ \mathbf{R}_{S_1} = \mathscr{L}_{\mathbf{V}_{31}} \circ \mathbf{R}_{S_1}. \tag{1.193}$$

This relation can be written in the symbolic form

$$\mathscr{L}_{\mathbf{V}_{32}} \circ \mathscr{L}_{\mathbf{V}_{21}} = \mathscr{L}_{\mathbf{V}_{31}}. \tag{1.194}$$

A particular case occurs if $S_3 = S_1$. The velocity $\mathbf{V}_{12}$ associated with the transformation from S_2 to S_1 is given by the Lorentz transformation as

$$\mathbf{V}_{12} = -\mathbf{V}_{21}; \tag{1.195}$$

hence,

$$\mathscr{L}_{-\mathbf{V}_{21}} \circ \mathscr{L}_{\mathbf{V}_{21}} = \mathscr{L}_0, \tag{1.196}$$

where $\mathscr{L}_0$ is that transformation that leaves the components unchanged.

The conditions given above are necessary to ensure that the transformations $\mathscr{L}_{\mathbf{V}}$ combine in a consistent manner, similar to that of the corresponding "displacements" in Figure 1.109. Furthermore, the velocity $\mathbf{V}_{31}$ in the combination law (1.194) must be determined from $\mathbf{V}_{32}$ and $\mathbf{V}_{21}$ by the Lorentz transformation law for velocities (1.127), in order that the transformations $\mathscr{L}_{\mathbf{V}}$, relating the components of $\mathbf{R}$ in different inertial systems, combine in a manner consistent with the Lorentz transformation law for event displacements in space-time.

SUMMARY The rule for transforming the components of an entity from one reference system must satisfy certain conditions, or the transformation rule is not consistent with the Lorentz transformation law.

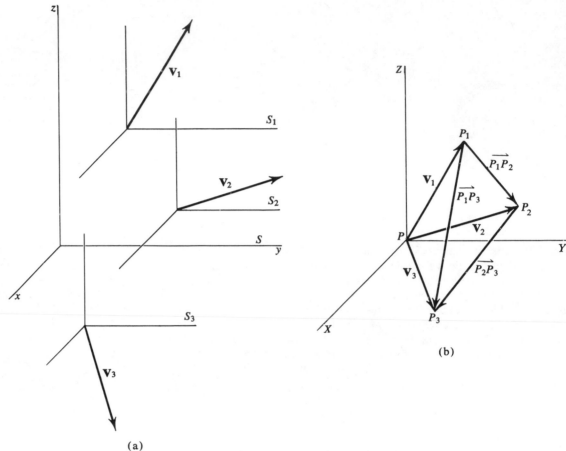

(a)

(b)

FIGURE 1.109   Since $\overrightarrow{P_1P_2} + \overrightarrow{P_2P_3} = \overrightarrow{P_1P_3}$, the transformation from S_1 to S_2 followed by the transformation from S_2 to S_3 must equal the transformation from S_1 to S_3.   (a) The instantaneous relation of the inertial reference frames S_1, S_2, S_3, to the reference frame S.   (b) The transformation from S to S_1 corresponds to the displacement $\mathbf{V}_1$, the transformation from S_1 to S_2 to the displacement from P_1 to P_2, $\overrightarrow{P_1P_2} = \mathbf{V}_2 - \mathbf{V}_1$, etc.   (Note that $\overrightarrow{P_1P_2}$ does not equal the velocity of S_2 relative to S_1, even though the displacement from P_1 to P_2 represents the transformation from S_1 to S_2.)

Problem 1.118

Discuss the rules that must be satisfied, according to the galilean transformation law, by the transformation equations of a set of components. What distinguishes the transformation equations that obey the galilean transformation law from those that obey the Lorentz transformation law?

Problem 1.119

Show that the reference frames denoted by the points inside a very small circle near P of Figure 1.108(a) satisfy the galilean transformation law.

Problem 1.120

A *group* is a set of elements, $A, B, C, \ldots$, for which there exists a law that defines the combination $A \circ B$ of any two elements of the set and that satisfies the following conditions:

(a) If A and B are elements of the set, then so is $A \circ B$.
(b) $A \circ (B \circ C) = (A \circ B) \circ C$.
(c) The set contains an element I (the identity element) that satisfies

$$I \circ A = A \circ I = A$$

for every member A of the set.
(d) To every element A in the set, there exists an element B such that

$$A \circ B = B \circ A = I.$$

Consider the set of transformations $\{\mathscr{L}_\mathbf{v}\}$ given above with the law of combination

$$\mathscr{L}_{\mathbf{v}_{32}} \circ \mathscr{L}_{\mathbf{v}_{21}} = \mathscr{L}_{\mathbf{v}_{31}}.$$

Show that the set forms a group.*

Problem 1.121

Consider the position vectors of the points in 3-space, $P_1, P_2, P_3, \ldots$, as the elements and the vector sum as the combination law; $P_1 \circ P_2$ corresponds to $P_1 + P_2$. Show that these elements with this combination law obey the group properties.

1.9.4   Tensors in space-time

The discussion of Section 1.9.3 shows that the transformation law for the components of an entity that is independent of the reference system is subject to certain restrictions. Armed with this knowledge, we are now in a position to examine entities other than scalars and 4-vectors that are defined relative to an inertial system by a set of components and that exist independent of the observer. As an introduction to this, we first show that the Lorentz transformations for 4-vectors do combine according to the conditions of Section 1.9.3.

An inertial observer S describes a 4-vector a by four components that we denote by (a_0, a_1, a_2, a_3), where a_0 is the time-like component and a_1, a_2, a_3 are the spatial components in the x, y, and z directions, respectively. One component is designated by a_μ with the Greek letter subscript μ equal to one of the numbers 0, 1, 2, or 3. The component $a_{\mu'}$ of a referred to another observer S' is determined from the components a_0, a_1, a_2, and a_3 by the linear and homogeneous transformation that relates the components of event displacements in the two inertial systems. We write this transformation as

$$a_{\mu'} = L_{\mu'0}a_0 - L_{\mu'1}a_1 - L_{\mu'2}a_2 - L_{\mu'3}a_3. \tag{1.197}$$

This can be written in a simpler form if we use this convention: Repeated Greek

---

* For another example of the use of group theory in physics, see F. J. Dyson, "Mathematics in the Physical Sciences," *Scientific American*, *211*: 129, September 1964.

letter indices in any expression denote the sum made up from that expression with the repeated indices set equal to zero and the negatives of that expression with the repeated indices set equal to 1, 2, and 3 in turn. Thus, with this convention, Equation (1.197) can be written as

$$a_{\mu'} = L_{\mu'\mu}a_{\mu}. \tag{1.198}$$

Also, the scalar product of two 4-vectors, $a_0b_0 - (a_1b_1 + a_2b_2 + a_3b_3)$, can be written within this convention* as

$$a \cdot b = a_{\mu}b_{\mu}. \tag{1.199}$$

The transformation coefficients $L_{\mu'\mu}$ depend on the relation between the observers S and S'. For example, if S' is traveling with the velocity $\beta c\hat{x}$ relative to S, these coefficients have the following values:

$$L_{0'0} = \gamma, \qquad L_{0'1} = \beta\gamma, \qquad L_{1'0} = -\gamma\beta, \qquad L_{1'1} = -\gamma,$$
$$L_{2'2} = L_{3'3} = -1, \tag{1.200}$$

all other $L_{\mu'\mu} = 0$. The transformation given by Equation (1.198) corresponds to the general form, Equation (1.192),

$$\mathbf{R}_{S_2} = \mathscr{L}_{\mathbf{v}_{21}} \circ \mathbf{R}_{S_1}. \tag{1.201}$$

The equation corresponding to the relation (1.194) can be deduced in the following way: Let $a_{\mu''}$ be the μth component of a relative to the inertial observer S''. Then,

$$a_{\mu''} = L_{\mu''\mu'}a_{\mu'} = L_{\mu''\mu'}L_{\mu'\mu}a_{\mu}; \tag{1.202}$$

but also,

$$a_{\mu''} = L_{\mu''\mu}a_{\mu}. \tag{1.203}$$

These relations are true for all a_{μ}, so

$$L_{\mu''\mu'}L_{\mu'\mu} = L_{\mu''\mu}, \tag{1.204}$$

which is a special case of

$$\mathscr{L}_{\mathbf{v}_{32}} \circ \mathscr{L}_{\mathbf{v}_{21}} = \mathscr{L}_{\mathbf{v}_{31}}. \tag{1.205}$$

These results show that Lorentz transformations of 4-vectors do indeed satisfy the conditions derived in Section 1.9.3. We now turn our attention to more complicated entities whose transformation laws satisfy these conditions.

Let a and b be two 4-vectors, and consider the entity $\mathbf{T}$ defined relative to an inertial observer S by the set of components

$$a_{\mu}b_{\nu} = T_{\mu\nu} \qquad (\mu = 0, 1, 2, 3 \qquad \text{and} \qquad \nu = 0, 1, 2, 3). \tag{1.206}$$

The components of $\mathbf{T}$ relative to another observer S' are defined by

$$T_{\mu'\nu'} = a_{\mu'}b_{\nu'} = L_{\mu'\mu}a_{\mu}L_{\nu'\nu}b_{\nu}$$
$$= L_{\mu'\mu}L_{\nu'\nu}T_{\mu\nu}. \tag{1.207}$$

---

* In some texts, the summation convention given above applies only to repeated indices of which one is a superscript and the other is a subscript, as in $a^{\mu}b_{\mu}$. Other texts introduce the time component as $x_4 = ict$ with $i^2 = -1$; in this notation, x_{μ} equals one of $x_1 = x$, $x_2 = y$, $x_3 = z$, or $x_4 = ict$, and the summation convention is

$$a_{\mu}b_{\mu} = \sum_{\mu=1}^{4} a_{\mu}b_{\mu} = a_1b_1 + a_2b_2 + a_3b_3 + a_4b_4$$
$$= a_1b_1 + a_2b_2 + a_3b_3 - a_0b_0.$$

Thus, with this notation, $a_{\mu}b_{\mu}$ is the negative of that given by our definition.

Relative to another observer S'', the components of **T** are

$$T_{\mu''\nu''} = L_{\mu''\mu'}L_{\nu''\nu'}T_{\mu'\nu'}, \tag{1.208}$$

and

$$T_{\mu''\nu''} = L_{\mu''\mu}L_{\nu''\nu}T_{\mu\nu}. \tag{1.209}$$

Hence, the condition

$$\mathscr{L}_{\mathbf{V}_{32}} \circ \mathscr{L}_{\mathbf{V}_{21}} = \mathscr{L}_{\mathbf{V}_{31}} \tag{1.210}$$

is satisfied, since [see Equation (1.204)]

$$(L_{\mu''\mu'}L_{\nu''\nu'})(L_{\mu'\mu}L_{\nu'\nu}) = (L_{\mu''\mu'}L_{\mu'\mu})(L_{\nu''\nu'}L_{\nu'\nu}) = L_{\mu''\mu}L_{\nu''\nu}. \tag{1.211}$$

Therefore, any entity **T** that is defined by the 16 components $T_{\mu\nu}$ that transform as

$$T_{\mu'\nu'} = L_{\mu'\mu}L_{\nu'\nu}T_{\mu\nu} \tag{1.212}$$

exists independent of the observer. We call **T** a *tensor of the second order*. (A 4-vector is a tensor of the first order.)

A tensor of the second order has 16 components defined relative to one inertial observer. Such tensors are important in the relativistic formulation of the laws of electromagnetism. Similar tensors play a fundamental role in Einstein's theory of gravitation, the general theory of relativity (Chapter 3). Tensors of other orders can also be defined.

S U M M A R Y   The Lorentz transformations of 4-vectors satisfy the conditions described in Section 1.9.3 for entities that exist independent of the observer. There are also other such entities, such as second-order tensors that are defined relative to an observer by 16 components.

Example 1.14

Q. The components of a tensor of the second order are represented by $T_{\mu\nu}$.

(a) Write out an explicit form for $T_{\mu\mu}$.
(b) Show that $T_{\mu\mu}$ is a scalar.

A. (a). According to our summation convention,

$$T_{\mu\mu} = T_{00} - T_{11} - T_{22} - T_{33}. \tag{1.213}$$

(b) The transformation law for $T_{\mu\mu}$ is given by Equation (1.212) to be

$$T_{\mu'\mu'} = L_{\mu'\mu}L_{\mu'\nu}T_{\mu\nu}. \tag{1.214}$$

The expression $L_{\mu'\mu}L_{\mu'\nu}$ appears in the transformation equation for the scalar product of two vectors,

$$a_{\mu'}b_{\mu'} = (L_{\mu'\mu}a_{\mu})(L_{\mu'\nu}b_{\nu}) = L_{\mu'\mu}L_{\mu'\nu}a_{\mu}b_{\nu}, \tag{1.215}$$

which, according to Equation (1.178), can be written as

$$a_{\mu'}b_{\mu'} = a_{\mu}b_{\mu}. \tag{1.216}$$

Therefore, the effect of $L_{\mu'\mu}L_{\mu'\nu}$ on the components $T_{\mu\nu}$ of a tensor of the second order is the same as a summation over the two indices:

$$L_{\mu'\mu}L_{\mu'\nu}T_{\mu\nu} = T_{\mu\mu}. \tag{1.217}$$

Hence,

$$T_{\mu'\mu'} = T_{\mu\mu}; \tag{1.218}$$

thus, $T_{\mu\mu}$ is a scalar.

Problem 1.122

The reference system S' is at rest with respect to S: The z and z' coordinate axes coincide, but the x' and y' coordinate axes are rotated through the angle θ about the positive z direction with respect to the x and y axes. Find the coefficients $L_{\mu'\mu}$ for the transformation law.

Problem 1.123

(a) If S' is traveling with the velocity $\beta c \hat{x}$ relative to S and S'' is traveling with the velocity $\beta' c \hat{x}'$ relative to S', find the velocity $\beta'' c \hat{x}$ of S'' relative to S.
(b) Show explicitly that

$$L_{\mu''\mu'}L_{\mu'\mu} = L_{\mu''\mu},$$

and describe the relation of this to Equation (1.194).

Problem 1.124

(a) Define a tensor of the third order.
(b) What is a tensor of the zeroth order?

Problem 1.125

One component of a tensor of the second order is $T_{\mu\nu}$. Show that the components (T_{01}, T_{02}, T_{03}) transform similar to the components of a vector in space under the rotation of Problem 1.122.

Problem 1.126

(a) A tensor of the second order is said to be symmetric if, relative to one reference system S, $T_{\mu\nu} = T_{\nu\mu}$. Show that, relative to any other reference system S', $T_{\mu'\nu'} = T_{\nu'\mu'}$. *Hint:* Consider $S_{\mu\nu} = T_{\mu\nu} - T_{\nu\mu}$. Show that the $S_{\mu\nu}$ are the components of a tensor of the second order and that each of these components is zero.
(b) A tensor of the second order is said to be skew-symmetric if, relative to one reference system S, $T_{\mu\nu} = -T_{\nu\mu}$. Show that, relative to any other reference system S', $T_{\mu'\nu'} = -T_{\nu'\mu'}$.

1.10   The Propagation 4-Vector for Waves

The theory of special relativity is required for the description of motions of particles in which relative speeds comparable to c are involved. This theory is also required for the description of some types of waves, phenomena, such as those of electrodynamics and the relativistic quantum theory, that are more complicated than are the motions of particles. Chapter 2 is devoted to the laws of dynamics of particles. In this section, we shall consider only part of the effects of relativity on wave phenomena.

Relative to an observer S, a wave is described by a wave function that varies over the position $\mathbf{r}$ in space and the time t relative to S. The wave function may be a scalar quantity, a vector, a tensor, or even some other entity that exists independent of the observer. Examples are wave functions that describe the change at time t in the pressure at $\mathbf{r}$ from its equilibrium value, the displacement at time t from the equilibrium position $\mathbf{r}$ of some material, or the variation at $\mathbf{r}$ and t of some other dynamical variable.

Harmonic plane waves are described relative to S by a wave function that can be factored into two parts, one that describes the amplitude and, for example, the polarization, and the other that describes the variation of the wave with position and time. For example, a linearly polarized harmonic plane wave can be described by the vector wave function

$$\mathbf{E}(\mathbf{r}, t) = E_0 \hat{e} \sin{(\mathbf{k} \cdot \mathbf{r} - \omega t)}, \tag{1.219}$$

where E_0 is the amplitude, the unit vector $\hat{e}$ describes the direction of polarization, and the sine factor describes the variation of $\mathbf{E}$ with position and time. The latter factor, with which we shall be concerned below, has a sinusoidal dependence on $\mathbf{r}$ and t relative to an inertial observer S, as in

$$\phi(\mathbf{r}, t) = \exp{[i(\mathbf{k} \cdot \mathbf{r} - \omega t + \alpha)]} \tag{1.220}$$

or in

$$\psi(\mathbf{r}, t) = \sin{(\mathbf{k} \cdot \mathbf{r} - \omega t + \alpha)}. \tag{1.221}$$

The propagation vector relative to S of this sinusoidal component $\mathbf{k}$ has magnitude, in terms of the wavelength λ,

$$k = \frac{2\pi}{\lambda} \tag{1.222}$$

and lies in the direction of propagation of this component. The circular frequency ω is related to the frequency ν and the speed v of the harmonic wave, both relative to S, by the equations

$$\omega = 2\pi\nu = \frac{2\pi v}{\lambda}. \tag{1.223}$$

Our interest in this section is in the manner in which $\mathbf{k}$ and ω depend on the observer.

1.10.1 The propagation vector and the doppler effect*

The wave function that represents a given wave to one observer S may be different from the wave function for another observer S'. The wave function can be analyzed in terms of plane-wave harmonic components. We need consider only the way in which the wave functions for harmonic plane waves transform, since, if we know this, it is possible to determine the way in which any wave function is transformed through use of a mathematical proposition known as Fourier's theorem.

One factor in the wave function, relative to S, of a harmonic plane wave depends on the character of the wave—that is, on whether it is a scalar or a tensor wave, or whatever. This factor transforms from one observer to another as its character would indicate, and will not concern us here. We need only note that the transformation equations for such entities as vectors are linear and homogeneous in the components, and thus the other factor, the time- and

---

* Feynman, Leighton, and Sands (vol. 1), Addison-Wesley, Sec. 34–6, p. 34–7; Sec. 34–7, p. 34–9.
Kacser, Prentice-Hall, Sec. 4.7, p. 85.
Kittel, Knight, and Ruderman, McGraw-Hill, pp. 361–362.
Resnick and Halliday (Part 2), John Wiley, Sec. 40–5, p. 1006.
Resnick, John Wiley, Sec. 2.7, p. 84.

position-dependent factor, appears in the wave function relative to every observer if it appears in the wave function of one.

This other factor, which is common to the wave functions relative to all observers, depends on the position $\mathbf{r}$ and the time t of the event at which it is evaluated only in the argument $(\mathbf{k} \cdot \mathbf{r} - \omega t)$ of a sinusoidal function such as $\exp [i(\mathbf{k} \cdot \mathbf{r} - \omega t + \alpha)]$ or $\sin (\mathbf{k} \cdot \mathbf{r} - \omega t + \alpha)$. Thus, if the wave function obeys a linear and homogeneous transformation law, this phase factor is the same for all observers; it is a scalar quantity. Therefore, the phase

$$p = \mathbf{k} \cdot \mathbf{r} - \omega t \qquad (1.224)$$

at the event $(ct, \mathbf{r})$ relative to that at the event $(0, 0)$ is the same for all observers to within an additive integral multiple of 2π, since, for example, $\sin (\theta + 2\pi n) = \sin \theta$. However, the transformation law for p cannot be discontinuous as the relationship between the observers is changed continuously, or else the one inertial system at which p jumps would be distinguished from others; hence, the additive integral multiple of 2π cannot appear. Thus, *the relative phase p is a scalar* if the wave function obeys a linear and homogeneous transformation law. The wave functions that describe the behavior of all types of waves with which we will be concerned, such as light and sound, do satisfy this condition of a linear and homogeneous transformation law.

We can give a physical interpretation of the scalar nature of p in the following way: Suppose we calculate the values of the function

$$\psi(\mathbf{r}, t) = \sin (\mathbf{k} \cdot \mathbf{r} - \omega t + \alpha) \qquad (1.225)$$

at each event on the line in Minkowski space from the event O, with coordinates $(0, 0)$ relative to S, to the event P, with coordinates $(ct, \mathbf{r})$ relative to S. Then the function ψ will have N maxima along the line from O to P (Figure 1.110), with

$$2\pi N = |\mathbf{k} \cdot \mathbf{r} - \omega t| = |p|. \qquad (1.226)$$

Each maximum corresponds to a wave crest, so N is the number of wave crests between the event O and the event P. Since p is a scalar in space-time, every inertial observer would count the same number of crests between these two events if there existed an experimental technique by which this counting could be done.

The coordinates $(ct', \mathbf{r}')$ relative to an inertial observer S' of the event displacement of P relative to O are linear and homogeneous functions of the coordinates $(ct, \mathbf{r})$ of that event displacement relative to S. Hence, the phase $p = \mathbf{k} \cdot \mathbf{r} - \omega t$, which is linear and homogeneous in the coordinates relative to S, is also linear and homogeneous in the coordinates relative to S':

$$p = \alpha x' + \beta y' + \gamma z' + \delta t', \qquad (1.227)$$

where the coefficients α, β, γ, and δ are determined by the transformation equations relating $(ct, \mathbf{r})$ and $(ct', \mathbf{r}')$ and are independent of the coordinates $(ct', \mathbf{r}')$. We can rewrite this in the form

$$p = \mathbf{k}' \cdot \mathbf{r}' - \omega' t', \qquad (1.228)$$

which defines $\mathbf{k}'$ and ω' in terms of $\mathbf{k}$ and ω and the transformation equations.

The space- and time-dependent factor in the wave function relative to S' has the form $\exp [i(\mathbf{k}' \cdot \mathbf{r}' - \omega' t' + \alpha)]$ or $\sin (\mathbf{k}' \cdot \mathbf{r}' - \omega' t' + \alpha)$, so $\mathbf{k}'$ and ω' are the propagation vector and circular frequency of the wave relative to S'.

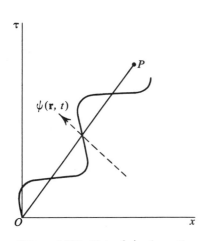

FIGURE 1.110 Plot of ψ along the event separation from O to P. The value of ψ at each point on the straight world line from O to P is given by the directed distance perpendicular to that line from the point to the curve.

Now we consider the transformation law relating $\mathbf{k}'$ and ω' to $\mathbf{k}$ and ω. The invariance of the form

$$\mathbf{k} \cdot \mathbf{r} - \omega t = -\left(\frac{\omega}{c}\tau - \mathbf{k} \cdot \mathbf{r}\right) \tag{1.229}$$

suggests that $(\omega/c, \mathbf{k})$ are the components relative to S of a 4-vector and that the form $(\omega/c)\tau - \mathbf{k} \cdot \mathbf{r}$ is the scalar product of that 4-vector and the event 4-vector x. That this suggestion is valid is shown by the following argument:

We introduce a 4-vector ℓ that is described relative to S by the components $(\omega/c, \mathbf{k})$ and inquire if the components of ℓ relative to S' are $(\omega'/c, \mathbf{k}')$. Since ℓ is a 4-vector,

$$\ell \cdot x = b_0 \tau - \mathbf{b} \cdot \mathbf{r} \tag{1.230}$$

is invariant in form, so that

$$\frac{\omega}{c}\tau - \mathbf{k} \cdot \mathbf{r} = b_{0'}\tau' - \mathbf{b}' \cdot \mathbf{r}'. \tag{1.231}$$

However,

$$\frac{\omega}{c}\tau - \mathbf{k} \cdot \mathbf{r} = \frac{\omega'}{c}\tau' - \mathbf{k}' \cdot \mathbf{r}', \tag{1.232}$$

so that

$$\left(b_{0'} - \frac{\omega'}{c}\right)\tau' - (\mathbf{b}' - \mathbf{k}') \cdot \mathbf{r}' = 0; \tag{1.233}$$

since this is true for all values of τ' and $\mathbf{r}'$, we must have

$$b_{0'} = \frac{\omega'}{c}, \qquad \mathbf{b}' = \mathbf{k}'. \tag{1.234}$$

Hence, $(\omega'/c, \mathbf{k}')$ are the components relative to S' of the 4-vector with components $(\omega/c, \mathbf{k})$ relative to S. We call this the *propagation 4-vector* and denote it by ℓ.

Since ℓ is a 4-vector, the transformation from the components $(\omega/c, \mathbf{k})$ to $(\omega'/c, \mathbf{k}')$ is the same as that from $(\tau, \mathbf{r})$ to $(\tau', \mathbf{r}')$. In particular, if S' moves with the constant velocity $\mathbf{V} = \beta c\hat{x}$ with respect to S, then

$$k_{x'} = \gamma\left(k_x - \beta\frac{\omega}{c}\right), \qquad k_{y'} = k_y, \qquad k_{z'} = k_z,$$

$$\frac{\omega'}{c} = \gamma\left(\frac{\omega}{c} - \beta k_x\right) = \frac{\gamma}{c}(\omega - \mathbf{V} \cdot \mathbf{k}). \tag{1.235}$$

The last expression results from the fact that $\beta k_x = V k_x/c = \mathbf{V} \cdot \mathbf{k}/c$.

The relation between the frequencies $\nu = \omega/2\pi$ and $\nu' = \omega'/2\pi$ measured by the two observers is given by

$$\nu' = \gamma\left(\nu - \frac{\mathbf{V} \cdot \mathbf{k}}{2\pi}\right). \tag{1.236}$$

If $V \ll c$, then $\gamma = [1 - (V^2/c^2)]^{-1/2} \approx 1$, and the doppler shift is given by

$$\nu' = \nu - \frac{\mathbf{V} \cdot \mathbf{k}}{2\pi} = \nu\left(1 - \frac{V}{v}\cos\theta\right) \qquad (V \ll c), \tag{1.237}$$

where v is the speed of the harmonic wave and θ is the angle between $\mathbf{V}$ and the direction of propagation of the wave. This is the formula for the doppler shift that we deduced in Problem 1.4 on the basis of our notions of space and time applicable to small relative speeds ($V/c \ll 1$).

The right-hand side of

$$\nu' = \nu \, \frac{1 - (V/v) \cos \theta}{\sqrt{1 - (V^2/c^2)}} \qquad (1.238)$$

differs by the factor $[1 - (V^2/c^2)]^{-\frac{1}{2}}$ from the right-hand side of Equation (1.237), which is valid for small V/c. This factor enters into the doppler shift for all angles θ, and in particular, it gives a doppler shift for $\theta = \pi/2$:

$$\nu = \nu' \sqrt{1 - (V^2/c^2)}, \qquad \mathbf{V} \cdot \mathbf{k} = 0. \qquad (1.239)$$

This effect, called the *transverse doppler effect*, is a direct result of time dilatation and hence is unobservable for small relative speeds ($V/c \ll 1$). The transverse doppler effect has been verified experimentally to high accuracy by spectroscopic measurements of the frequency of light emitted by rapidly moving ions.

SUMMARY   The relative phase of a plane wave $\mathbf{k} \cdot \mathbf{r} - \omega t$, at an event $P(ct, \mathbf{r})$ relative to that at the event $O(0, 0)$, is a scalar quantity if the wave function satisfies a linear and homogeneous transformation law. For such a wave, the components $(\omega/c, \mathbf{k})$ relative to an observer S are the components of a 4-vector.

Example 1.15

The doppler effect is the basis for radar measurements of a vehicle's speed. This use of the doppler effect comes about in the following way:

There are two types of radar* (radio direction and ranging) systems, pulsed radar, which measures distances, and doppler radar, which measures speeds. In all radar systems, electromagnetic radiation is emitted by the radar transmitter and is reflected by the object whose position or speed is to be determined (Figure 1.111). The reflected signal is then detected by the radar receiver. The two types of radar, pulsed and doppler, are distinguished by the different physical principles upon which they operate.

In pulsed radar, the electromagnetic radiation transmitted is in the form of very short pulses. The time interval between emission and detection of the pulse is measured (Figure 1.112), giving the distance from the radar set to the reflector. The time interval between transmission and reception for each mile of distance between the radar set and the reflector is

$$2 \text{ mi} \times (186,000 \text{ mi/sec})^{-1} \approx 10 \text{ } \mu\text{sec}.$$

Pulsed radar can be used to measure the speed of approach or recession of a reflector, since for that, we need to measure only the distances to the reflector at two different times. However, it would require very delicate apparatus to measure time intervals to an accuracy ($\sim 1 \text{ } \mu\text{sec}$) sufficient to distinguish between speeds of approach, say, of 60 and 70 mph (see Problem 1.132).

On the other hand, doppler radar measures speed directly, and the apparatus required to determine differences in speeds of even one mile per hour is simple

---

* See R. M. Page, *The Origin of Radar*, Anchor Books, Garden City, N.Y., 1962.

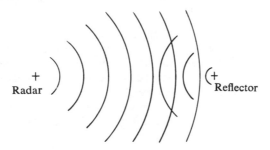

FIGURE 1.111   Transmitted and reflected waves.

and trustworthy. In doppler radar, a continuous harmonic wave, and not a pulse, of electromagnetic radiation is transmitted. One doppler radar device in use transmits such radiation at 2,455 MHz.

This continuous wave is partially reflected by the object whose speed is being measured, but the reflected radiation differs from that transmitted through a change in frequency as a result of the doppler effect. The process of reflection can be considered to take place in two parts. First, the reflector absorbs the transmitted wave at a frequency that is greater than the frequency emitted by the transmitter for an approaching reflector and less for a receding reflector. Second, the reflector emits the reflected wave that the radar receiver detects at an even higher frequency for approaching reflectors or at an even lower frequency for those receding (Problem 1.133). The difference in frequency between that emitted by a radar transmitter at 2,455 MHz and that detected by the radar receiver is 73 Hz for each mile per hour of approach or recession of the reflector (Problem 1.134).

It would be extremely difficult to measure frequencies around 2,455 MHz to such a degree of accuracy that differences of a few tens of hertz could be detected. However, the difference in frequencies between two such waves can be measured directly by use of the phenomenon of beats. The beats produced in a transmitted frequency of 2,455 MHz have frequencies of 73 Hz per mile per hour of reflector speed and are easy to detect. Indeed, the beat frequencies are in the audio range and could be detected as a pure tone in an audio generator, the tone having a higher pitch for higher reflector speeds.

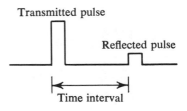

FIGURE 1.112   Shape of oscilloscope trace for pulsed radar system.

Problem 1.127

Show that if $(ct, \mathbf{r})$ is time-like or null, then relative to S, the N of Equation (1.226) is the number of wavelengths that leave the spatial origin $\mathbf{r} = 0$ after the time $t = 0$ and arrive at the position $\mathbf{r}$ in space before time t.

Problem 1.128

A star is moving directly away from the earth with a speed of $0.80c$. The star emits light of frequency 5.09×10^{14} Hz. Calculate the frequency of this light observed at the earth.

Problem 1.129

Calculate, in miles per hour, how fast you would have to travel toward a red light ($\lambda = 6.40 \times 10^{-7}$ m) to see it as green ($\lambda = 5.20 \times 10^{-7}$ m).

Problem 1.130

(a) Consider starlight striking the earth from the direction perpendicular to the plane of the earth's orbit. Calculate the angle of inclination, with respect to that direction for a long narrow terrestrial telescope, necessary in order that the starlight reach the bottom of the telescope.

(b) Compare your answer to (a) with the results derived earlier in Example 1.1. Explain why that previous derivation, based on the galilean transformation law, proved adequate to describe the aberration of starlight.

Problem 1.131

A wave moves with the velocity $v\hat{x}$ relative to S. The observer S' moves with the velocity $V\hat{x}$ relative to S. Find the velocity $\mathbf{v}'$ of the wave relative to S' from the transformation equations (1.235).

Problem 1.132

Two cars, one traveling 60 mi/hr (the legal speed limit) and one at 70 mi/hr, are observed for 30 sec by pulsed radar.

(a) Calculate the time interval measured by the pulsed radar set that corresponds to the distances traveled by these cars in the 30 sec.

(b) What time difference must be detected if the pulsed radar is to be used to determine the difference in speeds in order to obtain a conviction of the driver of the faster car?

Problem 1.133

(a) Show that the frequency of harmonic radiation absorbed by an approaching vehicle is greater than the frequency emitted by the stationary transmitter.

(b) Show that the frequency of harmonic radiation absorbed by a stationary receiver is greater than the frequency emitted by an approaching transmitter.

(c) State the problems corresponding to (a) and (b) for a receding vehicle, and answer them.

Problem 1.134

A doppler radar transmitter emits at 2,455 MHz. Calculate the beat frequency detected by the radar set from a vehicle that is

(a) approaching at 1 mi/hr,
(b) receding at 70 mi/hr.

Problem 1.135

A transverse wave described relative to S by a wave function of the form

$$\boldsymbol{\psi}(x, t) = A \cos (kx - \omega t)\hat{y} \pm A \sin (kx - \omega t)\hat{z}$$

is said to be circularly polarized [see Problem 1.3(f)]. The polarization is said to be right-handed if $\boldsymbol{\psi}(x, t)$, at a given point x, appears to rotate clockwise to an observer in S looking toward the source in the direction opposite to the direction of propagation (Figure 1.113).

(a) Define left-handed polarization.

Direction of propagation

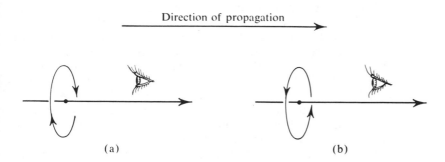

(a)   (b)

FIGURE 1.113 The two cases of circular polarization are shown in (a) and (b).

(b) Which of the signs, + or −, applies to right-handed polarization?
(c) What is the polarization of the wave of (b) as seen by an observer S' moving along the x axis of S? Discuss the polarization as seen by an observer S'' moving in another direction relative to S.

Additional Problems

Problem A1.1

The first evaluation of the speed of light was made by the Danish astronomer O. Roemer (1644–1710). He determined that the innermost satellite of Jupiter, called Io and about the same size as the moon, underwent a regular variation in its period of revolution; as the earth traveled away from Jupiter (for example, at point A of Figure A1.1), the period of Io increased, and as the earth traveled toward Jupiter, the period decreased. The average period of Io is 42.5 hr.

Roemer argued that the real period of the satellite was constant and that the apparent variation resulted from differences in the time delay in the light from Io arriving at the earth.

$$\text{Radius of Jupiter's orbit} = 7.78 \times 10^8 \text{ km,}$$

$$\text{Radius of Io's orbit} = 4.22 \times 10^5 \text{ km.}$$

(a) Calculate the distance that the earth travels in its orbit about the sun during the time of one period of Io.

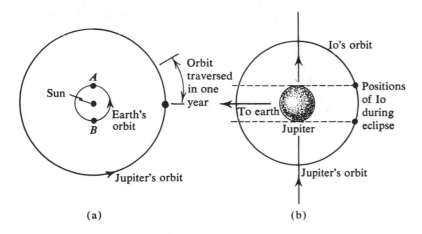

(a)   (b)

FIGURE A1.1 Orbits of Earth, Jupiter, and Io, Jupiter's innermost satellite. (a) The orbits of Jupiter and Earth. (b) The orbit of Io around Jupiter.

(b) Let $t = 0$ correspond to the time at which the earth is nearest Jupiter. Calculate, at time t, the delay or advance in the reappearance of Io after being eclipsed by Jupiter owing to the light traveling the change in the separation distance between the earth and Jupiter from the previous eclipse. Neglect the orbital motion of Jupiter and express your answer in terms of c, the speed of light.

(c) Calculate, in terms of c, the accumulated apparent delays over the six months in which the earth moves from the point nearest Jupiter to its farthest position from Jupiter. Show that this equals the time required for light to cross the diameter of the earth's orbit.

(d) Measurements of the eclipses of Io give 16 min 38 sec for the time required for light to cross a diameter of the earth's orbit. Calculate the speed of light.

(e) Plot the apparent variation of the period of Io's orbital motion as observed at the earth against time.

Problem A1.2

The direction of a star relative to the earth may vary during the course of a year if the direction of the star relative to the sun is fixed, as shown in Figure A1.2. The maximum angle subtended at the star by the radius of the earth's orbit is called the heliocentric or *annual parallax p* of the star. The values of the parallax in seconds of arc for a few stars are shown in the table.

(a) Let D be the distance from the sun to a star with parallax p measured in seconds of arc. Show that $D = 2.063 \times 10^5\ R/p$, where $R = 1.495 \times 10^8$ km is the radius of the earth's orbit.

(b) Calculate the distance, in kilometers and miles, from the sun to each of the stars listed above.

(c) A convenient unit of distance for the study of planetary motions is the *astronomical unit* (AU), equal to the mean earth-sun distance, 1.495×10^{11} m. Calculate, in astronomical units, the distances between each of the stars listed in the preceding table and the sun.

Star	Parallax, arcsec
Sirius	0.37
Vega	0.125
Centauri	0.75
Betelgeuse	0.012
Polaris	0.008

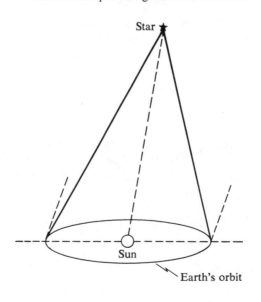
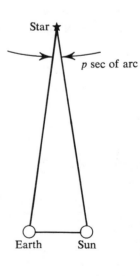

FIGURE A1.2   Parallax.

Star

Star

p sec of arc

Sun

Earth's orbit

Earth   Sun

(d) The *light-year* is the distance that light will travel in one year. Show that

$$1 \text{ light-year} = 9.461 \times 10^{15} \text{ m}.$$

(e) Calculate the distance in light-years between each of the stars listed above and the sun.

(f) How long before now was the light that you can see tonight emitted from each of the stars listed above?

(g) A convenient unit of distance for expressing stellar distances is the *parsec*. One parsec is the distance at which a star would have a *par*allax of one *sec*ond of arc. Show that

$$D = \frac{1}{p} \text{ parsec} = \frac{3.258}{p} \text{ light-years}.$$

(h) Calculate the distance, in parsecs, between each of the stars listed above and the sun.

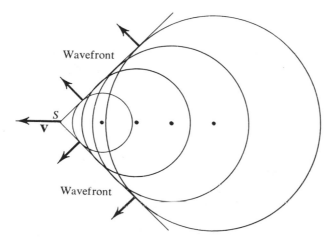

FIGURE A1.3 Wavefronts from a source moving with a velocity **V** relative to a medium in which the speed with which disturbances are transmitted is $v < V$.

Problem A1.3

Shock or bow waves are produced in a material when the source of a disturbance moves faster than the speed at which the waves from the disturbance are carried away from the source. The wave appears as a conical wavefront spreading out from the line of motion of the source (Figure A1.3).

The equivalent of shock waves for light is *Čerenkov radiation*,* which is produced by fast charged particles traversing a transparent medium. The charged particles must have a speed v greater than c/n, the speed of light in that medium. Most of the energy is produced in the visible region of the spectrum and accounts for the observation of Madame Curie of the glowing in the dark of glass bottles containing strong concentrations of radium salts.

(a) Show that the light travels along a cone of apex angle θ to the direction of the moving charged particles (Figure A1.4), where

$$\cos \theta = \frac{1}{\beta n}, \qquad \beta = \frac{v}{c}.$$

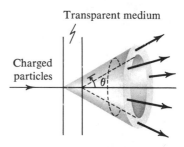

FIGURE A1.4   Cone of Čerenkov radiation.

---

* See J. V. Jelley, "Čerenkov Radiation: Its Origin, Properties and Applications," *Contemporary Physics*, 3: 45 (1961).

(b) Calculate the angle θ for Čerenkov radiation from particles moving with the speed $v \approx c$ through each of the materials listed in Problem 1.18.

Problem A1.4

(a) Show that, for relative speeds v near $c[|(c - v)/c| \ll 1]$, the formulas for Lorentz contraction and time dilatation take the form

$$l = l'\sqrt{2[1 - (v/c)]} \qquad \text{and} \qquad t = \frac{t'}{\sqrt{2[1 - (v/c)]}}.$$

(b) Calculate the Lorentz contraction factor and the time-dilatation factor for each of the following speeds:
   (i) $v = 0.99c$, (ii) $v = 0.999c$, (iii) $v = 0.9999c$.
(c) Find the speeds v for which the Lorentz contraction factor has the following values:
   (i) 10^{-3}, (ii) 10^{-6}, (iii) 10^{-9}.

Problem A1.5

(a) Derive the formula for time dilatation from the special Lorentz transformation equations in the following way: Use the formula

$$t' = \frac{t - (Vx/c^2)}{\sqrt{1 - (V^2/c^2)}}$$

to find relations between the times t_1 and t_2 and the positions x_1 and x_2 of a clock in S' that reads t_1' and t_2', respectively, at the two events. Then use the relation between $x_2 - x_1$ and $t_2 - t_1$ to derive the time-dilatation formula.
(b) Derive the length as determined by S of a measuring rod at rest in S' from the formulas

$$x = \frac{x' + Vt'}{\sqrt{1 - (V^2/c^2)}}, \qquad t = \frac{t' + (Vx'/c^2)}{\sqrt{1 - (V^2/c^2)}}.$$

Hint: Use the last formula to find $t_2' - t_1'$ in terms of the times t_2 and t_1 at which the positions of the ends of the rod are determined by S and the length $x_2' - x_1'$.

Problem A1.6

There is a set of synchronized clocks in each of two inertial systems, S and S'. The clocks in each frame are 1 m apart; they read time in meters. The system S' moves with the velocity $V = 4c\hat{x}/5$ relative to S. Draw a diagram showing the positions, according to S, of a few of the S' clocks at time $t = 0$ and the times t' that they read.

Problem A1.7

Determine the visual appearance of a rapidly moving sphere that is sufficiently far away that the light received by the eye is essentially a parallel beam. Determine the appearance only for that time interval in which the velocity of the sphere is essentially perpendicular to the line joining the sphere to the eye.

Problem A1.8

An inertial observer S draws a circle $x^2 + y^2 = 1$ in his xy plane.

(a) Show this spatial curve on the space-time diagram of S.
(b) An inertial observer S' moves with the velocity $4c\hat{x}/5$ relative to S. Draw his coordinate axes on the space-time diagram of S and mark the scales on these axes.
(c) What curve does S' observe as the circle of S?
(d) Explain your answer to (c) in terms of the space-time diagrams of (a) and (b).

Problem A1.9

The world line of a particle is given by the equation $x = \sqrt{4 + \tau^2} - 2$ relative to an inertial observer S.

(a) Draw the world line of the particle on the space-time diagram of S.
(b) Draw the 4-velocity on the world line at the points where the world line crosses the coordinate lines $\tau = 0$, $\tau = 1$, $\tau = 2$, and $\tau = 3$.
(c) Calculate the proper time t_0 of the particle at the time t relative to S.

Problem A1.10

A meter stick travels along the x axis of S with velocity $3c\hat{x}/5$ relative to the inertial observer S. A flat table, parallel to the x axis and containing a hole of length 1 m along the x direction, moves with velocity $c\hat{y}/5$ relative to S (Figure A1.5). The positions of the center of the meter stick, the center of the hole, and the origin $x = y = 0$ of the spatial coordinate axes of S coincide at $t = 0$. According to S, the meter stick is contracted in its direction of motion and therefore should pass through the hole. At first glance, it might be argued that, relative to the meter stick, the hole is Lorentz contracted and the meter stick cannot pass through the hole. Resolve this paradox* by determining the positions of the edges of the hole as viewed at one instant of time from the reference system of the meter stick.

Problem A1.11

Relative to an inertial observer S, a thin thread 1 m long is stretched to its breaking point along the x axis between the points $x = 0$ and $x = 1$. At time $\tau = 0$, the ends of the thread are accelerated almost instantaneously to a velocity of $3c\hat{x}/5$.

(a) Draw the world lines of the ends of the thread on the space-time diagram of S.
(b) Draw the world lines of the ends of the thread on the space-time diagram of an inertial observer S' moving with velocity $3c\hat{x}/5$ relative to S.
(c) Does the thread break when it undergoes the acceleration?† Explain why from the point of view of S and also from the point of view of S'.

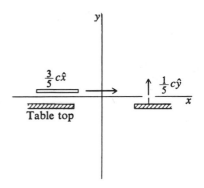

FIGURE A1.5 Does the stick pass through the hole?

---

* See R. Shaw, "Length Contraction Paradox," *American Journal of Physics, 30*: 72 (1962).
† See E. Dewan and M. Beran, "Note on Stress Effects Due to Relativistic Contraction," *American Journal of Physics, 27*: 517 (1959), A. A. Evett and R. K. Wangness, "Note on the Separation of Relativistically Moving Rockets," *American Journal of Physics, 28*: 566 (1960), and J. E. Romain, "A Geometrical Approach to Relativistic Paradoxes," *American Journal of Physics, 31*: 576 (1963).

Problem A1.12

A rod moves horizontally with velocity $\beta c\hat{x}$ over a flat table in which there is a hole equal in size to the proper length of the rod. It is argued from the point of view of the table that the rod is shorter than the hole is and that, therefore, the rod will undergo an acceleration g, directed downward, when it is over the hole and thus hit the far side. On the other hand, it is argued from the point of view of the rod that the hole is shorter and the straight rod will pass over the hole.

(a) Explain why the visual appearance of objects—for example, the rotation of an object considered in Section 1.6, plays no role in the resolution of a paradox like this.
(b) Let S be the reference system of the table and S' be that system that moves with velocity $\beta c\hat{x}$ relative to S. Relative to S, we assume that the rod undergoes the acceleration $-g\hat{y}$. Suppose the rod is straight according to S. Find its shape relative to S' and resolve the paradox.*
(c) The forces that the table exerts to support the rod disappear as the rod moves over the hole. These forces disappear along the rod, because of its motion, faster than the speed of sound, which is the speed with which disturbances propagate along the rod. Can the rod adjust to the loss of support as it moves across the edge of the hole? Discuss the assumption that the rod is straight according to S.

Problem A1.13

The speed of light in a transparent medium with index of refraction n is c/n relative to that medium. Show that the speed of light in the x direction in the transparent medium relative to an observer moving with the velocity $v\hat{x}$ relative to the medium is,† to first order in v/c,

$$v' = \frac{c}{n} \pm v\left(1 - \frac{1}{n^2}\right).$$

Problem A1.14

The transverse doppler effect has been measured indirectly by H. E. Ives and G. R. Stilwell.‡ They used the apparatus shown schematically in Figure A1.6 to determine the wavelengths of light in a particular spectral line of moving hydrogen atoms emitted in the backward and forward directions. The average of these wavelengths is shifted from the wavelength of the same spectral line emitted by stationary atoms by an amount given by the formula for the transverse doppler effect, as shown in the following problem.

(a) Let λ_0 be the wavelength of the spectral line relative to that reference system

---

* See W. Rindler, "Length Contraction Paradox," *American Journal of Physics*, 29: 365 (1961), and W. H. Wells, "Length Contraction Paradox," *American Journal of Physics*, 29: 858 (1961).
† This result agrees with experiments on the speed of light in moving liquids carried out by Fizeau. Since the velocity v′ is not simply the sum of the velocity of light in the medium and the velocity of the medium, it was argued that a material medium, such as the earth's atmosphere, is not the "preferred" frame of reference in which the ether is at rest. See P. G. Bergmann, *Introduction to the Theory of Relativity*, Prentice-Hall, Englewood Cliffs, N.J., 1942, pp. 20–21.
‡ H. E. Ives and G. R. Stilwell, "An Experimental Study of the Rate of a Moving Atomic Clock," *Journal of the Optical Society of America*, 28: 215 (1938).

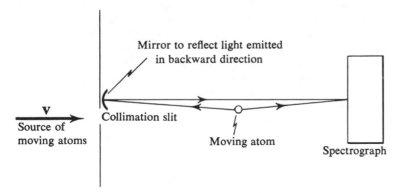

FIGURE A1.6 The Ives–Stilwell experiment.

in which the atom is at rest. Show that, in the reference system S in which the atom moves with velocity $\mathbf{V}$, the wavelength λ is given by

$$\lambda = \frac{\lambda_0(1 - \beta \cos \theta)}{\sqrt{1 - \beta^2}},$$

where $\beta = V/c$ and θ is the angle between $\mathbf{V}$ and the direction of propagation of the wave relative to S.

(b) The transverse doppler shift appears for $\theta = \pi/2$. Ives and Stilwell did not make their measurements at a right angle to the direction of $\mathbf{V}$, since it only required a very small deviation from this direction to produce an additional doppler shift comparable to the transverse shift expected. Determine the angle α relative to $\theta = \pi/2$ at which the transverse shift and the additional doppler shift are equal for $\beta = 0.004$, the order of speeds used by Ives and Stilwell. Determine the angle α' relative to $\theta = 0$ and the angle α'' relative to $\theta = \pi$ for which the angle dependent term of the equation in (a) gives a contribution equal to the $\sqrt{1 - \beta^2}$ factor.

(c) Show that the average of $\lambda_{\theta=0}$ and $\lambda_{\theta=\pi}$ is equal to the wavelength $\lambda_{\theta=\pi/2}$ that results from the transverse doppler effect.

Problem A1.15

The German astronomer Heinrich Wilhelm Olbers (1758–1840) published in 1826 a paper on the problem of the darkness of the night sky, a problem involving a calculation of the total light that should reach the earth from the stars if a few reasonable assumptions are satisfied. The calculation in this problem will show, on the basis of these assumptions, that the temperature of the earth should be a few thousand degrees, obviously in contradiction with reality. *Olbers' paradox,** the disagreement between the plausible calculation and actuality, was resolved by the work of the American astronomer Edwin P. Hubble (1889–1953), who, in 1929, announced that distant galaxies of stars are receding from us with a speed proportional to their distance. Part (a) of this problem concerns Olbers' paradox, and part (b), Hubble's law.

---

* The origin of this name for the problem of the darkness of the night sky is discussed in S. L. Jaki, "Olbers', Halley's, or Whose Paradox?," *American Journal of Physics*, *35*: 200 (1967).

(a) Assume that the whole universe is much like the part that we can see from the earth—that is, assume that everywhere in the universe the distribution of stars over large regions is uniform and that the stars, on the average, are similar to those in the neighborhood of the sun. Also, since the light that reaches us now from the distant stars was emitted long ago, assume that the average distribution and brightness have not changed appreciably with time. Furthermore, make the assumption that there are no systematic motions, so that the universe is static.

(i) Perform the following calculation of the light that travels toward us from all the stars. Suppose there are N stars per unit volume on the average. Show that, in a shell of radius R, thickness ΔR, centered at the earth, there are $4\pi R^2 N \Delta R$ stars. The light energy from an average star that travels toward us per unit time is proportional to $1/R^2$, say equal to κ/R^2; show that the light from the shell that travels toward us is $\kappa 4\pi N \Delta R$. Show that the light that travels toward us from all stars is infinite, on the basis of the above assumptions. Is this equal to the light that reaches the earth? Explain your answer.

(ii) The light that reaches us from a star, if the line of sight from the earth to the star is not broken by another star or interstellar dust, is proportional to $1/D^2$, where D is the distance to the star. Consider each star to be of average brightness. A measure of the portion of the sky cut out by the star is provided by the solid angle Ω subtended at the earth by the star (Figure A1.7). The solid angle Ω is equal to the ratio of the area A of a sphere of radius R centered at the earth, cut out by the cone with vertex at the earth and surrounding the star, to the square of the radius of the sphere R^2: $\Omega = A/R^2$. The unit of solid angle is the steradian.

Show that the light that reaches us from the star is proportional to the solid angle Ω, subtended by that part of the star in a direct line of sight to the earth. We denote the proportionality constant by α; thus the light reaching us from the star is $\alpha\Omega$. Under the above assumption of uniformity, each line of sight would end in a star. Show that the total light that should reach us, on the basis of the above assumptions, is $4\pi\alpha$, the same as if the earth were surrounded by a sphere of any radius at the temperature of the average star, some thousands of degrees.

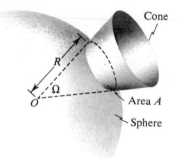

FIGURE A1.7 $\Omega = A/R^2$ sterad.

(iii) The temperature of the sun is $6,000°K$. Explain, as you would to a high school student, the relation of the above calculation to daylight and the darkness at night.

(b) Hubble showed that lines in the spectra of galaxies were shifted toward the red or long wavelengths*; the red shift $\Delta\lambda$ in wavelength λ per unit wavelength is directly proportional to the distance r from the earth†: $\Delta\lambda/\lambda = (v_r/c)r$, where v_r is the Hubble constant whose present accepted value is $3 \times 10^{-18}/\text{sec}$.

---

* See A. R. Sandage, "The Red-Shift," *Scientific American*, *195*: 170, September 1956.
† The distances were calculated partly on the basis of a relation between the brightness and the periods of some variable stars. See, for example, O. Struve, B. Lynds, and H. Pillans, *Elementary Astronomy*, Oxford Univ. Press, New York, 1959; R. P. Kraft, "Pulsating Stars and Cosmic Distances," *Scientific American*, *201*: 48, July 1959; and O. C. Wilson, "A New Scale of Stellar Distances," *Scientific American*, *204*: 107, January 1961.

(i) Express v_r in kilometers per second·megaparsecs.

(ii) Show that the red shift can be interpreted as a motion of the galaxies away from the earth with recession speeds $V = v_r r$, for $V/c \ll 1$. *Hint:* Consider the doppler shift in wavelength for $V/c \ll 1$.

(iii) Hubble observed red shifts up to $\Delta\lambda/\lambda \sim 1/8$. Calculate the speed of recession and the distance of a galaxy showing this red shift.

(iv) The quasars [6] (quasistellar radio sources)* 3C273 and 3C48 show red shifts of 0.16 and 0.37. Find their speeds of recession and distances from Hubble's relationship, and show in each case that the neglect of V/c compared to unity provides a good approximation.

(v) The motion of galaxies away from the earth appears to place the earth in a unique position, the center of the expansion. Show that if, in fact, the universe is expanding, every point in the universe can be considered equally well as the center of expansion. *Hint:* Consider a reference frame scaffolding, such as that shown in Figure 1.38, in which the distances between points of juncture of the scaffolding are increasing linearly with time. Choose an arbitrary point as origin and show that each juncture point is receding from that origin with a speed proportional to its distance from the origin.

(vi) Assume that the universe was created by an explosion of condensed matter that is and has been expanding at the rate given by Hubble's relationship. Calculate the age of the universe on the basis of this assumption.†

(vii) Explain how the expansion of the universe can be used as a basis for the resolution of Olbers' paradox. *Hint:* Consider the light energy emitted by a star per unit time and its relation to the energy per unit volume, as the volume expands.

Problem A1.16

(This problem should be attempted only if you are familiar with the principle of equivalence and the formulas for the potential energies due to gravity. You may wish to do this Problem after you have studied Section 3.2.)

The principle of equivalence states that an inertial reference frame at rest in a region of a gravitational force is equivalent to a noninertial reference frame in which no gravitational forces are present but which is undergoing an acceleration relative to the fixed stars. Thus, the principle of equivalence can be combined with the special theory of relativity to provide information on the behavior of light under gravitation.‡

(a) Consider an emitter of light of frequency ν_0 at a height h above a light receiver near the earth's surface. Determine the frequency ν of the light that strikes the receiver in the following way: Replace the system of emitter and receiver at rest in the region of gravitational force by a noninertial system in a force-free region, undergoing an acceleration g upward relative to an inertial frame. Let Δv be the difference in the speed of the receiver when

---

* See J. L. Greenstein, "Quasi-Stellar Radio Sources," *Scientific American, 209*: 54, December 1963.
† See, for example, H. Bondi, *The Universe at Large*, Anchor Books, Garden City, N.Y., 1960, for a description of cosmological theories that can be used to explain the expansion.
‡ See L. I. Schiff, "On Experimental Tests of the General Theory of Relativity," *American Journal of Physics, 28*: 340 (1960).

the light is received and the speed of the emitter when the light was emitted. Show that $\Delta v = gh/c$ and that $v \approx v_0[1 + (\Delta v/c)] = v_0[1 + (gh/c^2)]$.

  (i) R. V. Pound and G. A. Rebka, Jr.,* measured $(v - v_0)/v_0$ for a fall of $h = 74$ ft and obtained $(2.56 \pm 0.26) \times 10^{-15}$. Compare their result with the prediction of (a).

(b) Consider two clocks A and B separated in a region of gravitational force by a distance h along the direction of gravitational force. Replace the system as in (a), and let the clocks move past an inertial clock C. Let v_A be the speed of A past C and v_B that of B. Show that the periods T_A, T_B, and T_C of the clocks at A, B, and C, respectively, are related by

$$\Delta T_A = \Delta T_C[1 - (v_A^2/2c^2)], \quad \Delta T_B = \Delta T_C[1 - (v_B^2/2c^2)],$$

so that

$$\Delta T_B = \Delta T_A\left(1 - \frac{[v_B^2 - v_A^2]}{2c^2}\right) = \Delta T_A\left(1 - \frac{gh}{c^2}\right).$$

Hint: Assume that v_A^2/c^2 and v_B^2/c^2 are both much smaller than unity.

  (i) Show that $\Delta T_B = \Delta T_A\{1 + [(V_A - V_B)/c^2]\}$, where V_A and V_B are the gravitational potential energies per unit mass at A and B, respectively.

  (ii) Light of frequency v_0 is emitted at A and absorbed as light of frequency v at B. Show from (b) that

$$v = v_0\left(1 + \frac{gh}{c^2}\right).$$

  (iii) Compare a series of clocks at various heights above a star of mass M and radius R to show that light of frequency v_0 emitted at the surface of the star is determined to have frequency v a large distance from the star, with

$$v = v_0\left(1 - \frac{GM}{Rc^2}\right).$$

This shift in frequency toward the red end of the visible light spectrum is called the gravitational red shift and must be distinguished from the recessional red shifts of distant galaxies (Problem A1.15).

Problem A1.17

(a) The components relative to an observer S of a skew-symmetric tensor of the second order, are $T_{\mu\nu}$.

  (i) Show that $T_{\mu\mu} = 0$.

  (ii) Show that $T_{\mu\nu}$ has only six independent nonvanishing components, and list these.

(b) Two 4-vectors with components a_μ and b_μ relative to S are a and ℓ.

  (i) Show that $a_\mu b_\nu - a_\nu b_\mu$ is a skew-symmetric tensor of the second order.

  (ii) The cross or *vector product* of two 3-vectors $\mathbf{a}$ and $\mathbf{b}$ is defined as a vector written as $\mathbf{a} \times \mathbf{b}$ (or $\mathbf{a} \wedge \mathbf{b}$) (Figure A1.8) with magnitude $|\mathbf{a}|\,|\mathbf{b}|\sin\theta$, direction perpendicular to the plane of $\mathbf{a}$ and $\mathbf{b}$, and with sense as determined by the right-hand rule shown in Figure A1.9:

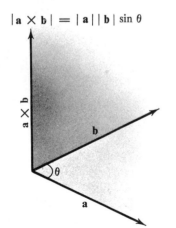

$$|\mathbf{a} \times \mathbf{b}| = |\mathbf{a}|\,|\mathbf{b}|\sin\theta$$

FIGURE A1.8 The definition of the vector product $\mathbf{a} \times \mathbf{b}$.

---

* R. V. Pound and G. A. Rebka, Jr., "Apparent Weight of Photons," *Physical Review Letters, 4:* 337 (1960).

curl the fingers of the right hand from **a** to **b** and the thumb designates the sense. Show that the magnitude of **a** × **b** is equal to the area of the parallelogram with sides **a** and **b**. Show that **a** × **b** = −**b** × **a**.

(iii) Show that the magnitude of **a** × **b** is equal to |**a**| times the magnitude of the component of **b** perpendicular to **a**.

(iv) Use the result of (iii) to show that

$$(\mathbf{a} + \mathbf{b}) \times \mathbf{c} = \mathbf{a} \times \mathbf{c} + \mathbf{b} \times \mathbf{c}.$$

(v) Show that $\hat{x} \times \hat{x} = 0$ and $\hat{x} \times \hat{y} = \hat{z}$. Find the vector products of all other combinations of unit vectors along the directions of the coordinate axes.

(vi) Use the results of (iv) and (v) to show that

$$\mathbf{a} \times \mathbf{b} = (a_2 b_3 - a_3 b_2)\hat{x} + (a_3 b_1 - a_1 b_3)\hat{y} + (a_1 b_2 - a_2 b_1)\hat{z}.$$

(vii) Show that **a** × **b** determined by the use of a left-hand rule is opposite in sense to **a** × **b** determined with a right-hand rule. A vector that changes sign under the exchange of a left for a right hand is called a pseudovector or a polar vector to distinguish it from a true or axial vector.

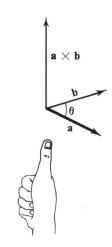

FIGURE A1.9   Right-hand rule for the determination of the sense of **a** × **b**.

(c) Consider the transformation

$$x' = x \cos \theta - y \sin \theta, \qquad y' = x \sin \theta + y \cos \theta, \qquad z' = z, \qquad \tau' = \tau.$$

(i) Show that the components (P_1, P_2, P_3), where $P_1 = T_{23}$, $P_2 = T_{31}$, and $P_3 = T_{12}$, of any skew-symmetric tensor of the second order behave like the components of a 3-vector under that transformation.

(ii) Show that the components (V_1, V_2, V_3), where $V_1 = T_{01}$, $V_2 = T_{02}$, and $V_3 = T_{03}$, of any skew-symmetric tensor of the second order behave like the components of a 3-vector under that transformation.

(d) Consider the transformation

$$x' = -x, \qquad y' = -y, \qquad z' = -z, \qquad \tau' = \tau.$$

(i) Show that the components (a_1, a_2, a_3) of a 3-vector become $(-a_1, -a_2, -a_3)$ under that transformation.

(ii) Show that the components (d_1, d_2, d_3) of a pseudo 3-vector become (d_1, d_2, d_3) under that transformation.

(iii) Show that **V** is a true 3-vector and **P** is a pseudo 3-vector.

(e) Let S' be a reference system that is moving with the velocity $\mathbf{v} = \beta c \hat{x}$ relative to S.

(i) Suppose that $V_1 = V_2 = V_3 = 0$. Find $(V_{1'}, V_{2'}, V_{3'})$ in terms of the 3-vector **P** with components (P_1, P_2, P_3). Show that

$$\mathbf{V'} = \frac{\mathbf{v} \times \mathbf{P}}{c\sqrt{1 - (v^2/c^2)}}.$$

(ii) Suppose that $P_1 = P_2 = P_3 = 0$. Find the components $(P_{1'}, P_{2'}, P_{3'})$ in terms of $\mathbf{V} = (V_1, V_2, V_3)$. Show that

$$\mathbf{P'} = \frac{-\mathbf{v} \times \mathbf{V}}{c\sqrt{1 - (v^2/c^2)}}.$$

(iii) Find **P'** and **V'** in terms of **P** and **V** in the case in which neither **P** nor **V** vanishes.

Problem A1.18

(This problem is for students who have some familiarity with matrices.)

(a) Show that the components $T_{\mu\nu}$ of a tensor of the second order can be written in matrix form.

(b) Show that the components $T_{\mu\nu}S_{\nu\rho}$, calculated according to our summation convention, can be calculated from the matrix product of the following three matrices:
   (i) that corresponding to $T_{\mu\nu}$,
   (ii) that with elements $g_{00} = -g_{11} = -g_{22} = -g_{33} = 1$ and $g_{\mu\nu} = 0$, $\mu \neq \nu$,
   (iii) that corresponding to $S_{\nu\rho}$.

(c) Write the components b_μ as the elements of a column matrix with one element per row and the components a_μ as the elements of a row matrix with one element per column. Show that $a \cdot b$ is equal to the matrix product of the following three matrices:
   (i) that corresponding to a_μ,
   (ii) that of (b)(ii) above,
   (iii) that corresponding to b_μ.

(d) Show that $T_{\mu\mu}$, calculated according to our summation convention, is the trace of the matrix formed by the matrix product of
   (i) the matrix corresponding to $T_{\mu\nu}$,
   (ii) the matrix of (b)(ii) above.

ADVANCED REFERENCES

1. Other methods of measuring c are described in E. Bergstrand, "Determination of the Velocity of Light," *Encyclopedia of Physics*, S. Flügge (Ed.), vol. 24, Springer-Verlag, Berlin, 1956, in J. H. Sanders, *The Velocity of Light*, Pergamon Press, New York, 1965, and in J. F. Mulligan and D. F. McDonald, "Some Recent Determinations of the Velocity of Light II," *American Journal of Physics*, *25*: 180 (1957).

2. A list of repetitions of the Michelson–Morley experiment is given in Table 1 of R. S. Shankland, S. W. McCuskey, F. C. Leone, and G. Kuerti, "New Analysis of the Interferometer Observations of Dayton C. Miller," *Reviews of Modern Physics*, *27*: 167 (1955). See also J. P. Cedarholm, G. F. Bland, B. L. Havens, and C. H. Townes, "New Experimental Test of Special Relativity," *Physical Review Letters*, *1*: 342 (1958), reprinted in *Special Relativity Theory, Selected Reprints*, American Institute of Physics, New York, 1963.

3. The contribution of H. Poincaré to the theory of relativity is discussed in C. Scribner, Jr., "Henri Poincaré and the Principle of Relativity," *American Journal of Physics*, *32*: 672 (1964).

4. A direct verification of this postulate is described in D. Sadeh, "Experimental Evidence for the Constancy of the Velocity of Gamma Rays, Using Annihilation in Flight," *Physical Review Letters*, *10*: 271 (1963). See also T. A. Filippas and J. G. Fox, "Velocity of Gamma Rays from a Moving Source," *Physical Review*, *135*: B1071 (1964).

5. J. Terrell, "Invisibility of the Lorentz Contraction," *Physical Review*, *116*, 1041 (1959).

6. See H. Y. Chiu, "Gravitational Collapse," *Physics Today*, *17*: 21, May 1964, and also G. C. McVittie, *Physics Today*, *17*: 70, July 1964.

Special Relativity Theory: Introductory Dynamics

2

Newton's laws of motion have been remarkably successful in describing the motions of macroscopic objects, but the kinematics of newtonian mechanics, and thus perhaps the range of validity of Newton's laws, are limited to circumstances in which *every one* of the relative speeds involved is very small compared to c. This limitation is generally not restrictive in those problems involving the motions of macroscopic objects only; the upper limit in speed is imposed, in practice, by the fact that tremendous energies would be required to accelerate macroscopic objects to speeds near that of light.

Atomic and subatomic particles can be accelerated to such speeds, however, so we turn now to the problem of describing the motions of objects under circumstances in which at least one of the relative speeds involved is comparable to c. We must discover whether Newton's laws are valid even if some of the relative speeds involved are large. If they are not valid, we must determine what modifications are required in these laws of motion. Answers to these problems form the subject matter of this chapter.

There are two theories, each correct within a restricted range of application, that we will need to consider in our discussions of these problems. The first of these theories is the kinematics of special relativity that we studied in Chapter 1. The second is newtonian mechanics, which is so useful in describing the motions of familiar objects in those circumstances in which newtonian and relativistic kinematics coincide. Any relativistic theory of dynamics must agree with newtonian mechanics, at least in that limit, and so it is worthwhile at this point to review the structure of newtonian mechanics. The problems at the end of this introduction provide material for further study in the field of newtonian mechanics.

One class of motions, force-free motions, is distinguished above all others in newtonian mechanics: by definition, a particle that moves with a constant velocity relative to any inertial frame of reference experiences no force. According to newtonian mechanics, deviations from this class of motions are attributable to the presence of objects within the environment of the given particle. The objects are said to interact with the particle.

The interaction between only two particles results in each experiencing an acceleration (Figure 2.1), and the two accelerations are related: the ratio of the magnitudes of the accelerations, defined as the inverse ratio of the masses,

$$\frac{m_1}{m_2} = \frac{|\mathbf{a}_2|}{|\mathbf{a}_1|},$$

(2.1)

is the same for any two given particles independent of the means by which they interact. If the interaction results in one particle experiencing a smaller acceleration than the other, its mass is larger than the other's, so the ratio of the masses is a measure of the relative inertia of the two particles. The masses of objects are given in terms of the mass of an arbitrarily chosen standard. In addition, mass is conserved in all reactions that are described by newtonian mechanics.

The deviation from motion under no force,

$$\mathbf{v}(t) = \text{a constant vector}, \tag{2.2}$$

is measured by the acceleration with respect to an inertial frame

$$\mathbf{a} = \frac{d\mathbf{v}}{dt} = \frac{d^2\mathbf{r}}{dt^2} \tag{2.3}$$

or, more conveniently, by the product mass times acceleration. This is defined as the force experienced by the particle:

$$\mathbf{F} = m\mathbf{a}. \tag{2.4}$$

The force experienced by a particle is determined, according to Newton's second law of motion, by the objects in the environment of the particle. Thus, the differential equation that describes the motion of the particle $\mathbf{r}(t)$ in terms of the environment and the (arbitrarily) given initial position and velocity is Newton's equation of motion

$$m\frac{d^2\mathbf{r}}{dt^2} = \mathbf{F}(\text{environment}). \tag{2.5}$$

Moreover, the force is a vectorially additive function of the environment; the force due to two objects in the environment is the vector sum of the forces that would result from the presence of each of the objects in the absence of the other.

This principle of the additivity of environments means that all forces may be reduced to a sum of fundamental forces, forces between two particles that depend on the intrinsic properties of the particles. Newtonian mechanics depends very critically on the properties of these fundamental forces. There must be few of these in nature, and they must have a simple mathematical form or physical phenomena would be too complicated to describe within the newtonian formalism.

There do exist two fundamental forces that are simple in mathematical form: the electrostatic force and the gravitational force. The electrostatic force is the electric force between two stationary particles, each of which possesses the property of electric charge, given by Coulomb's law as

$$\mathbf{F}_{2(1)} = K\frac{q_1q_2}{r^2}\,\hat{r}. \tag{2.6}$$

Here $\mathbf{F}_{2(1)}$ is the electrostatic force experienced by particle 2 due to the presence of particle 1, $\mathbf{r}$ is the position vector of particle 2 relative to particle 1 (Figure 2.2), q_1 and q_2 are the respective charges of the particles, and K is a positive universal constant. In the system of units in which forces are measured in newtons, distances in meters, and charges in coulombs,

$$K = 8.988 \times 10^9 \text{ N} \cdot \text{m}^2/\text{C}^2. \tag{2.7}$$

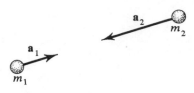

FIGURE 2.1   Accelerations produced in two interacting particles.

The gravitational force is given by Newton's law of gravitation as

$$\mathbf{F}_{2(1)} = -G\frac{m_1 m_2}{r^2}\hat{r}, \qquad (2.8)$$

where $\mathbf{F}_{2(1)}$ is the gravitational force, m_1 and m_2 are the inertial masses,* and G is a positive universal constant given by

$$G = 6.670 \times 10^{-11}\,\mathrm{N\cdot m^2/kg^2}. \qquad (2.9)$$

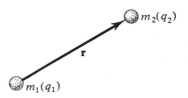

FIGURE 2.2   Separation vector **r**.

The electrostatic and gravitational forces do not describe all the interactions experienced by particles. For example, there are also velocity-dependent interactions called magnetic forces that come into play between two charged particles in relative motion. The problems involved in describing magnetic interactions in terms of the relative position vector and the relative velocity between two particles led in the nineteenth century to the development of another type of theory, field theory, for the description of electromagnetic phenomena. We shall not consider those problems, but we shall use the formula for the magnetic force experienced by one particle of charge q and velocity $\mathbf{v}$ due to the presence of other moving charges:

$$\mathbf{F} = q\mathbf{v} \times \mathbf{B}, \qquad (2.10)$$

where $\mathbf{B} = \mathbf{B}(\mathbf{r}, t)$ is a vector field called the magnetic induction, which represents the effects of the other charges and will be given where needed in the work to follow.†

There are other forms for forces, such as friction, that describe in a phenomenological manner the effects of many interactions. These forces are not fundamental in nature, however, and so will be excluded from our discussions.

Finally, the forces between pairs of particles in a system are not all independent. A system composed of particles, each moving according to the equation of motion

$$m_i \frac{d^2\mathbf{r}_i}{dt^2} = m_i\mathbf{a}_i = \mathbf{F}_i, \qquad i = 1, 2, 3, \ldots, \qquad (2.11)$$

appears as one object whose motion, according to Newton's laws, satisfies the equation

$$M \frac{d^2\mathbf{R}}{dt^2} = M\mathbf{A} = \mathbf{F}. \qquad (2.12)$$

This behavior is explained in newtonian mechanics by Newton's third law, which states that if the force experienced by the ith particle due to the presence of the jth particle is $\mathbf{F}_{i(j)}$, then

$$\mathbf{F}_{i(j)} = -\mathbf{F}_{j(i)}. \qquad (2.13)$$

For any system of particles, the equations of motion and knowledge of the forces involved completely determine the motions in terms of the initial

---

* The masses thus correspond to gravitational charges. The fact that the coefficients in the formula for the gravitational force can be determined as a measure of inertia in a collision experiment that does not depend on gravitational interactions is important in modern theories of gravitation (see Chapter 3).

† In units in which forces are measured in newtons, charges in coulombs, and velocities in meters per second, $\mathbf{B}$ is measured in webers per square meter (Wb/m²) or teslas (T).

positions and velocities. Therefore, armed with this knowledge we can find the time variation of any entity that can be measured and is associated with the particle or its motion. Such an entity is called a dynamical variable. Dynamical variables, in general, vary with the time in a manner dependent on the motions under consideration. There do exist, however, some dynamical variables that do not change with time if the interactions are described by fundamental forces. That these dynamical variables are conserved is a very general feature of motions, and provides partial information on the behavior of a system even in those cases where we do not or cannot solve the equations of motion. The discussion in the rest of this chapter will involve only two of these conservation laws.

An isolated system of particles may be viewed as an object, and within the framework of newtonian mechanics the behavior of that object is subject to Newton's equation of motion. Therefore, since by definition there are no external forces acting on an isolated system, some average position within the object moves with constant velocity. It is a consequence of Newton's laws that this average position is described by a vector equal to the mass-weighted average position vector of all the particles in the system:

$$\mathbf{R} = \frac{m_1 \mathbf{r}_1 + m_2 \mathbf{r}_2 + \cdots}{m_1 + m_2 + \cdots} \tag{2.14}$$

is the position vector of the center of mass. We introduce the (linear) momentum of the ith particle as

$$\mathbf{p}_i = m_i \mathbf{v}_i \tag{2.15}$$

and obtain from Newton's laws that $\mathbf{P} = M \, d\mathbf{R}/dt$, where

$$M = m_1 + m_2 + \cdots \tag{2.16}$$

and

$$\mathbf{P} = m_1 \mathbf{v}_1 + m_2 \mathbf{v}_2 + \cdots = \frac{d}{dt}(m_1 \mathbf{r}_1 + m_2 \mathbf{r}_2 + \cdots), \tag{2.17}$$

is conserved. This conservation law of linear momentum is independent of the form of the forces of interaction between the particles of the system.

The other conservation law that will be of interest to us directly involves the form of these forces. The law depends on them through a scalar function of the position vectors $\mathbf{r}_1, \mathbf{r}_2, \ldots$ of the particles of the system called the potential energy $V(\mathbf{r}_1, \mathbf{r}_2, \ldots)$. The potential energy is defined* in terms of the forces $\mathbf{F}_1, \mathbf{F}_2, \ldots$ experienced by the particles through the equation

$$V(\mathbf{r}_1 + d\mathbf{r}_1, \mathbf{r}_2 + d\mathbf{r}_2, \ldots) - V(\mathbf{r}_1, \mathbf{r}_2, \ldots) = -\mathbf{F}_1 \cdot d\mathbf{r}_1 - \mathbf{F}_2 \cdot d\mathbf{r}_2 \ldots, \tag{2.18}$$

and the value, say $V = 0$, for some specified reference configuration such as that for which the particles are sufficiently far apart that the forces can be neglected. The dynamical variable E defined by

$$E = \tfrac{1}{2} m_1 v_1^2 + \tfrac{1}{2} m_2 v_2^2 + \cdots + V(\mathbf{r}_1, \mathbf{r}_2, \ldots) \tag{2.19}$$

is the energy of the system, and is a conserved dynamical variable. The energy of the system includes the kinetic energy of each particle, $\tfrac{1}{2} m_i v_i^2$, that is equal to the work done on the particle between a state of rest and a state of motion with velocity $\mathbf{v}_i$:

---

* It is not possible to define a potential energy function for every type of force—the force of friction, for example—but we have excluded such nonfundamental forces from consideration here.

$$\tfrac{1}{2}m_iv_i^2 = \int_{v_i'=0}^{v_i'=v_i} d(\tfrac{1}{2}m_iv_i'^2) = \int_0^{v_i} m_i\mathbf{v}_i \cdot d\mathbf{v}_i = \int_0^{v_i} m_i\mathbf{a}_i \cdot \mathbf{v}_i \, dt$$
$$= \int_0^{v_i} \mathbf{F}_i \cdot d\mathbf{r}_i, \tag{2.20}$$

where $\mathbf{F} \cdot d\mathbf{r}$ is the work done by the force $\mathbf{F}$ on the particle during its motion through the displacement $d\mathbf{r}$.

The conservation law of energy sometimes can be reduced to a form that does not depend on the form of the force functions. Consider a system of particles that are initially moving freely (without interacting), then interact and finally move away from each other to become free again. The energy E is conserved throughout the motion so the initial energy is equal to the final energy. The relation given by the conservation law

$$\left(\sum \frac{1}{2} m_iv_i^2 \right)_{\text{initial}} = \left(\sum \frac{1}{2} m_iv_i^2 \right)_{\text{final}} \tag{2.21}$$

does not depend on the form of the potential energy. Thus we conclude our review of newtonian mechanics.

Section 2.1 is concerned with a critical examination of the role of the newtonian concepts of space and time in the formulation of Newton's laws and the statement of a principle that permits our extending these laws to yield descriptions of motions involving large relative speeds. We shall find that Newton's laws involve the concept of absolute simultaneity in a critical manner and these laws must be modified in some way. Where can we look for this modification? Since the problem arises from the fact that the relation between various inertial frames is different from what we had believed previously, we will investigate that part of mechanics that involves the relation between the description of motions relative to different frames, the principle of relativity.

This principle, which may be deduced from Newton's laws of motion, remains valid even in those circumstances under which Newton's laws no longer apply. The principle of relativity provides us with a starting point for our investigation of laws of mechanics valid in circumstances involving relative speeds comparable to c. Section 2.1 contains a statement of this principle and a discussion of its application to the problem of finding the appropriate laws of motion.

Laws of physics that are valid regardless of the relative speeds involved are called *relativistic laws*. Those laws whose validity is limited to circumstances not involving *any* relative speeds comparable to c are called *nonrelativistic laws*.* Relativistic laws of motion must reduce, under the appropriate circumstances, to nonrelativistic laws. Furthermore, the basic entities that appear in the simplest relativistic laws possess analogues in their nonrelativistic limit, the newtonian formulation of mechanics. We can arrive, therefore, at these relativistic laws by suitable generalizations of the corresponding newtonian laws.

Two problems are involved in generalizing laws from newtonian mechanics to the corresponding relativistic laws. In the first place, we must find dynamical variables that are relevant to a description of the behavior of interacting objects and that, in addition, are defined in a manner consistent with the Lorentz transformation law. The dynamical variables in point are the 4-momentum introduced in Section 2.2 and the Minkowski force introduced in Section 2.3.

---

* Some equations appearing in this chapter are valid only under nonrelativistic conditions. Nonrelativistic equations included in the text of this chapter (but not those that appear in the Examples or Problems) are designated by the abbreviation N.R. after the equation.

Second, we must find relations between these dynamical variables that restrict their values to those satisfied in actual motions. This aim cannot be achieved completely within the framework of the concepts that are simple generalizations of the concepts of newtonian mechanics. Therefore, the studies in this chapter are restricted to two types of equations between dynamical variables—conservation laws and the simplest equations of motion—that can be introduced within that framework.

We begin in Section 2.2 with those dynamical relations in which the interactions do not appear explicitly, namely, the generalizations of the nonrelativistic conservation laws of momentum and mass and, under the appropriate restrictions, the conservation law of energy. The corresponding relativistic conservation laws do not provide a complete description of the motions of objects, but they do provide an introduction to the study of the effects of interactions between objects. Furthermore, these conservation laws apply under analogous conditions even at the atomic and subatomic levels, where many of our ordinary concepts break down.

The simplest equation of motion is that for an object that experiences a given external interaction. In this case, the motion of the object does not modify the environment appreciably; the motion can be described by a force function that does not depend on the motion of the object itself. In Section 2.3, we formulate a simple generalization of Newton's equation of motion, which is compatible with the principle of relativity and which describes the motion of an object under a given external interaction.

Most of the practical applications of relativistic mechanics involve the fundamental particles of physics, which are described in Section 2.4.* The masses of these particles are sufficiently small that they can be accelerated in the laboratory to relativistic speeds; furthermore, these particles often are observed moving at such speeds when they strike the earth's atmosphere from outer space. A study of these particles shows conclusively the necessity for the relativistic generalizations of Newton's laws of motion and the corresponding conservation laws. The behavior of these particles cannot be described in detail without the use of the concepts of quantum mechanics; nevertheless, this behavior is restricted by the relativistic conservation laws.

Section 2.4 contains many problems involving experimentally determined numbers that provide useful examples of the developments given in Sections 2.2 and 2.3.

Problem 2.1

Consider two inertial systems S and S' related by the galilean transformation law.

(a) An observer in S measures the mass ratio m_1/m_2 of two particles in a collision experiment. Show from the galilean transformation law that an observer in S' who makes the appropriate measurements in that experiment also obtains the number m_1/m_2 for the ratio.

(b) Consider the fundamental force $\mathbf{F}_{i(j)}$, experienced by particle i due to the

---

* Some problems preceding Section 2.4 involve the particles known as the electron and the proton. If these particles are not familiar to you from your previous studies in physics or chemistry, you need only view them in those problems as objects having the masses given on the endpapers of this text and the electric charges $-e$ and e, respectively.

presence of particle j, given relative to S by

$$\mathbf{F}_{i(j)} = k \frac{\mathbf{r}_i - \mathbf{r}_j}{|\mathbf{r}_i - \mathbf{r}_j|^3}.$$

Show from the galilean transformation law that, relative to S',

$$\mathbf{F}'_{i(j)} = k \frac{\mathbf{r}'_i - \mathbf{r}'_j}{|\mathbf{r}'_i - \mathbf{r}'_j|^3}.$$

(c) Consider a system of particles $1, 2, \ldots, N$ interacting among themselves through the fundamental forces of part (b), and experiencing no other forces.
   (i) Show that $\mathbf{P} = m_1\mathbf{v}_1 + m_2\mathbf{v}_2 + \cdots + m_N\mathbf{v}_N$ is conserved as a consequence of Newton's laws.
   (ii) Let $\mathbf{F}_i$ be the total force experienced by the ith particle. Show from the galilean transformation law that the equation of motion $m_i\mathbf{a}_i = \mathbf{F}_i$ relative to S is equivalent to $m'_i\mathbf{a}'_i = \mathbf{F}'_i$ relative to S'. Show that the galilean principle of relativity follows from this result.

Problem 2.2

(a) The total energy of a system depends on the position $\mathbf{r}_i$ and the velocity $\mathbf{v}_i$ of each particle:

$$E(\mathbf{r}_1, \mathbf{r}_2, \ldots; \mathbf{v}_1, \mathbf{v}_2, \ldots) = \sum_i \frac{1}{2} m_i v_i^2 + V(\mathbf{r}_1, \mathbf{r}_2, \ldots).$$

Show that if

$$\Delta\mathbf{v}_i = \mathbf{a}_i \Delta t = \mathbf{F}_i \frac{\Delta t}{m_i} \quad \text{and} \quad \Delta\mathbf{r}_i = \mathbf{v}_i \Delta t$$

for sufficiently small Δt, the energy is conserved. *Hint:* Expand $E(\mathbf{r} + \Delta\mathbf{r}, \mathbf{v} + \Delta\mathbf{v})$ and neglect the terms in $(\Delta t)^2$.
(b) Show that the rate at which work is done on a particle is given by $\mathbf{F}\cdot\mathbf{v}$, the power P, and that $P = d(\frac{1}{2}mv^2)/dt$.
(c) The electric work done on a charged particle is proportional to the charge of the particle. The work done in joules per coulomb is the potential difference in volts experienced by the particle. The work done on an electron when it is accelerated through one volt is called an electron volt (eV). Show that $1 \text{ eV} = 1.602 \times 10^{-19}$ J.
(d) An electron is accelerated from rest to motion with a kinetic energy of 10 MeV. The acceleration takes place over 10^{-7} sec. Calculate the average power in watts (J/sec) required for this acceleration.
(e) Calculate the speed of a 10-MeV electron if the kinetic energy were given by the nonrelativistic formula $E = \frac{1}{2}mv^2$.
(f) The radius of a proton is about one femtometer (1 fm = 10^{-15} m). Calculate the radius in angströms (1 Å = 10^{-10} m).
(g) The masses of some particles are conveniently expressed in the (unified) atomic mass unit, u, equal to 1/12 the mass of an atom of C^{12}: 1 u = 1.660 $\times 10^{-27}$ kg. Calculate the mass of the proton in unified atomic mass units.
(h) Calculate the approximate density of the proton and the density of water in kilograms per cubic meter and unified atomic mass units per cubic femtometer.

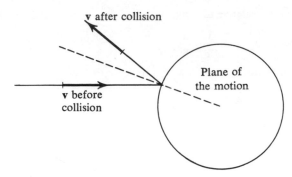

FIGURE 2.3 Collision with a fixed hard sphere.

Problem 2.3

Use nonrelativistic mechanics in this problem. A point particle incident on a fixed hard sphere experiences a force at the surface of the sphere that reverses the component of momentum normal to the sphere at the point of contact, and leaves the corresponding tangential component unchanged.

(a) Show that the incident and scattered motions make equal angles with the radius of the sphere drawn to the point of contact, and that the incident motion, the scattered motion, and that radius lie in one plane (Figure 2.3).

(b) The angle Θ through which the particle's motion is deflected is called the scattering angle. The distance S between the line of the particle's motion and a parallel line through the center of the force center is called the impact parameter. Show that for hard-sphere scattering by a sphere of radius a (Figure 2.4) $S = a \cos (\Theta/2)$.

(c) Consider the collision of a hard sphere of radius r with an identical hard sphere (Figure 2.5). Show that $S = 2r \cos (\Theta/2)$.

Problem 2.4

Use nonrelativistic mechanics in this problem. Consider a particle incident on a spherical system of particles for which the attractive force between each pair of the particles is exerted only over a very short distance. On approaching

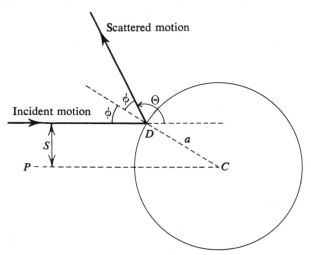

FIGURE 2.4 Diagram for the calculation of the relation between Θ and S for hard-sphere scattering.

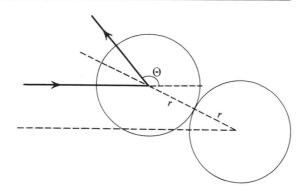

FIGURE 2.5   Collision of identical hard spheres.

the system, the particle experiences a brief attraction, but once inside the net force is zero (Figure 2.6).

(a) Show that the effects of such a force are approximated by the square-well potential (Figure 2.7)

$$V(r) = -V_0, \qquad r < a$$
$$= 0, \qquad r > a.$$

Hint: Consider the force for which the potential varies smoothly from $-V_0$ at $r = a - \varepsilon$ to 0 at $r = a + \varepsilon$.

Particle undergoing deflection

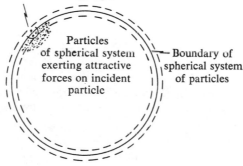

Particles of spherical system exerting attractive forces on incident particle

Boundary of spherical system of particles

FIGURE 2.6   This force is approximated by a square-well force. Inside the dashed region the particle experiences a net force toward the center of the sphere.

(b) Use conservation laws to show that the normal (n) and tangential (t) components of the velocity inside and outside the sphere are related by

$$(v_t)_{\text{outside}} = (v_t)_{\text{inside}}$$
$$(v_n)_{\text{inside}} = \sqrt{(v_n)^2_{\text{outside}} + 2V_0/m}.$$

Show that (Figure 2.8) if E = the incident energy,

$$\frac{\sin i}{\sin r} = n = (1 + V_0/E)^{1/2}.$$

(c) Consider a particle with energy $E = V_0$ that is scattered by the square well of part (a). Show that (Figure 2.9)

$$S = a \sin i, \qquad \Theta = 2(i - r),$$

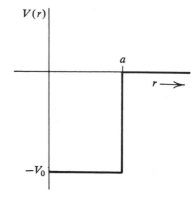

FIGURE 2.7   The square-well potential.

and

$$S^2 = \frac{2a^2 \sin^2 (\Theta/2)}{3 - 2\sqrt{2} \cos (\Theta/2)}.$$

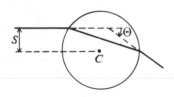

FIGURE 2.8   The angle of incidence and the angle of refraction r.

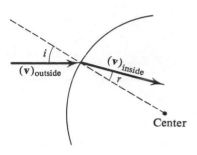

FIGURE 2.9   The scattering angle and the impact parameter for square-well scattering.

Problem 2.5

A particle moves toward a force center with the nonrelativistic speed v_i before it experiences any appreciable acceleration. The repulsive force is given by $\mathbf{F} = (k/r^2)\hat{r}$.

(a) Show directly from Newton's equation of motion that the angular momentum relative to the force center $\mathbf{L} = m\mathbf{r} \times \mathbf{v}$ is conserved. Show that this implies that the motion takes place in a plane.

(b) Set up a polar-coordinate system in the plane of the motion with the origin at the force center. Show that $mr^2(d\theta/dt) = L$ is a constant. *Hint:* Show that the component of velocity perpendicular to $\mathbf{r}$ is $r(d\theta/dt)$.

(c) Show from the energy conservation law that $\frac{1}{2}mv^2 + k/r = \frac{1}{2}mv_i^2$. Show that $v^2 = (dr/dt)^2 + r^2(d\theta/dt)^2$.

(d) Show that

$$\frac{L^2}{2mr^4}\left(\frac{dr}{d\theta}\right)^2 + \frac{L^2}{2mr^2} + \frac{k}{r} = \frac{1}{2} mv_i^2.$$

Hint: Use $dr/dt = (dr/d\theta)(d\theta/dt)$.

(e) Set $u = 1/r$ and show that

$$\left(\frac{du}{d\theta}\right)^2 + u^2 + \frac{2mk}{L^2} u = \frac{m^2v_i^2}{L^2}.$$

(f) Show that

$$u = \frac{1}{r} = \frac{mk}{L^2}\left(\varepsilon \cos \theta - 1\right), \qquad \varepsilon = \sqrt{1 + (L^2v_i^2/k^2)}$$

is a solution of the differential equations of (d) or (e).

(g) Show that (Figure 2.10) $L = mSv_i$ and that $1/r \rightarrow 0$ for $\theta = \theta_i$.

(h) Show that the scattering angle Θ is given by $k \cot (\Theta/2) = mSv_i^2$.

Problem 2.6

A particle of mass m and electric charge q moves relative to an inertial system S with the nonrelativistic speed v in a region of constant magnetic induction $\mathbf{B}$.

FIGURE 2.10   Relation between orbit and scattering parameters S and Θ for scattering by an inverse-square central force.

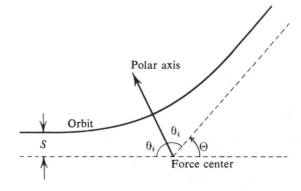

Polar axis

Orbit

S

θ_i

θ_i

Θ

Force center

(a) The particle moves relative to S in a plane perpendicular to **B**. Show that the particle travels on a circle of radius $R = mv/|q|B$.

(b) The component parallel to **B** of the velocity relative to S of the particle is $\mathbf{v}_\parallel$ and that perpendicular to **B** is $\mathbf{v}_\perp$. Show that, relative to S, the particle moves on a circle of radius $R = mv_\perp/|q|B$, whose center travels with the velocity $\mathbf{v}_\parallel$.

Problem 2.7

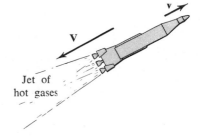

FIGURE 2.11   Thrust of rocket.

Consider a rocketship in free space. Hot gases are ejected from the tail of the rocket at the constant rate of μ kg/sec with the constant velocity **V** relative to the rocket (Figure 2.11). Let $M(t)$ denote the mass of the rocket and $\mathbf{v}(t)$ its velocity at time t.

(a) Show that as a consequence of Newton's laws the velocity **v** changes in the short time interval dt according to the relation $M\,d\mathbf{v} + (\mu\,dt)\,\mathbf{V} = 0$ or $M(d\mathbf{v}/dt) = -\mu\mathbf{V}$.

(b) Show that $M(t) = M_0 - \mu t$, where $M_0 = M(0)$.

(c) Show that if the rocket were at rest at time $t = 0$, the velocity of the rocket at time t would be given by $\mathbf{v}(t) = \mathbf{V}\ln(1 - \mu t/M)$.

2.1 The Dynamic Postulate of Einstein's Special Relativity Theory

It is a common experience when traveling by boat or plane that you are unable to detect any effects of the "motion." The behavior of objects relative to the "moving" boat or plane is indistinguishable from their behavior, under similar circumstances, relative to the earth. More striking is our inability to detect any effects of the earth's orbital speed of 30 km/sec, about 70,000 mi/hr, upon the course of any physical process. These facts corroborate the principle of relativity, which states, in effect, that *the laws of nature do not differentiate one particular inertial reference system from any other*. The question of the validity of this principle is the main topic of this section.

2.1.1 *The incompatibility of Newton's laws and the principle of relativity* *

The galilean form of the principle of relativity is a consequence of newtonian mechanics (see Problem 2.1). Thus, our inability to detect any effects of steady motions involving only small relative speeds follows as a consequence of Newton's laws of motion and the newtonian ideas of space and time. We now know, from our critical examination of the properties of space and time (relativistic kinematics) in Chapter 1, that the newtonian ideas of space and time are untenable in circumstances involving large relative speeds. Therefore in the light of insight, provided by relativistic kinematics, into the properties of space and time, we must examine the validity of Newton's laws—particularly their compatibility with the principle of relativity as they apply to circumstances involving relative speeds comparable to c.

Our main concern lies in the relation between the laws of mechanics as

* Kacser, Prentice-Hall, Sec. 2.4, p. 14; Sec. 4.9, p. 101; Sec. 5.2, p. 144.
Resnick, John Wiley, Sec. 3.2, p. 111.

referred to various inertial systems. Therefore let us assume for the moment that Newton's laws of motion are valid for all motions relative to one particular inertial system S, say that in which the sun is (almost) at rest.

Newton's equation of motion for a particle relates two vectors, the force and the acceleration. The force depends on the relation of the particle to its environment and is thus independent of the reference frame relative to which the motion is measured. On the other hand, the acceleration is measured relative to the reference frame. As a result, the equation of motion (2.5) takes on the same form relative to all inertial frames connected to the inertial system S by the galilean transformation equations

$$\mathbf{r} = \mathbf{r}' + \mathbf{V}t, \qquad \mathbf{v} = \mathbf{v}' + \mathbf{V}, \qquad \mathbf{a} = \mathbf{a}', \qquad t = t'. \qquad \text{N.R.} \qquad (2.22)$$

Furthermore, since $t = t'$, the concept of simultaneity and related concepts like instantaneous relative displacement are the same for all such inertial systems. We conclude that if Newton's laws of motion are valid in the one particular inertial reference system S, the same statements of the laws are valid in all other inertial reference systems connected to S by galilean transformations.

Newton's laws of motion are consistent with the principle of relativity for all inertial systems moving with speeds, relative to the particular inertial system S, sufficiently small that the Lorentz transformation reduces to the galilean transformation. However, Newton's laws are not compatible with the principle of relativity for circumstances involving relative speeds of the order of c. For one thing, acceleration plays an important role in Newton's laws, and yet the acceleration $d\mathbf{v}/dt$ measured relative to one inertial frame does not transform under a Lorentz transformation into the acceleration relative to another inertial frame (Problem 2.8). That is, if the acceleration of the particle relative to the inertial system S is determined by the environment, then relative to another inertial system S' the environment determines a more complicated entity that depends on the velocity of the particle relative to S' and the velocity of S' relative to S. This result is not compatible with the principle of relativity, since, from an examination of the behavior of the particle, we could decide whether we are referring the motion to the "privileged" inertial system S or to the "moving" inertial system S'.

Furthermore, the concept of simultaneity plays an important part in the formulation of Newton's laws. For instance, his second law states that the acceleration of a particle at a given instant t is determined by the properties of the environment, objects at other points in space, at that same time t. Consider, for example, the motion of two particles that interact through a repulsive force. If the two particles undergo straight-line motion along the line joining their positions at any instant, say the x axis, then their world lines appear as shown on the space-time diagram in Figure 2.12. This diagram depicts the particles approaching each other, undergoing a repulsive interaction, and then receding from each other. Suppose now that particle 1 experiences a force that depends on the instantaneous separation distance from particle 2. Then one inertial observer S determines that separation distance relative to the event E_1 as $\sqrt{-(E_1 E_2)^2}$, while another inertial observer S' determines that separation distance as $\sqrt{-(E_1 E_3)^2}$. Thus, the instantaneous force depends on the observer and, moreover, on the motion of the object, since the event separation $\underline{E_1 E_3}$ depends on what happens to the particle, that is, on where the world line goes, after E_2. We conclude that the notion of force as a simple function of the instantaneous environment, as implied by newtonian mechanics, is not compatible with the principles of relativity and the invariance of the speed of light.

It is possible that the newtonian notion of force can be modified suitably to fit into the framework of the principle of relativity. In this event, we must inquire next into the validity of Newton's third law. This states that a component force experienced by one particle at the time t is equal in magnitude and opposite in direction to the corresponding component force experienced by another particle, spatially separated from the first, at that same instant t. This would mean, in the circumstances shown in Figure 2.12, that the force experienced by particle 1 at E_1 is equal and opposite to the force experienced by particle 2 at E_2 according to S and at E_3 according to S'. This comes about because of the fact, shown in our study of relativistic kinematics, that spatially separated events that are simultaneous relative to one inertial system are not simultaneous in all other inertial systems. Therefore, if Newton's third law, which depends on the use of the concept of simultaneity, is valid in one inertial system S, this law is not valid in the same form in all inertial systems. We see again that Newton's laws are not compatible with the principle of relativity for circumstances involving relative speeds of the order of c.

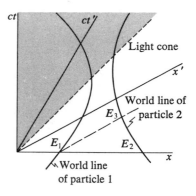

FIGURE 2.12 The distance between particles 1 and 2 relative to E_1 is $\sqrt{-(E_1E_2)^2}$ relative to S and $\sqrt{-(E_1E_3)^2}$ relative to S'.

SUMMARY   Newton's laws of motion are incompatible with the principle of relativity because of the properties of space-time revealed by the invariance of the speed of light. For one thing, the acceleration measured relative to one inertial frame does not determine, by itself, the acceleration relative to another inertial frame. Also, the relative character of simultaneity shows that, in all but the simplest cases, the instantaneous force experienced by an object cannot depend on the simultaneous configuration of the environment relative to all observers. Furthermore, the instantaneous action of one object on another in general cannot be equal and opposite to the reaction simultaneously relative to all inertial observers.

Problem 2.8

An inertial system S' moves with velocity $\beta c \hat{x}$ relative to another inertial system S. Let $\mathbf{a} = d^2\mathbf{r}/dt^2$ be the acceleration of a particle relative to S and $\mathbf{a}' = d^2\mathbf{r}'/dt'^2$ be the acceleration of the same particle relative to S'.

(a) Show that

$$a_x' = a_x \frac{(1 - \beta^2)^{3/2}}{[1 - (\beta v_x/c)]^3},$$

$$a_y' = a_y \frac{1 - \beta^2}{[1 - (\beta v_x/c)]^2} + \frac{a_x \beta v_y}{c} \frac{1 - \beta^2}{[1 - (\beta v_x/c)]^3},$$

$$a_z' = a_z \frac{1 - \beta^2}{[1 - (\beta v_x/c)]^2} + \frac{a_x \beta v_z}{c} \frac{1 - \beta^2}{[1 - (\beta v_x/c)]^3}.$$

(b) Show, from the results of (a), that the acceleration $\mathbf{a}$ relative to S does not determine, by itself, the acceleration $\mathbf{a}'$ relative to S'. Compare this with the discussion in Section 1.9.

2.1.2   *The dynamic postulate of special relativity**

The incompatibility of Newton's laws of motion and the principle of relativity presents us with several alternatives. Is there one special inertial reference system, one that we might call "absolute," in which Newton's laws of motion retain the form that we have considered previously? Or must we modify our laws of physics, in particular Newton's laws, so that they take on the same form relative to every inertial system and thus satisfy the principle of relativity? Or is there some middle course that we must follow?

There is no point in debating about which must be done—the answer rests with the results of experiments. Newton's laws are outstandingly successful in the description of physical phenomena over a wide range of validity, from the motions of objects as small as dust particles to the motions of planets in the solar system. However, *a careful analysis shows that the successes of newtonian mechanics are limited to circumstances involving relative speeds much less than c.* Therefore, it is possible that the principle of relativity is valid, together with the necessary modification in Newton's laws, without any contradiction from that great body of experience that corroborates newtonian mechanics within its range of validity.

On the other hand, we cannot point to one particular experiment that demonstrates conclusively that all inertial reference systems are equivalent for our description of the laws of nature. Indeed, some reflection will show that the principle of relativity can be based only on agreement between *every one* of a very large number of experiments and consequences of that principle. Experiments that involve relative speeds of the order of c are being performed every day in scientific laboratories around the world. *There is no experimental evidence to show the existence of an absolute reference system.* All experimental results are consistent with the principle of relativity. Therefore, we must accept the dynamic postulate of special relativity put forward by Einstein.

▶ *Principle of Relativity: All inertial reference systems are completely equivalent as regards our description of all laws of nature.*

We now face the task of formulating laws of motion that are consistent with the principle of relativity and the relation between different inertial systems, the Lorentz transformation law. These laws also must reduce to Newton's laws for those circumstances in which all speeds involved are much less than c.

The formulation of relativistic laws of motion involves two issues. One of these concerns the determination of an equation of motion that takes on the same form relative to all inertial reference systems so that the equation of motion does not distinguish one inertial system from all others. The other difficulty concerns the concept of simultaneity; this concept, applied to spatially separated events, cannot enter into any relativistic formulation of a physical law. These issues have been resolved for some types of interactions, but within a framework of ideas different from that we have encountered thus far. A relativistic theory of gravitation, including laws that describe the motions of interacting particles,

---

* Feynman, Leighton, and Sands (vol. 1), Addison-Wesley, Sec. 15–1, p. 15–1.
Kacser, Prentice-Hall, Sec. 5.1, p. 142.
Resnick, John Wiley, Sec. 1.9, p. 35.
Taylor and Wheeler, W. H. Freeman, Sec. 3, p. 11.

has been proposed by Einstein and is discussed in Chapter 3. The motions of interacting particles with electric charges are described by another theory that involves the field concept. The behavior of particles interacting in other ways can be described only within the framework of the concepts of quantum mechanics.

The rest of this chapter is devoted to laying a foundation for a study of relativistic interactions while, at the same time, restricting our considerations in order to avoid the major problems involved in an investigation of these interactions. The difficulty arising from the relative nature of simultaneity appears in the problem of describing the interactions between spatially separated particles. A partial description of the effects of these interactions is given by conservation laws that, as in newtonian mechanics, are satisfied regardless of the nature of the interaction. In Section 2.2, we explore the nature of these conservation laws in relativity theory. In Section 2.3 we establish a relativistic formulation of the equation of motion for a particle experiencing a given external interaction. By restricting our investigation thus to circumstances involving interactions that are given independent of the motion of the particle under consideration, we resolve one of the issues without encountering the other.

SUMMARY   The principle of relativity has withstood the test of experiment, whereas the range of validity of the newtonian formulation of mechanics is restricted to circumstances involving relative speeds that are much less than c. Therefore, we require a reformulation of the laws of motion consistent with the dynamic postulate of special relativity.

Problem 2.9

A student reads the section above and describes to a friend the reasons that Newton's laws of motion are not universally valid. The friend asks "Then why did you study and learn something that is wrong?" State the answer that you would give to this question.

Problem 2.10

What is the difference between Einstein's postulate and the galilean principle of relativity?

2.1.3   Covariant equations

In the next two sections, we will be concerned with generalizing nonrelativistic equations, such as Newton's equation of motion

$$m\frac{d^2\mathbf{r}}{dt^2} = m\mathbf{a} = \mathbf{F}, \qquad \text{N.R.} \tag{2.23}$$

to relativistic equations that are valid under circumstances in which relative speeds of the order of c are manifest. We must find, for this purpose, equations that are compatible both with the principle of relativity and with the principle of the invariance of the speed of light. Therefore, we first examine the conditions that an equation, expressing a physical law, must satisfy in order that it be compatible with these principles.

A law of physics is a statement of a relationship between different physical entities, each of which can be experimentally measured, at least in principle. If it is not possible to state this relationship in an identical manner with respect to every inertial system, we would be able to differentiate between inertial systems, a result contrary to the principle of relativity. Therefore, the principle of relativity implies that it is possible to state a physical law in such a manner that the statement takes on the same form relative to every inertial system.

The principle of the invariance of the speed of light is taken into account by the requirement that the different inertial systems be related through the Lorentz transformation law.

The statement of a physical law may consist of, or have as an integral part, an equation analogous to (2.23), which relates the behavior of two different entities. For example, Equation (2.23) gives a relation between the acceleration $\mathbf{a} = d^2\mathbf{r}/dt^2$ that is defined by the motion of the particle and the force $\mathbf{F}$ that is determined by the environment. In order that an equation be compatible with the principle of relativity, it must be possible to express that equation in the same form relative to every inertial reference system.

Consider the relativistic formulation of an equation that consists of a relation between two different entities, each of which is described by a set of components. In the first place, each of these entities must exist independent of the inertial reference system, as described in Section 1.9. Furthermore, the components of the two entities must transform in the same way under a Lorentz transformation from one inertial system to another in order that the form of the equation does not change under that transformation. Equations that satisfy these conditions are said to be *covariant* under the Lorentz transformation law. Hence, the principle of relativity implies that it must be possible to express any equation that is an integral part of a law of physics in a form covariant under the Lorentz transformation law.

SUMMARY   An equation relating two entities is said to be covariant if the entities can be defined independent of any inertial reference system and if the components of the two entities transform in the same way under a Lorentz transformation. Any equation that is a statement of a law of physics must be covariant under Lorentz transformations, or the equation would violate the principle of relativity.

Problem 2.11

The components of a 4-vector ℓ are denoted by b_μ and the components of a tensor of the second order **T** are denoted by $T_{\mu\nu}$. Is the equation $b_\mu = T_{\mu 3}$ a covariant equation? State the reasons for your answer.

Problem 2.12

This problem provides an example of finding a covariant equation, which describes the motion of a particle, in terms of information that is not presented in a covariant form. The motion of a particle relative to an inertial system S is described by the equation $\mathbf{r} = \mathbf{v}_0 t$, where $\mathbf{v}_0$ is a constant vector..

(a) Show that this is not a covariant equation.

(b) Let t_0 be the proper time of the particle measured relative to the event $t = 0$, $\mathbf{r} = 0$. Show that, relative to any inertial system, the motion is described by the covariant equation $v(t_0) = v_0$, where v_0 is the 4-vector whose components relative to S are

$$\left(\frac{c}{\sqrt{1 - (v_0^2/c^2)}}, \frac{\mathbf{v}_0}{\sqrt{1 - (v_0^2/c^2)}} \right).$$

Problem 2.13

Show that the equation

$$m \frac{d^2 \mathbf{r}}{dt^2} = \mathbf{F}$$

is an equation that is not covariant under the Lorentz transformation law (see Problem 2.8).

2.2 The Conservation of Momentum and Energy

The problem of formulating relativistic laws for the motions of interacting particles is complicated by consequences of the principle of the invariance of the speed of light, such as the relative nature of simultaneity. The restricted problem of finding the covariant equation of motion, without consideration being given to the forms of the fundamental interactions between particles, will be discussed in Section 2.3. In this section, we shall investigate those results that can be studied without even inquiring into the covariant form of the equation of motion.

We will be concerned with a partial description of the results of interactions that take place in a circumscribed region of space-time (Figure 2.13). That is, we consider a system of particles, initially noninteracting, that, over a finite period of time, undergo interactions in a restricted region of space, after which there appear particles that are again noninteracting. We impose the restriction in the initial part of this section that no form of energy or momentum other than that associated with the particles appears in space-time outside the region of interaction.

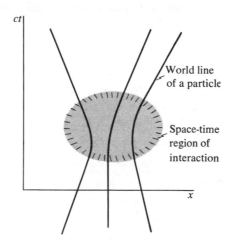

FIGURE 2.13 The interaction between the particles occurs in a limited region of space-time.

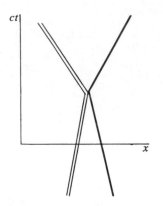

FIGURE 2.14   Space-time diagram of a point interaction between two particles with world lines represented by a single and a double line, respectively.

The simplest examples of interactions that take place in a restricted region of space and time are point forces, forces that act between particles only when they are coincident (Figure 2.14). Since the point of action of the interaction and the source of the interaction are one and the same world point in space-time, there are no problems with simultaneity involved in describing the interaction. Since point interactions are localized at a world point, they often are called *local interactions*.

We will arrive at the relativistic conservation laws by generalizing the appropriate nonrelativistic laws in the following way: We first consider the description of the collision under investigation according to nonrelativistic mechanics. We will argue that the relativistic laws must reproduce this description in the appropriate circumstances. Therefore, we consider possible covariant generalizations of the nonrelativistic laws and, in this way, arrive at dynamical laws that are compatible with the principle of relativity.*

The simplest circumstance in which a conservation law holds is in the force-free motion of a single particle. We consider this case first, since it provides us with an introduction to the dynamical variables that appear under more general circumstances.

It must be kept in mind that, after we have obtained a plausible set of laws, we must turn to the results of experiments for the decision as to whether or not these laws are valid.

2.2.1   The covariant form of the conservation law for the force-free motion of a single particle †

We first consider the simplest motion, namely, that of a particle that experiences no force. In order to obtain a covariant equation that describes such a motion, we consider the characteristics of the motion relative to one inertial system and then determine how we can describe these characteristics independent of that reference system.

A particle that experiences no force moves, relative to any inertial system S, with a velocity $\mathbf{v}(t)$ that is constant in time t. At any instant of time, the velocity vector $\mathbf{v}(t)$ determines in a unique manner a 4-vector, the 4-velocity v, which is an entity that exists independent of the reference system S (see Section 1.9.2). Relative to the inertial system S, the 4-velocity v has the components

$$v = \left[\frac{c}{\sqrt{1 - (v^2/c^2)}}, \frac{\mathbf{v}}{\sqrt{1 - (v^2/c^2)}}\right] = (\gamma c, \gamma \mathbf{v}), \qquad (2.24)$$

where $\gamma = [1 - (v^2/c^2)]^{-1/2}$.

In the general case of accelerated motion, the 4-velocity v varies from event to event along the world line (Figure 2.15), and it can be represented as a 4-vector

* There are methods of deducing possible forms of the relativistic dynamical laws other than the one presented in the text. Our method uses the 4-vector concept, whereas the others, which are more complicated in their mathematical development, start directly from the Lorentz transformation law. These other methods are described in Example 2.7 and in Problem 2.34 [1].
† Feynman, Leighton, and Sands (vol. 1), Addison-Wesley, Sec. 17–4, p. 17–5.
Kittel, Knight, and Ruderman, McGraw-Hill, pp. 385–388.
Resnick, John Wiley, Sec. 3.7, p. 143.
Taylor and Wheeler, W. H. Freeman, Sec. 12, p. 111.

function of one parameter that specifies the events along the world line. The transformation properties of this parameter are fixed by the principle of relativity. Since we wish to describe the characteristics of the motion independent of any inertial system, the parameter that specifies events along the world line must be a scalar. The appropriate parameter is, therefore, the proper time t_0 of the particle (Section 1.8); thus, we write the 4-velocity as $v(t_0)$.

In the case of force-free motion, the velocity $\mathbf{v}(t)$ does not change during the motion and the associated 4-velocity $v(t_0)$ is also constant throughout the motion (Figure 2.16). Let v_0 be a constant 4-vector equal to the 4-velocity $v(t_0)$ at any event, say the event $t_0 = 0$, during the particle's motion. Then the condition that the particle experiences no force can be expressed by the covariant equation

$$v(t_0) = v_0. \tag{2.25}$$

The condition for free motion can be expressed also by any *scalar* multiple of this equation. In fact, the corresponding nonrelativistic equation,

$$\mathbf{v}(t) = \mathbf{v}_0 \qquad \text{N.R.} \tag{2.26}$$

is a consequence of a general law for systems of particles, that of conservation of momentum, for the special case of a one-particle system:

$$\mathbf{p}(t) = \mathbf{p}_0 \qquad \text{N.R.} \tag{2.27}$$

where $\mathbf{p} = m\mathbf{v}$ and m is the scalar called mass. The special case [Equation (2.27)] of this general law reduces to Equation (2.26) for a single-particle system, because each term in the equation is proportional to m. This feature does not appear in the general case, so no such reduction of the law for a many-particle system to an equation between velocities and independent of the masses is possible in general. This suggests, but does not prove, that the form (2.27) of the equation for single-particle force-free motion is more relevant to our problem than is the form (2.26). This conjecture will be confirmed in Sections 2.2.2 through 2.2.4; in anticipation of this, it is worthwhile for us to examine, at this point, the dynamical variables that appear in the appropriate relativistic generalization of Equation (2.27).

The spatial components of the covariant equation $v(t_0) = v_0$ that describes single-particle force-free motion give the nonrelativistic conservation law $\mathbf{v}(t) = \mathbf{v}_0$ in the limiting cases for which $v/c \ll 1$. Therefore, that covariant equation can be modified so that the spatial components reduce to the nonrelativistic law of momentum conservation by multiplication with a scalar that reduces to the mass m in the nonrelativistic limit. However, a scalar has the same value relative to all inertial systems, and the value of the relevant scalar factor is m when referred to a system in which $v/c \ll 1$. Therefore, the appropriate scalar multiplicative factor is m; *the covariant form of the single-particle law of momentum conservation is*

$$mv(t_0) = mv_0. \tag{2.28}$$

The mass coefficient m may be measured, according to its definition, in a collision in which the relative speeds involved are sufficiently small that relativistic effects play no part. Sometimes m is called the *rest mass*, or *proper mass*, of the particle to indicate that it is measured in terms of acceleration ratios by an experiment in which nonrelativistic mechanics ($v/c \approx 0$) is applicable. As shown

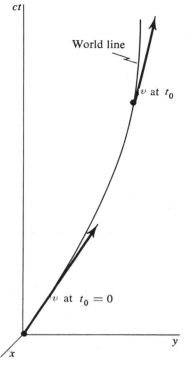

FIGURE 2.15   The 4-velocity v varies from event to event if the world line is not straight; that is, if the motion is accelerated.

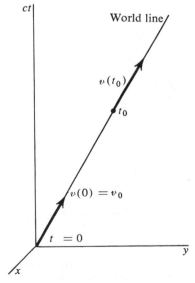

FIGURE 2.16   The world line of a particle moving with a constant velocity.

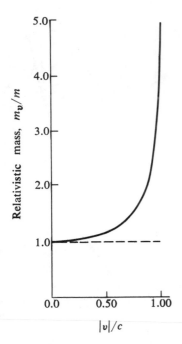

FIGURE 2.17   The relativistic mass m_v.

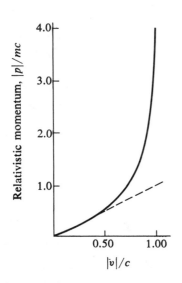

FIGURE 2.18   The relativistic momentum. The dashed line is the curve for mv, the nonrelativistic expression for the momentum.

by its definition, m is a scalar quantity. We shall assume throughout this chapter, except where it is stated explicitly otherwise, that the magnitude of the (rest) mass does not change throughout the motion, as occurs, for example, if the object is an evaporating liquid drop. Under this assumption, m is an *invariant scalar*.

The 4-vector mv is called the 4-momentum p or, for reasons that will be apparent shortly, the *energy-momentum vector*:

$$p \equiv mv. \tag{2.29}$$

The spatial component of the 4-momentum is a 3-vector,

$$\mathbf{p} = m\gamma\mathbf{v} = \frac{m}{\sqrt{1 - (v^2/c^2)}}\,\mathbf{v}, \tag{2.30}$$

called the *relativistic momentum*. This 3-vector can be written $\mathbf{p} = m_v\mathbf{v}$, a form identical to that for the nonrelativistic momentum of a particle of mass m_v, and so

$$m_v = \frac{m}{\sqrt{1 - (v^2/c^2)}} \tag{2.31}$$

is called the *relativistic mass*. Figures 2.17 and 2.18 show the dependence of the relativistic mass and the magnitude of the relativistic momentum $m_v v$ on the speed $v = |\mathbf{v}|$, and Figure 2.19 shows the speed v of a particle in terms of its relativistic momentum.

The time-like component relative to S of the 4-momentum is

$$p_0 = m\gamma c = m_v c = \frac{mc}{\sqrt{1 - (v^2/c^2)}}. \tag{2.32}$$

We obtain some insight into the significance of this term by considering the familiar case of the nonrelativistic limit, $v/c \ll 1$. In this circumstance, we can expand the factor $[1 - (v^2/c^2)]^{-1/2}$ in powers of v/c to obtain

$$p_0 = mc\left(1 - \frac{v^2}{c^2}\right)^{-1/2} = mc\left(1 + \frac{v^2}{2c^2}\right). \qquad \text{N.R.} \tag{2.33}$$

This can be rewritten as

$$p_0 c = mc^2 + \tfrac{1}{2}mv^2 \qquad \text{N.R.} \tag{2.34}$$

from which we see that, in the nonrelativistic limit, $p_0 c - mc^2$ is the kinetic energy. We define that quantity to be the *relativistic kinetic energy* T with respect to S for all possible values of v:

$$p_0 c = mc^2 + T. \tag{2.35}$$

Figure 2.20 shows T as a function of v.

The term mc^2 that is added to T in the equation

$$p_0 = \frac{mc^2 + T}{c} \tag{2.36}$$

is an energy that depends only on the rest mass of the particle and that vanishes for zero rest mass. As we shall see shortly, it is possible under some circumstances to convert this amount of energy to other forms through the annihilation of the rest mass m. The term mc^2 is called the *rest energy* E_0:

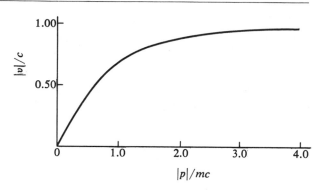

FIGURE 2.19   The speed $|v|$ of a particle in terms of its relativistic momentum p.

$$E_0 = mc^2. \tag{2.37}$$

The sum of the rest energy and the relativistic kinetic energy is called the *relativistic energy* E of the particle (Figure 2.21):

$$E = E_0 + T = mc^2 + (m_v c^2 - mc^2)$$
$$= m_v c^2. \tag{2.38}$$

It must be kept in mind in the following that merely defining $(m_v c^2 - mc^2)$ as the relativistic kinetic energy and $m_v c^2$ as the relativistic energy does not bestow on these quantities all the properties possessed by the similarly named expressions of nonrelativistic mechanics. The properties of T and E require further investigation, as presented in Sections 2.2.2 through 2.2.6.

The 4-momentum has components

$$\not{p} = mv = \left(\frac{E}{c}, \mathbf{p}\right); \tag{2.39}$$

for this reason, $\not{p}$ is sometimes called the energy-momentum vector. The norm of the energy-momentum vector is given by

$$\not{p} \cdot \not{p} = \frac{E^2}{c^2} - \mathbf{p} \cdot \mathbf{p}$$
$$= \gamma^2(m^2 c^2 - m^2 v^2) \tag{2.40}$$
$$= m^2 c^2,$$

and is thus an invariant quantity. This relation, illustrated in Figure 2.22, is frequently useful in calculations involving the energy and momentum of a particle. This relation also provides a definition of the (rest) mass that is valid for relativistic and nonrelativistic speeds. The relation between the relativistic energy $m_v c^2$ and the rest mass m can be stated as follows:

▶ *The relativistic energy is the time-like component of a 4-vector $\not{p}c$ whose norm is proportional to the square of the rest mass.*

In relativistic mechanics, the relativistic momentum $\mathbf{p}$ is not equal to a constant times the velocity $\mathbf{v}$, as in nonrelativistic mechanics, but equals $m\gamma\mathbf{v}$. The additional factor γ also appears in the relativistic energy $E = m\gamma c^2$, and hence the velocity is related to the momentum and energy by the relation

$$\mathbf{v} = \frac{\mathbf{p}c^2}{E}. \tag{2.41}$$

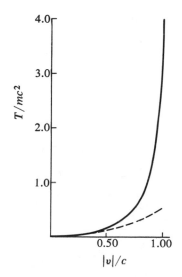

FIGURE 2.20   The kinetic energy T. The dashed line shows the nonrelativistic form, $T = \frac{1}{2}mv^2$.

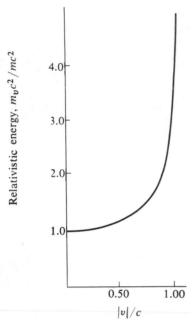

FIGURE 2.21 The relativistic energy $E = m_v c^2$.

SUMMARY The covariant equation that describes single-particle force-free motion and that reduces to the nonrelativistic law of momentum conservation for $v/c \ll 1$ is

$$\not{p}(t_0) = \not{p}_0,$$

where $\not{p} = m v$ is the 4-momentum or the energy-momentum vector, and m is the rest mass. The components of $\not{p}$ relative to a reference system are written as $(E/c, \mathbf{p})$, where E is the relativistic energy, $\mathbf{p} = m_v \mathbf{v}$ is the relativistic momentum, and m_v is the relativistic mass; $T = E - mc^2$ is called the relativistic kinetic energy.

Example 2.1

Q. The components of the energy-momentum vector $\not{p}$ are $(E/c, \mathbf{p})$ relative to an inertial system S and $(E'/c, \mathbf{p}')$ relative to another inertial system S'; S' moves with the velocity $\beta c \hat{x}$ relative to S. Find the relations between the components $(E/c, p_x, p_y, p_z)$ and $(E'/c, p_{x'}, p_{y'}, p_{z'})$.

A. The energy-momentum vector $\not{p}$ is a 4-vector; its components are related by the Lorentz transformation law Equation (1.173). Hence,

$$E' = \gamma(E - \beta c p_x), \qquad p_{x'} = \gamma\left(p_x - \frac{\beta E}{c}\right), \qquad p_{y'} = p_y, \qquad p_{z'} = p_z.$$

$$(2.42)$$

Example 2.2

In the physics of particles of high energies, it is frequently convenient to express the masses and relativistic momentums as energies. This is achieved if we multiply the usual entities, in the customary notation m and p, by the appropriate powers of c, namely c^2 and c, respectively, and call the products the (rest) mass and momentum. These entities, expressed as energies, can be denoted by the conventional symbols m and $\mathbf{p}$, so that we have, for example, the relations

$$m^2 = E^2 - \mathbf{p}\cdot\mathbf{p}, \qquad (2.43)$$

$$E = m + T, \qquad (2.44)$$

and

$$\beta = \frac{\mathbf{p}}{E}. \qquad (2.45)$$

Most physicists working with particles at high energies are familiar with the masses, in megaelectron volts, of the particles of interest. Therefore, to avoid confusion between the two remaining quantities, the energy and the magnitude of the momentum, we write units of the momentum expressed in megaelectron volts as MeV/c.

Q. (a) The mass of a proton is 1.67×10^{-27} kg. Find the mass of a proton in megaelectron volts.
(b) A proton has a relative kinetic energy of 2.00 GeV. Find the energy, momentum, and speed of the proton.

A. (a) The energy corresponding to a rest mass of 1.67×10^{-27} kg is

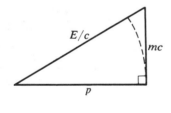

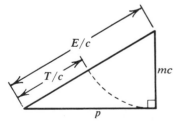

FIGURE 2.22 $E^2/c^2 = p^2 + (mc)^2$. Note that if either p or E/c is much larger than mc, $p \approx E/c$.

$$m = 1.67 \times 10^{-27} \text{ kg} \times (3.00 \times 10^8 \text{ m/sec})^2$$
$$= 1.50 \times 10^{-11} \text{ J}$$
$$= 1.50 \times 10^{-11} \text{ J} \times \frac{1 \text{ MeV}}{1.602 \times 10^{-13} \text{ J}} \qquad (2.46)$$
$$= 938 \text{ MeV}.$$

(b) The energy of the proton E is given by

$$E = m + T = 938 \text{ MeV} + 2.00 \times 10^3$$
$$= 2.94 \times 10^3 \text{ MeV}. \qquad (2.47)$$

The momentum p of the proton is given by

$$p = \sqrt{E^2 - m^2} = \sqrt{(2.94 \times 10^3)^2 - (938)^2}$$
$$= 7.76 \times 10^6 = 2.78 \times 10^3 \text{ MeV}. \qquad (2.48)$$

In the terminology of high-energy particle physics, p is equal to 2.78×10^3 MeV/c, or 2.78 GeV/c. The speed of the proton is given by

$$\beta = \frac{p}{E} = 0.946. \qquad (2.49)$$

Problem 2.14

Calculate the kinetic energy, in megaelectron volts, and the momentum in megaelectron volts per c of an 80-kg man walking 4 mi/hr.

Problem 2.15

Show that 1 u = 931 MeV.

Problem 2.16

(a) Show that, for a particle traveling with speed v relative to an inertial system S,

$$\frac{v}{c} = \sqrt{1 - (m/m_v)^2} = \sqrt{1 - (E_0/E)^2}.$$

(b) A particle moves relative to an inertial system S with relativistic kinetic energy much greater than the particle's rest energy. Show that

$$\frac{c - v}{c} = \frac{1}{2}\left(\frac{E_0}{T}\right)^2.$$

Problem 2.17

(a) Find the rest energy of an electron, in megaelectron volts.
(b) Find the relativistic masses of electrons with the following relativistic kinetic energies:
   (i) 1 eV,
   (ii) 1 keV,
   (iii) 1 MeV,
   (iv) 1 GeV (or, as it is often written, 1 BeV).

(c) Calculate the ratio of relativistic mass to rest mass of each of the electrons of (b).
(d) Calculate $(c - v)/c$ for each of the electrons of (b).

Problem 2.18

An electron has a momentum of 78 MeV/c. Find the following:

(a) the energy,
(b) the kinetic energy,
(c) the speed of the electron.

Problem 2.19

(a) Show that the relativistic energies E' and E of Equation (2.42) are related by

$$E' = \frac{[1 - (v_x\beta/c)]E}{\sqrt{1 - \beta^2}}.$$

(b) Deduce the transformation equation for the relativistic mass,

$$m_v = m_v' \frac{[1 + (v_x\beta/c)]}{\sqrt{1 - \beta^2}}.$$

Problem 2.20

The world line of a particle is described relative to an inertial observer S by the equation

$$x = 3 \sin\left(\frac{\pi ct}{10}\right).$$

(a) Draw the world line of the particle on the space-time diagram of S.
(b) Draw, on this space-time diagram, the energy-momentum vectors, with an appropriate scale, at the points on the world line for which
    (i) $ct = 0$,
    (ii) $ct = 4$,
    (iii) $ct = 5$.

Problem 2.21

(a) What is the mass of a π^+ meson, in kilograms?

$$\text{Mass of } \pi^+ \text{ meson} = 139.6 \text{ MeV}.$$

(b) Find the relativistic energies and relativistic momentums of π^+ mesons moving with the following speeds:
    (i) $\beta = 0.01$,
    (ii) $\beta = 0.1$,
    (iii) $\beta = 0.50$,
    (iv) $\beta = 0.90$,
    (v) $\beta = 0.99$,
    (vi) $\beta = 0.999$,
    (vii) $\beta = 0.9999$.

Problem 2.22

Show that $v = pc/\sqrt{p^2 + m^2c^2}$.

Problem 2.23

A plane wave is specified in part by the 4-propagation vector k with components $(\omega/c, \mathbf{k})$ relative to an observer (Section 1.10). The free motion of a particle is specified by the 4-momentum p with components $(E/c, \mathbf{p})$ relative to an observer. Quantum mechanics attributes a particle-like behavior to waves and wave-like behavior to particles.

(a) What is the form of the covariant equation that relates the two descriptions?
(b) What is the equivalent rest mass of a light wave?

Problem 2.24

It was tacitly assumed in the above analysis that motion with constant velocity $\mathbf{v}$ relative to one inertial system would be described as unaccelerated motion relative to any other inertial system. Show in the following way that this result follows from the fact that the Lorentz transformation equations are linear. Write a general Lorentz transformation as

$$x_{\mu'} = L_{\mu'\mu}x_\mu + a_{\mu'},$$

where the $L_{\mu'\mu}$ and $a_{\mu'}$ are constants. Write out the explicit forms of this equation for $\mu' = 0$ and, say, $\mu' = 1$; from these explicit expressions, show that if dx/dt, dy/dt, and dz/dt are all constant, then dx'/dt' is also constant.

2.2.2   Conservation laws for elastic collisions *

We now turn to the problem of finding relativistic conservation laws for systems of particles. We begin with the simplest case by restricting our considerations here to collisions in which none of the basic properties, such as rest mass and charge, of any of the particles involved is changed by the collision. A collision in which this restriction is satisfied is called an *elastic collision*. A collision in which this restriction is not satisfied is illustrated in Figure 2.23. Keep in mind that our concern is with the values of the dynamical variables outside the region of space-time in which the interactions take place (see Figure 2.13).

Let us consider first the description of an elastic collision according to non-relativistic mechanics. This description should be valid if all relative speeds involved are much less than c. We consider the effects of a collision between N particles, each of whose dynamical variables before the collision is labeled by a superscript (a), with $a = 1, 2, \ldots, N$. For example, the velocity of particle a is denoted by $\mathbf{v}^{(a)}$ (Figure 2.24).

The effect of the collision of the particles on their motions is described in part by standard conservation laws. A nonrelativistic collision of particles

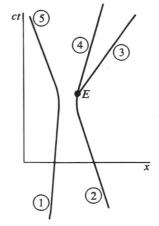

FIGURE 2.23   Space-time diagram of an inelastic collision: Particle 2 disintegrates into particles 3 and 4 at event E.

* Feynman, Leighton, and Sands (vol. 1), Addison-Wesley, Sec. 16–4, p. 16–6.
Kacser, Prentice-Hall, Sec. A.5, p. 121; Sec. 5.4, p. 148.
Kittel, Knight, and Ruderman, McGraw-Hill, pp. 382–385.
Resnick, John Wiley, Sec. 3.3, p. 114.
Taylor and Wheeler, W. H. Freeman, Sec. 11, p. 103.

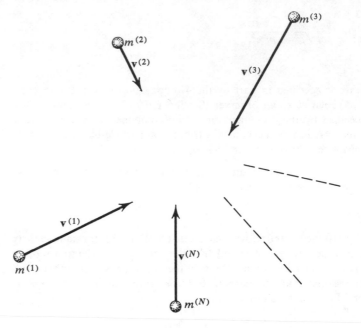

FIGURE 2.24 The dynamical variables of the particles are labeled by a super-script.

that interact with forces derivable from a potential energy satisfy, in general, a number of conservation laws—those of momentum, energy, and angular momentum. (For the time being, we shall omit consideration of the conservation law of angular momentum; the relativistic form of this law will be discussed in Problem A2.17.) The nonrelativistic conservation laws of momentum and energy have the following forms:

$$\sum_{a=1}^{N} m^{(a)}\mathbf{v}_i^{(a)} = \sum_{a=1}^{N} m^{(a)}\mathbf{v}_f^{(a)} \qquad \text{N.R.} \qquad (2.50)$$

and

$$\sum_{a=1}^{N} \frac{1}{2} m^{(a)}[v_i^{(a)}]^2 = \sum_{a=1}^{N} \frac{1}{2} m^{(a)}[v_f^{(a)}]^2 \qquad \text{N.R.} \qquad (2.51)$$

where the subscripts i (for initial) and f (for final) denote the labeled values of the dynamical variables before the collision and after the collision, respectively.

These conservation laws do not provide a complete description of the motions, since, from these conservation laws alone, we cannot obtain a unique specification of the result of the collision. For example, in a collision of two identical particles interacting with a known force, it is necessary also to know the impact parameter and the plane of the motion before the unique angle of scatter can be determined (Figure 2.25). Moreover, in order to obtain the complete specification of the motion, we must know the force; we wish at this time to obtain as much information as possible about relativistic dynamics while avoiding the problems associated with a determination of the interactions between particles. Therefore, we restrict our considerations to the relativistic generalizations of the standard conservation laws, Equations (2.50) and (2.51).

The nonrelativistic laws of conservation of momentum and energy are valid under circumstances in which all relative speeds are much less than c. The conservation laws of relativistic mechanics that we seek must reduce to the corresponding laws of newtonian mechanics under these circumstances; thus, we are led to expect that the fundamental concepts and laws of nonrelativistic mechanics correspond to analogous concepts and laws in relativistic mechanics. Therefore, we anticipate that in relativistic mechanics there exist conservation laws of the relativistic analogues of momentum and energy.

There is one simplifying feature in relativity theory, namely, that the relativistic mass and the relativistic energy (and hence, essentially, the relativistic kinetic energy) are one and the same entity aside from the immaterial factor of c^2. The fact that two apparently dissimilar nonrelativistic concepts, energy and mass, correspond to one relativistic concept, relativistic energy or relativistic mass, arises in the following manner: For a relative speed $v \ll c$, the relativistic mass $m_v = m/\sqrt{1 - (v^2/c^2)}$ is essentially the constant m, and, in fact, a measurement of m_v using a conventional technique for measuring masses is certain to give m within the limits of experimental errors. Thus, the difference $m_v - m$ cannot be determined by a mass measurement, but, on the other hand, it can be obtained by a measurement of the kinetic energy $\frac{1}{2}mv^2$, since $m_v - m = \frac{1}{2}mv^2/c^2$ for $v \ll c$. Therefore, the nonrelativistic mass is the substantial part of the relativistic mass, and the nonrelativistic kinetic energy is, apart from the

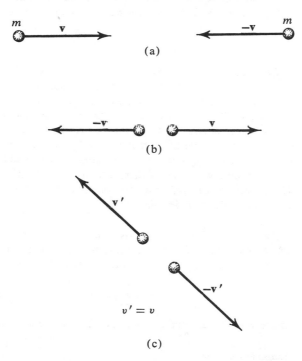

FIGURE 2.25 The conservation laws of momentum and energy are satisfied by different motions. (a) Before the collision. (b) After a head-on collision. (c) After a glancing collision.

factor of c^2, the (relatively small) difference between the relativistic mass and the rest mass. The nonrelativistic conservation law of energy (2.51) will thus follow from (1) the fact that the collision is elastic, so that the rest masses are unchanged, and (2) a conservation law of relativistic energy.

The nonrelativistic equations (2.50) and (2.51) for an elastic collision therefore are equivalent to the set

$$\sum_{a=1}^{N} m^{(a)} \mathbf{v}_i^{(a)} = \sum_{a=1}^{N} m^{(a)} \mathbf{v}_f^{(a)} \qquad \text{N.R.} \tag{2.52}$$

and

$$\sum_{a=1}^{N} m_{vi}^{(a)} = \sum_{a=1}^{N} m_{vf}^{(a)} \quad \text{or} \quad \sum_{a=1}^{N} m_{vi}^{(a)} c^2 = \sum_{a=1}^{N} m_{vf}^{(a)} c^2 \qquad \text{N.R.} \tag{2.53}$$

where $m_v^{(a)}$ denotes the relativistic mass of particle (a).

At this point, we can proceed directly to the relativistic form of these conservation laws. We note in the first place that, since $m_v = m$ for $|v|/c \ll 1$ Equation (2.52) is the nonrelativistic limit of

$$\sum_{a=1}^{N} m_{vi}^{(a)} \mathbf{v}_i^{(a)} = \sum_{a=1}^{N} m_{vf}^{(a)} \mathbf{v}_f^{(a)} \tag{2.54}$$

and in the second place that $[m_v^{(a)} c, m_v^{(a)} \mathbf{v}^{(a)}] = m^{(a)} v^{(a)}$ is a 4-vector and hence can be defined independent of the observer. As a result of these facts, we see that

$$\sum_{a=1}^{N} m^{(a)} v_i^{(a)} = \sum_{a=1}^{N} m^{(a)} v_f^{(a)} \tag{2.55}$$

is a covariant equation that reduces in the nonrelativistic limit to Equations (2.52) and (2.53). This is the *relativistic law of 4-momentum conservation*:

$$\mathscr{P}_i \equiv \sum_{a=1}^{N} \not p_i^{(a)} = \sum_{a=1}^{N} \not p_f^{(a)} \equiv \mathscr{P}_f. \tag{2.56}$$

We obtained this law by generalizing the nonrelativistic conservation laws, so we cannot maintain that we derived it. Indeed, the validity of this law must be based on its agreement with experiments and, in order that they support this law and not its nonrelativistic limit, these experiments must pertain to a range of experience outside of which the nonrelativistic laws are valid. Experiments with fast ($v \approx c$) particles that are performed day after day in high-energy laboratories throughout the world support the validity of the 4-momentum conservation law. Some of these experiments are discussed in the problems. The process of generalization by which we arrived at this conservation law is examined in detail in Section 2.2.3.

SUMMARY   The nonrelativistic conservation laws of mass, momentum, and kinetic energy for elastic collisions may be generalized to one covariant 4-vector law. This is the conservation law of 4-momentum

$$\sum_{a=1}^{N} \not p_i^{(a)} = \sum_{a=1}^{N} \not p_f^{(a)},$$

whose validity has been corroborated by many experiments.

Example 2.3

> **Q.** A particle of rest mass m_1 and kinetic energy T_1 relative to an inertial system S is incident on a particle of rest mass m_2 stationary with respect to S.
>
> (a) Find the components of the total 4-momentum vector relative to S.
> (b) Find the kinetic energy T_2 of the particle m_2 relative to that inertial system S' in which the particle m_1 is at rest.

A. (a) The energy relative to S of particle m_1 is given by

$$E_1 = m_1 c^2 + T_1, \qquad (2.57)$$

and the magnitude of its momentum relative to S is given by the relation

$$\frac{E_1^2}{c^2} - p_1^2 = m_1^2 c^2 \qquad (2.58)$$

to be

$$p_1 = \sqrt{(E_1^2/c^2) - m_1^2 c^2} = \sqrt{T_1[2m_1 + (T_1/c^2)]}. \qquad (2.59)$$

Let $\hat{x}$ designate the direction of $\mathbf{p}_1$. Then the total 4-momentum has components, relative to S, given by

$$\mathscr{P} = \left(\frac{E_1}{c} + m_2 c, \mathbf{p}_1\right) \qquad (2.60)$$

with

$$E_1 = m_1 c^2 + T_1, \qquad \mathbf{p}_1 = \sqrt{T_1[2m_1 + (T_1/c^2)]}\hat{x}. \qquad (2.61)$$

(b) Let the components of $\mathscr{P}$ be $[(E_2/c) + m_1 c, \mathbf{p}_2]$ relative to S'. Since the 4-momentum with components $(E_2/c, \mathbf{p}_2)$ describes the motion of m_2,

$$\frac{E_2^2}{c^2} - \mathbf{p}_2^2 = m_2^2 c^2. \qquad (2.62)$$

The components of $\mathscr{P}$ relative to S and S' involve the given data and the unknown T_2. Therefore, the one relation necessary to calculate the one unknown can be obtained from the fact that $\mathscr{P}^2$ is an invariant. We equate the norm of $\mathscr{P}$ calculated relative to S to the norm calculated relative to S':

$$\left(\frac{E_1}{c} + m_2 c\right)^2 - \mathbf{p}_1^2 = \left(\frac{E_2}{c} + m_1 c\right)^2 - \mathbf{p}_2^2. \qquad (2.63)$$

Cancellation of $m_1^2 c^2 + m_2^2 c^2$ from each side gives the relation

$$E_1 m_2 = m_1 E_2. \qquad (2.64)$$

Therefore,

$$T_2 = E_2 - m_2 c^2 = (m_1 c^2 + T_1) \frac{m_2}{m_1} - m_2 c^2$$
$$= \frac{m_2 T_1}{m_1}. \qquad (2.65)$$

Example 2.4

Q. Two particles travel with 4-momentums $\mathscr{P}_1$ and $\mathscr{P}_2$, respectively. The *center-of-momentum system* is that inertial system in which the total relativistic momentum is zero (Figure 2.26). Label the components of 4-vectors relative to the center-of-momentum system by a superscript asterisk.

(a) Find the total (relativistic) energy in the center-of-momentum system.

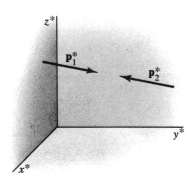

FIGURE 2.26 The center-of-momentum system: The total relativistic momentum $\mathbf{P}^* = \mathbf{p}_1^* + \mathbf{p}_2^*$ is zero relative to that system.

(b) Find the velocity of the center-of-momentum system relative to the inertial system S in which the total 4-momentum $\mathscr{P} = \not{p}_1 + \not{p}_2$ has components $(E/c, \mathbf{P})$.

A. (a) Relative to the center-of-momentum system, the components of the total 4-momentum $\mathscr{P}$ are $(E^*/c, 0)$, since $\mathbf{P}^* = 0$. Therefore,

$$\left(\frac{E^*}{c}\right)^2 = \mathscr{P}\cdot\mathscr{P} \quad \text{or} \quad E^* = c\sqrt{\mathscr{P}\cdot\mathscr{P}} = c\sqrt{(\not{p}_1 + \not{p}_2)\cdot(\not{p}_1 + \not{p}_2)}. \tag{2.66}$$

(b) The components of $\mathscr{P}$ are $(E^*/c, 0)$ relative to the center-of-momentum system and $(E/c, \mathbf{P})$ relative to S. Let $\mathbf{V}$ be the velocity of the center-of-momentum system with respect to S. Then the transformation law that determines the components $(E/c, \mathbf{P})$ from the components $(E^*/c, 0)$ is identical to that that determines the components $\not{p} = (\varepsilon/c, \mathbf{p})$ relative to S of the 4-momentum of a single particle of mass m in terms of the components $\not{p} = (mc, 0)$ relative to the rest system of the particle The velocity of the particle relative to S is $c^2\mathbf{p}/\varepsilon$; hence,

$$\mathbf{V} = \frac{c^2\mathbf{P}}{E}. \tag{2.67}$$

This can also be expressed in terms of the components $(E_1/c, \mathbf{p}_1)$ and $(E_2/c, \mathbf{p}_2)$ of $\not{p}_1$ and $\not{p}_2$, respectively, relative to S. Since $E = E_1 + E_2$ and $\mathbf{P} = \mathbf{p}_1 + \mathbf{p}_2$, we have

$$\mathbf{V} = \frac{c^2(\mathbf{p}_1 + \mathbf{p}_2)}{E_1 + E_2}. \tag{2.68}$$

Example 2.5

Q. A particle of rest mass m undergoes an elastic collision in which its 4-momentum is changed from $\not{p}_i$ to $\not{p}_f$ (Figure 2.27). Show that the 4-momentum transferred to the particle, $\not{q} = \not{p}_f - \not{p}_i$, is space-like unless $\not{q} = 0$.

A. The norm of the 4-momentum transfer is given by

$$\begin{aligned}
\not{q}^2 &= (\not{p}_f - \not{p}_i)\cdot(\not{p}_f - \not{p}_i) = \not{p}_f^2 + \not{p}_i^2 - 2\not{p}_f\cdot\not{p}_i \\
&= m^2c^2 + m^2c^2 - 2\not{p}_f\cdot\not{p}_i = 2(m^2c^2 - \not{p}_f\cdot\not{p}_i).
\end{aligned} \tag{2.69}$$

We can evaluate $\not{p}_f\cdot\not{p}_i$ relative to a particular inertial system, and, since $\not{p}_f\cdot\not{p}_i$ is a scalar, we can thus obtain the value of $\not{p}_f\cdot\not{p}_i$ relative to any other inertial system. Since

$$\not{p}_f\cdot\not{p}_i = p_{0f}p_{0i} - \mathbf{p}_f\cdot\mathbf{p}_i, \tag{2.70}$$

it is most convenient to evaluate this relative to an inertial system, say S', in which $\mathbf{p}_f' = 0$ (or, alternatively, in that system in which $\mathbf{p}_i'' = 0$). Relative to S',

$$\not{p}_f\cdot\not{p}_i = p_{0'f}p_{0'i} \tag{2.71}$$

and

$$p_{0'f} = mc \quad \text{and} \quad p_{0'i} = \sqrt{m^2c^2 + \mathbf{p}_i'^2}, \tag{2.72}$$

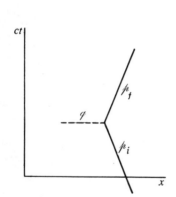

FIGURE 2.27   The 4-momentum transfer to the particle: $\not{q} = \not{p}_f - \not{p}_i$.

where $\mathbf{p}'_i$ is the initial momentum relative to S'. Hence,

$$\not{p}_f \cdot \not{p}_i = mc\sqrt{m^2c^2 + \mathbf{p}'^2_i}$$
$$= m^2c^2\sqrt{1 + (\mathbf{p}'^2_i/m^2c^2)} \qquad (2.73)$$

and

$$q^2 = 2m^2c^2\left[1 - \sqrt{1 + (\mathbf{p}'^2_i/m^2c^2)}\right] \leq 0; \qquad (2.74)$$

$q^2 = 0$ only if $\mathbf{p}'_i = 0$, in which case the particle is at rest relative to S' both before and after the collision, so $q = 0$. If $q \neq 0$, then $q^2 < 0$, and the 4-momentum transfer is space-like.

Example 2.6

Q. A particle of mass m and kinetic energy T_i relative to an inertial system S collides elastically with a particle of mass M initially at rest with respect to S (Figure 2.28).

(a) Find a relation between the kinetic energy T_i of the incident projectile, the energy $\Delta E = T_i - T_f$ transferred from the projectile to the target, and the angle θ through which the projectile is scattered.

(b) Find the relation of (a) for the extreme-relativistic case in which $mc^2 \ll T_i$ and in which the terms in mc^2/T_i can be neglected compared to unity.

A. (a) Let $\not{p}$ and $\not{p}'$ be the initial and final 4-momentums, respectively, of the particle of mass m, and let $\mathscr{P}$ and $\mathscr{P}'$ be those of M. Then, relative to S,

$$\not{p} = \left(mc + \frac{T_i}{c}, \mathbf{p}\right), \qquad \not{p}' = \left(mc + \frac{T_f}{c}, \mathbf{p}'\right) \qquad (2.75)$$

$$\mathscr{P} = (Mc, 0) \qquad (2.76)$$

and

$$\mathbf{p} \cdot \mathbf{p}' = |\mathbf{p}|\,|\mathbf{p}'|\cos\theta. \qquad (2.77)$$

The conservation law of energy-momentum is

$$\not{p} + \mathscr{P} = \not{p}' + \mathscr{P}'. \qquad (2.78)$$

Since $\mathbf{p} \cdot \mathbf{p}'$ depends on the angle θ, we obtain a relation involving θ from the equality of the norms of the vectors $\not{p} - \not{p}'$ and $\mathscr{P}' - \mathscr{P}$:

$$\not{p} - \not{p}' = \mathscr{P}' - \mathscr{P}, \qquad (2.79)$$

so that

$$\not{p}^2 + \not{p}'^2 - 2\not{p} \cdot \not{p}' = \mathscr{P}'^2 + \mathscr{P}^2 - 2\mathscr{P}' \cdot \mathscr{P}. \qquad (2.80)$$

Since

$$\not{p}^2 = \not{p}'^2 = m^2c^2, \qquad \mathscr{P}'^2 = \mathscr{P}^2 = M^2c^2,$$
$$\not{p} \cdot \not{p}' = \not{p}_0\not{p}'_0 - \mathbf{p} \cdot \mathbf{p}', \qquad \text{and} \qquad \mathscr{P}' \cdot \mathscr{P} = P'_0Mc, \qquad (2.81)$$

we obtain

$$m^2c^2 - \left(mc + \frac{T_i}{c}\right)\left(mc + \frac{T_i}{c} - \frac{\Delta E}{c}\right) + |\mathbf{p}|\,|\mathbf{p}'|\cos\theta$$
$$= M^2c^2 - P'_0Mc. \qquad (2.82)$$

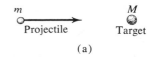

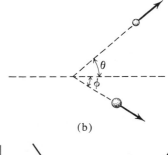

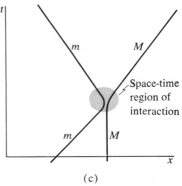

FIGURE 2.28 An elastic collision with a target that was initially stationary relative to S. (a) Before the collision. (b) After the collision. (c) Space-time diagram of the collision.

The time-like component of the energy-momentum conservation law gives the relation

$$
\begin{aligned}
P_0' c &= p_0 + P_0 - p_0' \\
&= (mc^2 + T_i) + Mc^2 - (mc^2 + T_i - \Delta E) \quad\quad (2.83) \\
&= Mc^2 + \Delta E.
\end{aligned}
$$

Also,

$$
|\mathbf{p}| = \sqrt{p_0^2 - m^2 c^2} = \sqrt{2mT_i + (T_i^2/c^2)}, \quad\quad (2.84)
$$

$$
|\mathbf{p}'| = \sqrt{2m(T_i - \Delta E) + [(T_i - \Delta E)^2/c^2]}. \quad\quad (2.85)
$$

Therefore,

$$
\begin{aligned}
\cos\theta &= \frac{M^2 c^2 - P_0' Mc + m(2T_i - \Delta E) + (T_i/c^2)(T_i - \Delta E)}{|\mathbf{p}|\,|\mathbf{p}'|} \\
&= \frac{-\Delta E M + m(2T_i - \Delta E) + (T_i/c^2)(T_i - \Delta E)}{\sqrt{2mT_i + (T_i^2/c^2)}\,\sqrt{2m(T_i - \Delta E) + [(T_i - \Delta E)^2/c^2]}}.
\end{aligned} \quad (2.86)
$$

(b) For $mc^2 \ll T_i$, we can neglect all the terms involving m in the above equation, since this can be written as

$$
\cos\theta = \frac{-\Delta E M + (T_i^2/c^2)[1 + (2mc^2/T_i)] - (T_i \Delta E/c^2)[1 + (mc^2/T_i)]}{[T_i(T_i - \Delta E)/c^2]\,\sqrt{1 + (2mc^2/T_i)}\,\sqrt{1 + [2mc^2/(T_i - \Delta E)]}}. \quad (2.87)
$$

Therefore, unless $\Delta E \sim T_i$,

$$
\cos\theta \approx \frac{-c^2 \Delta E M - T_i \Delta E + T_i^2}{T_i(T_i - \Delta E)} = 1 - \frac{c^2 \Delta E M}{T_i(T_i - \Delta E)}. \quad\quad (2.88)
$$

This can also be written in the convenient form

$$
\frac{1}{T_f} - \frac{1}{T_i} = \frac{1}{E_f} - \frac{1}{E_i} = \frac{1}{Mc^2}(1 - \cos\theta), \quad\quad (2.89)
$$

since $T_i = E_i - mc^2 \approx E_i$ and $T_f \approx E_f$.

Example 2.7

The form of a relativistic conservation law, that of energy-momentum, was conjectured in the text on the basis of the 4-vector concept. It is possible also to use the Lorentz transformation to arrive at the same form by more direct generalizations of the nonrelativistic laws. These direct generalizations involve much the same assumptions as given in the text and the use of the formula (1.191) for the transformation law of $\gamma = \sqrt{1 - (v^2/c^2)}$:

$$
\sqrt{1 - (v'^2/c^2)} = \frac{\sqrt{1 - (v^2/c^2)}\,\sqrt{1 - (V^2/c^2)}}{1 - (vV/c^2)} \quad\quad (2.90)
$$

In nonrelativistic mechanics, the effect of a collision is described in part by the conservation law of total momentum $\sum_{a=1}^{N} m^{(a)} \mathbf{v}^{(a)}$ and the conservation law of total mass $\sum_{a=1}^{N} m^{(a)}$. Furthermore, because of the form of the galilean transformation equations, these conservation laws take on the same form relative to all inertial systems (for which each relative speed involved is much less than c). We assume that these laws are valid also in relativistic mechanics and that their form is the same in all inertial reference systems. In order that this generalization be valid, it is necessary to give up the property that the $m^{(a)}$ in the above forms for the total momentum and total mass are constants, a condition that is valid in nonrelativistic mechanics, and to permit $m^{(a)}$ to depend on the speed $v^{(a)}$: $m^{(a)} = m^{(a)}(v^{(a)})$. The problem of finding the form of the conservation laws compatible with the principle of relativity and the principle of the invariance of the speed of light reduces to determining the form of the function $m^{(a)}(v^{(a)})$.

Q. (a) Relative to an inertial system S, two identical particles move along a straight line toward each other with equal speeds, collide elastically, and recoil along the same straight line. Describe the motion after the collision on the basis of the assumption that the total mass and total momentum are conserved. In particular, show that the total momentum before and after the collision is zero.

   (b) Describe the motion relative to an inertial system S' moving with velocity $V\hat{x}$ relative to S, where $\hat{x}$ is along the line of motion of the two particles.

   (c) Find the form of the function $m(v)$ if, relative to S', the total momentum is $M'[-V\hat{x}]$, where M' is the total mass relative to S'.

A. (a) Let the two particles be labeled 1 and 2 with $\mathbf{v}_1 = v\hat{x}$ and $\mathbf{v}_2 = -v\hat{x}$ before the collision. Denote the corresponding entities evaluated after the collision by a bar over the symbol for the entity. Then the conservation laws state that

$$m_1(v_1) + m_2(v_2) = m_1(\bar{v}_1) + m_2(\bar{v}_2) \tag{2.91}$$

and

$$m_1(v_1)\mathbf{v}_1 + m_2(v_2)\mathbf{v}_2 = m_1(\bar{v}_1)\bar{\mathbf{v}}_1 + m_2(\bar{v}_2)\bar{\mathbf{v}}_2. \tag{2.92}$$

One solution of these equations is $\bar{\mathbf{v}}_1 = \mathbf{v}_1$, $\bar{\mathbf{v}}_2 = \mathbf{v}_2$, which describes the case in which there is no collision. But, since the particles are identical,

$$m_1(v) = m_2(v) = m(v), \tag{2.93}$$

so that

$$m(v_1) + m(v_2) = m(\bar{v}_1) + m(\bar{v}_2), \tag{2.94}$$

$$m(v_1)\mathbf{v}_1 + m(v_2)\mathbf{v}_2 = m(\bar{v}_1)\bar{\mathbf{v}}_1 + m(\bar{v}_2)\bar{\mathbf{v}}_2. \tag{2.95}$$

Hence, there exists another solution with

$$\bar{\mathbf{v}}_1 = \mathbf{v}_2, \qquad \bar{\mathbf{v}}_2 = \mathbf{v}_1. \tag{2.96}$$

This solution describes an elastic collision, since, because of the identity of the particles, all dynamical properties of the system are the same after the collision as they were before. Hence, the motion after the collision is described by

$$\bar{\mathbf{v}}_1 = -v\hat{x}, \qquad \bar{\mathbf{v}}_2 = v\hat{x}. \tag{2.97}$$

The total momentum is

$$\mathbf{P} = m(v)v\hat{x} + m(v)(-v\hat{x}) = 0 \quad , \tag{2.98}$$

both before and after the collision.

(b) The velocity $v\hat{x}'$ relative to S' of a particle moving with the velocity $v\hat{x}$ relative to S is given by

$$v' = \frac{v - V}{1 - (vV/c^2)}. \tag{2.99}$$

Therefore,

$$v_1' = \bar{v}_2' = \frac{v - V}{1 - (vV/c^2)}, \qquad v_2' = \frac{-v - V}{1 + (vV/c^2)}. \tag{2.100}$$

(c) The relativistic masses of the particles with respect to S' are $m(v_1')$ and $m(v_2')$; by assumption, we have

$$m(v_1') + m(v_2') = M', \tag{2.101}$$

$$m(v_1')v_1' + m(v_2')v_2' = -M'V, \tag{2.102}$$

where M' is the total relativistic mass with respect to S'. We solve

$$m(v_1')v_1' + m(v_2')v_2' = -[m(v_1') + m(v_2')]V \tag{2.103}$$

for the ratio

$$\frac{m(v_2')}{m(v_1')} = \frac{v_1' + V}{-v_2' - V} = \frac{\{(v - V)/[1 - (vV/c^2)]\} + V}{\{(v + V)/[1 + (vV/c^2)]\} - V}$$

$$= \frac{1 + (vV/c^2)}{1 - (vV/c^2)} \tag{2.104}$$

$$= \frac{\sqrt{1 - (v^2/c^2)} \, \sqrt{1 - (V^2/c^2)} \, \sqrt{1 - (v_1'^2/c^2)}}{\sqrt{1 - (v^2/c^2)} \, \sqrt{1 - (V^2/c^2)} \, \sqrt{1 - (v_2'^2/c^2)}}.$$

Therefore,

$$m(v_1')^2 \sqrt{1 - (v_1'^2/c^2)} = m(v_2') \sqrt{1 - (v_2'^2/c^2)}. \tag{2.105}$$

We always can choose V such that $v_2' = 0$, in which case we obtain, setting $v_1' = v$,

$$m(v) = \frac{m(0)}{\sqrt{1 - (v^2/c^2)}} = \frac{m}{\sqrt{1 - (v^2/c^2)}}. \tag{2.106}$$

Example 2.8

Q. A particle of mass m is incident with kinetic energy T on an identical particle at rest relative to an inertial system S. The collision is elastic.

(a) Show that the angle between the direction of motion of the two particles after the collision is $90°$, according to nonrelativistic mechanics.

(b) Let θ be the angle in S through which the incident particle is scattered and ϕ be the angle in S between the final direction of motion of the recoiling target particle and the initial direction of motion of the incident particle (Figure 2.29). Show that, in the energy units introduced above,

$$\tan\theta \tan\phi = \frac{2m}{2m + T}. \tag{2.107}$$

(c) Show that $(\theta + \phi) < 90°$ for $T > 0$.

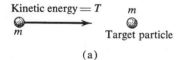

Kinetic energy $= T$ m

m Target particle

(a)

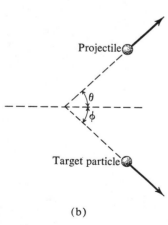

Projectile

θ

ϕ

Target particle

(b)

FIGURE 2.29 Definition of the angles θ and ϕ. (a) Before the collision. (b) After the collision.

A. Label the incident particle 1 and the target particle 2. An unprimed entity denotes the value of the entity before the collision; a primed entity, that after the collision.

(a) The nonrelativistic conservation laws for an elastic two-particle collision

$$m_1\mathbf{v}_1 + m_2\mathbf{v}_2 = m_1\mathbf{v}_1' + m_2\mathbf{v}_2',$$
$$\tfrac{1}{2}m_1v_1^2 + \tfrac{1}{2}m_2v_2^2 = \tfrac{1}{2}m_1v_1'^2 + \tfrac{1}{2}m_2v_2'^2 \qquad (2.108)$$

reduce, since $m_1 = m_2$ and $\mathbf{v}_2 = 0$, to

$$\mathbf{v}_1 = \mathbf{v}_1' + \mathbf{v}_2', \qquad v_1^2 = v_1'^2 + v_2'^2. \qquad (2.109)$$

These two equations are consistent only if $\mathbf{v}_1 \cdot \mathbf{v}_1 = v_1^2$ so that

$$(\mathbf{v}_1' + \mathbf{v}_2') \cdot (\mathbf{v}_1' + \mathbf{v}_2') = v_1'^2 + 2\mathbf{v}_1' \cdot \mathbf{v}_2' + v_2'^2 = v_1'^2 + v_2'^2; \qquad (2.110)$$

hence,

$$\mathbf{v}_1' \cdot \mathbf{v}_2' = 0. \qquad (2.111)$$

Therefore, the angle between the two recoiling particles is 90°.

(b) Let the momentum $\mathbf{p}_1$ lie along the x direction and the momentum $\mathbf{p}_1'$ lie in the xy plane. Then, because of the conservation of momentum, the momentum $\mathbf{p}_2'$ also lies in the xy plane. The angles θ and ϕ are given by the relations (see Figure 2.30)

$$\tan\theta = \frac{|p_{1y}'|}{|p_{1x}'|}, \qquad \tan\phi = \frac{|p_{2y}'|}{|p_{2x}'|}. \qquad (2.112)$$

We can evaluate the desired product, $|p_{1y}'p_{2y}'/p_{1x}'p_{2x}'|$, most easily by finding the momentum components involved in terms of the components $(E^*/c, \mathbf{p}^*)$ relative to the center-of-momentum system of the corresponding 4-momentums $\not{p}$.

In the center-of-momentum system

$$\mathbf{p}_1^{*\prime} + \mathbf{p}_2^{*\prime} = 0, \qquad E_1^{*\prime} = E_2^{*\prime} = \tfrac{1}{2}E^{*\prime}. \qquad (2.113)$$

The components of $\mathbf{p}_1'$ and $\mathbf{p}_2'$ are given in energy units by

$$p_{1x}' = \frac{p_{1x}^{*\prime} + VE_1^{*\prime}}{\sqrt{1-V^2}}, \qquad p_{1y}' = p_{1y}^{*\prime},$$

$$p_{2x}' = \frac{p_{2x}^{*\prime} + VE_2^{*\prime}}{\sqrt{1-V^2}} = \frac{-p_{1x}^{*\prime} + VE_1^{*\prime}}{\sqrt{1-V^2}}, \qquad p_{2y}' = p_{2y}^{*\prime} = -p_{1y}^{*\prime}. \quad (2.114)$$

Therefore,

$$\tan\theta\tan\phi = \left| \frac{(p_{1y}^{*\prime})^2(1-V^2)}{(p_{1x}^{*\prime} + VE_1^{*\prime})(-p_{1x}^{*\prime} + VE_1^{*\prime})} \right|$$
$$= \frac{(p_{1y}^{*\prime})^2(1-V^2)}{V^2 E_1^{*\prime 2} - (p_{1x}^{*\prime})^2}. \qquad (2.115)$$

The speed of the center-of-momentum system relative to S is V. Since particle 2 is initially at rest relative to S, V is also the initial speed of particle 2 relative to the center-of-momentum system. Since $|\mathbf{p}_1^*| = |\mathbf{p}_2^*|$, V is also the initial speed of particle 1 relative to the

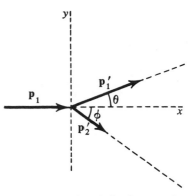

FIGURE 2.30 The relation between the angles θ and ϕ and the momenta $\mathbf{p}_1$, $\mathbf{p}_1'$, and $\mathbf{p}_2'$.

center-of-momentum system. Furthermore, according to the conservation law of energy-momentum, $\mathbf{p}_1^* + \mathbf{p}_2^* = 0$, $|\mathbf{p}_1^{*\prime}| = |\mathbf{p}_2^{*\prime}|$; hence, $E_1^* = E_2^* = \frac{1}{2}E^*$, $E_1^{*\prime} = E_2^{*\prime} = \frac{1}{2}E^{*\prime}$, and $E^* = E^{*\prime}$. Therefore, $v_1^{*\prime} = v_1^* = V$. We conclude that

$$V = \frac{|\mathbf{p}_1^{*\prime}|}{E_1^{*\prime}} = \frac{\sqrt{(p_{1x}^{*\prime})^2 + (p_{1y}^{*\prime})^2}}{E_1^{*\prime}}. \tag{2.116}$$

Hence,

$$\tan\theta \tan\phi = 1 - V^2. \tag{2.117}$$

The speed V of the center-of-momentum system relative to S is given by Equation (2.68) as

$$V = \frac{mv_1/\sqrt{1 - v_1^2}}{m/\sqrt{1 - v_1^2} + m} = \frac{v_1}{1 + \sqrt{1 - v_1^2}}. \tag{2.118}$$

Therefore,

$$\begin{aligned}
\tan\theta \tan\phi &= 1 - \frac{v_1^2}{(1 + \sqrt{1 - v_1^2})^2} \\
&= \frac{2\sqrt{1 - v_1^2}\,(1 + \sqrt{1 - v_1^2})}{(1 + \sqrt{1 - v_1^2})^2} \\
&= \frac{2}{1 + 1/\sqrt{1 - v_1^2}} = \frac{2m}{2m + T}.
\end{aligned} \tag{2.119}$$

(c) Since

$$\tan(\theta + \phi) = \frac{\tan\theta + \tan\phi}{1 - \tan\theta \tan\phi} = \frac{(\tan\theta + \tan\phi)(2m + T)}{T}, \tag{2.120}$$

$\tan(\theta + \phi) < \infty$ for all $T > 0$. Therefore, $\theta + \phi < 90°$ for $T > 0$.

Problem 2.25

A 20-MeV electron is incident on a stationary proton.

(a) Calculate the speed of the electron.
(b) Calculate the kinetic energy of the proton relative to the electron.
(c) Calculate the speed of the proton relative to the electron.
(d) Calculate the relativistic energy in the center-of-momentum system.
(e) Calculate the speed of the center-of-momentum system relative to the proton.
(f) Calculate the speed of the center-of-momentum system relative to the electron.

Problem 2.26

A 10-GeV electron is incident on a stationary proton. Answer parts (a) to (f) of Problem 2.25.

Problem 2.27

(a) The 4-momentums of two particles have components relative to the center-of-momentum reference system of

$$\left(\frac{e_i}{c}, \mathbf{p}_i\right) \quad \text{and} \quad \left(\frac{E_i}{c}, \mathbf{P}_i\right)$$

before the collision and

$$\left(\frac{e_f}{c}, \mathbf{p}_f\right) \quad \text{and} \quad \left(\frac{E_f}{c}, \mathbf{P}_f\right)$$

after the collision. Show that

$$e_i = e_f, \qquad E_i = E_f, \qquad |\mathbf{p}_i| = |\mathbf{p}_f|, \qquad |\mathbf{P}_i| = |\mathbf{P}_f|;$$

also show that the initial momentums lie along one line and that the final momentums lie along a line.

(b) A 20-MeV electron is incident on a stationary proton. The electron is scattered through 90° in the center-of-momentum system. Find the angle through which the electron is scattered in that reference system in which the proton was initially at rest.

Problem 2.28

An electron, traveling with velocity $\beta = 0.999\hat{x}$ undergoes a collision in which its velocity is changed to $\beta' = (0.999\hat{x} + 0.999\hat{y})/\sqrt{2}$.

(a) Calculate the relativistic momentum and the relativistic energy before the collision.
(b) Calculate the relativistic momentum and the relativistic energy after the collision.
(c) Calculate the energy-momentum transfer to the other particle.
(d) Calculate the norm of the energy-momentum transfer.

Problem 2.29

(a) Show that the total 4-momentum vector $\mathscr{P} = \sum_{a=1}^{N} \not{p}^{(a)}$ of a system of N noninteracting particles is time-like.
(b) Show that $\mathscr{P}$ lies in a direction in space-time that is inside the cone of the absolute future.
(c) Let the components of $\mathscr{P}$ relative to an inertial system S be $(E/c, \mathbf{P})$. Show that, relative to an inertial system S' moving with velocity $\mathbf{V}$ relative to S, the components of $\mathscr{P}$ are $(E_0/c, 0)$, if $\mathbf{V} = \mathbf{P}c^2/E$. Find E_0.
(d) Show that $V < c$.
(e) Since $\mathscr{P}$ is time-like and points into the absolute future and since $\mathscr{P}$ has zero spatial components relative to S', it is natural to consider the system as an object with rest mass M given by $M = E_0/c^2$ that is moving with the velocity $\mathbf{V}$ relative to S. Show that

$$E = \frac{Mc^2}{\sqrt{1 - (V^2/c^2)}}, \qquad \mathbf{P} = \frac{M}{\sqrt{1 - (V^2/c^2)}}\mathbf{V}.$$

(f) Show that $M \geq \sum_{a=1}^{N} m^{(a)}$. Explain why $M > \sum_{a=1}^{N} m^{(a)}$ unless every one of the particles is moving with velocity $\mathbf{V}$ relative to S.

Problem 2.30

A particle of mass m collides elastically with a particle of mass M. Initially,

m had a kinetic energy of T_i with respect to an inertial system S and M was at rest relative to S. Assume that $mc^2/T_i \ll 1$ and that $mc^2/T_f \ll 1$.

(a) Show that the magnitude of the final momentum $\mathbf{p}'$ of m is given by

$$p' = |\mathbf{p}'| = \frac{p}{1 + (2T_i/Mc^2) \sin^2 (\theta/2)},$$

where $p = |\mathbf{p}| = T_i/c$. *Hint:* Take the norm of $\not p + \mathscr{P} - \not p' = \mathscr{P}'$.

(b) Show that the kinetic energy T of the recoiling mass M is given by

$$T = \Delta E = \frac{T_i^2}{Mc^2} \frac{2 \sin^2 (\theta/2)}{1 + (2T_i/Mc^2) \sin^2 (\theta/2)}.$$

(c) Show that the magnitude of the 3-momentum transfer $\mathbf{q} = \mathbf{p} - \mathbf{p}' = \mathbf{P}' - \mathbf{P}$ is given by

$$q = |\mathbf{q}| = \frac{2T_i \sin (\theta/2)}{c[1 + (2T_i/Mc^2) \sin^2 (\theta/2)]^{\frac12}} \left[1 + \frac{(T_i^2/M^2c^4) \sin^2 (\theta/2)}{1 + (2T_i/Mc^2) \sin^2 (\theta/2)}\right]^{\frac12}.$$

(d) Show that the norm of the 4-momentum transfer $\not q = \not p - \not p' = \mathscr{P}' - \mathscr{P}$ is given by

$$q^2 = \left(\frac{\Delta E}{c}\right)^2 - q^2 = -4pp' \sin^2 \frac{\theta}{2},$$

or

$$\sqrt{-\not q^2} = \frac{(2T_i/c) \sin (\theta/2)}{[1 + (2T_i/Mc^2) \sin^2 (\theta/2)]^{\frac12}}.$$

(e) Let ϕ be the angle between $\mathbf{p}$ and $\mathbf{P}'$. Show that

$$\Delta E = T_i \frac{2(T_i/Mc^2) \cos^2 \phi}{[1 + (T_i/Mc^2)]^2 - (T_i^2/M^2c^4) \cos^2 \phi}.$$

Hint: Eliminate $|\mathbf{P}'|$ from the norm of $\not p + \mathscr{P} - \mathscr{P}' = \not p'$ through $\mathscr{P}'^2 = M^2c^2$.

Problem 2.31

An electron with 100-MeV kinetic energy strikes a proton at rest. The collision is elastic. Find the kinetic energy of the recoiling proton if the electron is scattered through

(a) 10°,   (b) 90°,   (c) 180°.

Problem 2.32

An electron with kinetic energy $T_i \gg m_e c^2$ strikes a proton at rest. The electron is scattered elastically through 180°. Plot the energy of the scattered electron versus the incident energy for the range $10 \text{ MeV} \le T_i \le 10^3 \text{ MeV}$.

Problem 2.33

A 187-MeV electron is incident on a stationary proton. Plot the energy of the elastically scattered electron as a function of θ, the angle of scatter.*

---

* These results can be compared with the measured values shown in Figure 4 of R. W. McAllister and R. Hofstadter, "Elastic Scattering of 188-MeV Electrons from the Proton and the Alpha Particle," *Physical Review, 102*: 851 (1956).

Problem 2.34

The principle of the invariance of the speed of light, as expressed by the Lorentz transformation law, is consistent with the conservation laws of momentum and mass only if the mass occurring in the conservation equations is the relativistic mass $m_v = m/\sqrt{1 - (v/c)^2}$. This result was demonstrated in Example 2.7 and also is a consequence of this problem.

Consider the elastic collision of two identical particles. Relative to the inertial system S, the particles have the velocities

$$\mathbf{v}_{1i} = v_x\hat{x} + v_y\hat{y}, \qquad \mathbf{v}_{2i} = -v_x\hat{x} - v_y\hat{y}$$

before the collision and

$$\mathbf{v}_{1f} = v_x\hat{x} - v_y\hat{y}, \qquad \mathbf{v}_{2f} = -v_x\hat{x} + v_y\hat{y}$$

after the collision, as shown in Figure 2.31.

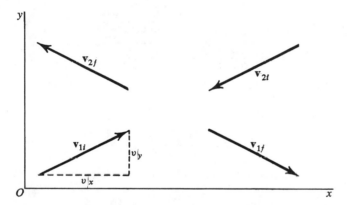

FIGURE 2.31   The initial and final velocities.

(a) Define the total momentum by $m_{v_1}\mathbf{v}_1 + m_{v_2}\mathbf{v}_2$ and the total mass by $m_{v_1} + m_{v_2}$, with the masses m_{v_1} and m_{v_2} being functions of the speed:

$$m_{v_1} = m(|\mathbf{v}_1|), \qquad m_{v_2} = m(|\mathbf{v}_2|).$$

Show that the total momentum and the total mass are conserved for the motions under consideration.

(b) Let $\mathbf{r}_1(t)$ denote the position of particle 1 relative to S at time t and $\mathbf{r}_2(t)$ the corresponding position of particle 2. Let $t = 0$ denote the instant of the collision and $\mathbf{r} = 0$ the position of the collision. Show that

$$\begin{aligned} \mathbf{r}_1(t) &= v_x t\hat{x} + v_y t\hat{y}, & t &< 0, \\ &= v_x t\hat{x} - v_y t\hat{y}, & t &> 0, \\ \mathbf{r}_2(t) &= -v_x t\hat{x} - v_y t\hat{y}, & t &< 0, \\ &= -v_x\hat{x}t + v_y t\hat{y}, & t &> 0. \end{aligned}$$

(c) Let S' be an inertial reference system that moves with velocity $v_x\hat{x}$ with respect to S. Let the origins of S and S' coincide at $t = t' = 0$. Show that

$$\mathbf{r}_1'(t') = \frac{v_y t'}{\sqrt{1 - (v_x^2/c^2)}} \, \hat{y}', \qquad t' < 0,$$

$$= \frac{-v_y t'}{\sqrt{1 - (v_x^2/c^2)}} \, \hat{y}', \qquad t' > 0,$$

$$\mathbf{r}_2'(t') = \frac{-2v_x}{1 + (v_x^2/c^2)} \, t'\hat{x}' - \frac{v_y \sqrt{1 - (v_x^2/c^2)}}{1 + (v_x^2/c^2)} \, t'\hat{y}', \qquad t' < 0,$$

$$= \frac{-2v_x}{1 + (v_x^2/c^2)} \, t'\hat{x}' + \frac{v_y \sqrt{1 - (v_x^2/c^2)}}{1 + (v_x^2/c^2)} \, t'\hat{y}', \qquad t' > 0.$$

(d) Show that

$$\mathbf{v}_1'(t') = \frac{v_y}{\sqrt{1 - (v_x^2/c^2)}} \, \hat{y}', \qquad t < 0,$$

$$= -\frac{v_y}{\sqrt{1 - (v_x^2/c^2)}} \, \hat{y}', \qquad t > 0,$$

$$\mathbf{v}_2'(t') = \frac{-2v_x}{1 + (v_x^2/c^2)} \, \hat{x}' - \frac{v_y \sqrt{1 - (v_x^2/c^2)}}{1 + (v_x^2/c^2)} \, \hat{y}', \qquad t' < 0,$$

$$= \frac{-2v_x}{1 + (v_x^2/c^2)} \, \hat{x}' + \frac{v_y \sqrt{1 - (v_x^2/c^2)}}{1 + (v_x^2/c^2)} \, \hat{y}', \qquad t' > 0,$$

and that

$$|\mathbf{v}_1'(t')| = \frac{v_y}{\sqrt{1 - (v_x^2/c^2)}},$$

$$|\mathbf{v}_2'(t')| = \frac{\sqrt{4v_x^2 + v_y^2[1 - (v_x^2/c^2)]}}{1 + (v_x^2/c^2)}.$$

(e) Show that the conservation law of momentum holds only if

$$m(|\mathbf{v}_1'|) - \frac{1 - (v_x^2/c^2)}{1 + (v_x^2/c^2)} \, m(|\mathbf{v}_2'|) = 0.$$

(f) Show that the result of (e) is equivalent to

$$m(v_1')\sqrt{1 - (v_1'^2/c^2)} = m(v_2')\sqrt{1 - (v_2'^2/c^2)}$$

and that

$$m(v) = \frac{m(0)}{\sqrt{1 - (v^2/c^2)}}.$$

Problem 2.35

When electrons were scattered elastically off stationary electrons, the following values of the scattering angles (Figure 2.29) were measured.*

(a) Calculate the values of $\theta + \phi$ expected for the given values of β_1 and θ.
(b) Why do these results corroborate relativistic mechanics and disagree with newtonian mechanics?

---

* F. C. Champion, "On Some Close Collisions of Fast β-Particles with Electrons, Photographed by the Expansion Method," *Proceedings of the Royal Society*, *A136*: 630 (1932). *Note:* β particles are electrons that have been emitted from a radioactive nucleus.

$\beta_1 = v_1/c$	θ, deg	$\theta + \phi$, deg
0.85 ± 0.01	20.0 ± 0.5	83.6 ± 1.0
0.83	26.6	81.2 ± 1.0
0.83	31.4	81.0 ± 1.0
0.82	22.0	84.1 ± 1.0
0.85	22.0	82.2 ± 1.0
0.83	22.4	82.1 ± 1.0
0.84	23.4	82.7 ± 1.0
0.90	24.5	79.6 ± 1.0
0.88	35.4	76.8 ± 0.5
0.85	21.1	82.7 ± 1.0
0.91	36.9	75.2 ± 0.5
0.93	29.6	72.5 ± 0.5
0.85	21.8	82.4 ± 1.0
0.82	36.9	80.6 ± 0.5

Problem 2.36

An inertial system S' moves with velocity $\mathbf{V}$ relative to another inertial system S.

(a) Find the velocity of an inertial system S'' such that the velocity of S' relative to S'' is equal and opposite that of S relative to S''.
(b) Find the velocity of S' relative to S''.

Problem 2.37

A particle of mass m moves with 4-momentum p, which has components $(E/c, \mathbf{p})$ relative to an inertial system S. Let $\mathscr{X}$ be a time-like 4-vector with components $(X_0, \mathbf{X})$ relative to S and components $(X_0', 0)$ relative to another inertial system S'.

(a) Find the energy E' of the particle relative to S'. *Hint:* Consider the product $p \cdot \mathscr{X}$.
(b) Find the magnitude of the relative momentum $\mathbf{p}'$ of the particle relative to S'. *Hint:* The quantity $(E'^2/c^2) - p'^2$ is an invariant.
(c) Find the speed v' of the particle relative to S'.

Problem 2.38

Two particles have masses m_1 and m_2 and 4-momentums p_1 and p_2, respectively.

(a) Find the energy, the magnitude of the momentum, and the speed of particle 1 relative to particle 2.
(b) Find the energy, the magnitude of the momentum, and the speed of particle 1 relative to the center-of-momentum system for the two-particle system.
(c) A proton of kinetic energy 6.2 GeV is incident on a stationary proton. Calculate the total energy in the center-of-momentum system. Calculate the kinetic energy and speed of each proton relative to the center-of-momentum system.

(d) A proton of kinetic energy 6.2 GeV is incident on a proton of kinetic energy 25 MeV. Assume that the protons are traveling toward each other along the line joining them. Calculate the total energy in the center-of-momentum system. Calculate the speed and kinetic energy of one of the protons relative to the other.

(e) Repeat (d) for the case in which the protons are separating along the line joining them.

2.2.3 *An examination of the process of generalization of the nonrelativistic conservation laws*

A greater understanding of the law of conservation of 4-momentum can be obtained by a more careful examination than that given above of the process of generalization from the nonrelativistic law. Therefore, we consider in greater detail the procedure by which we arrived at the covariant form of the conservation law. Note that we are still considering the values of the dynamical variables only outside the restricted region of space-time in which they interact (Figure 2.13).

There are two newtonian concepts, momentum and energy, with conservation laws that must result in the nonrelativistic limit from the laws of relativistic mechanics. One of these nonrelativistic conservation laws is a relation between 3-vectors (the momentums), and the other is a relation between scalars (the energies). Since, in the appropriate limit, the relativistic laws reduce to a nonrelativistic form, we can expect to have, in relativity theory, laws that involve relations between entities, one of which transforms under spatial rotations like a 3-vector and the other like a scalar. A nonrelativistic 3-vector equation and a nonrelativistic scalar equation together consist of four component relations, as does a 4-vector equation. Moreover, for a general 4-vector, three of the four components transform under rotations like a 3-vector and the remaining component transforms like a scalar. Therefore, it is gratifying that nonrelativistic mechanics involves a 3-vector law and a scalar law, since it is possible to obtain these forms as the appropriate limit from a 4-vector relation and, as we know, 4-vectors can be defined independent of the observer.

The desired conservation law must relate a 4-vector that depends on the circumstances of the particles before the collision to a similar 4-vector dependent on the circumstances of the particles after the collision. We can expect, by analogy with the nonrelativistic case, that the 4-vector in question is the sum of 4-vectors, each of which is an attribute of an individual particle and its motion. This expectation, which turns out to be justified, means also that we can assign such a 4-vector to a subsystem of free particles.

Consider now the pertinent 4-vector associated with the particle labeled (a). In the free motion that occurs either before or after the collision, the only 4-vector that depends solely on the properties of the particle and its motion is the 4-velocity $v^{(a)}$, or a scalar times the 4-velocity, $\kappa^{(a)}v^{(a)}$, where the scalar $\kappa^{(a)}$ may depend on the properties of the particle (a) (Figure 2.32).

This can be seen also from the following argument: The spatial 3-vector component of this 4-vector lies along the line of the direction of motion of the particle, and because of the isotropy of space, the particle and its (free) motion define no other direction in space. The only 4-vector whose spatial 3-vector lies along the direction of the motion has the form $\kappa^{(a)}v^{(a)}$, since, if there existed

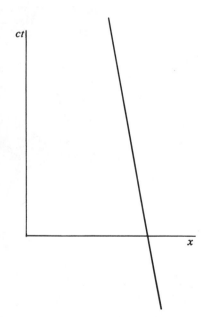

ct

x

FIGURE 2.32 The world line of a particle moving freely. The only 4-vector direction in space-time defined by the particle and its motion is along the world line of the particle.

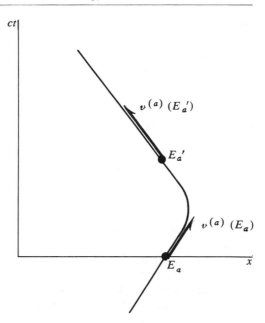

FIGURE 2.33   The value of $v^{(a)}$ depends on the event on the world line at which it is evaluated.

another such 4-vector $w^{(a)}$, then, for an appropriate choice of $\kappa^{(a)}$, $w^{(a)} - \kappa^{(a)}v^{(a)}$ would be a nonzero 4-vector with zero spatial components relative to all inertial systems, a circumstance that is impossible.

The 4-vector $\kappa^{(a)}v^{(a)}$ may vary during the interaction; thus, it depends on the event E_a on the world line at which it is evaluated (Figure 2.33):

$$\kappa^{(a)}v^{(a)} = \kappa^{(a)}v^{(a)}(E_a). \tag{2.121}$$

Consider the sum $\sum_{a=1}^{N} \kappa^{(a)}v^{(a)}(E_a)$ evaluated at events E_1, E_2, ..., that are simultaneous relative to one inertial observer S (Figure 2.34). These events are not simultaneous relative to other inertial observers, and so the sum $\sum_{a=1}^{N} \kappa^{(a)}v^{(a)}(E_a)$ is not necessarily equal to the sum $\sum_{a=1}^{N} \kappa^{(a)}v^{(a)}(E_a')$ evaluated at events E_1', E_2', ... simultaneous relative to another inertial observer S'. In general, $\sum_{a=1}^{N} \kappa^{(a)}v^{(a)}(E_a)$ is a 4-vector property of the system that depends on the events E_1, E_2, E_3, ...[*]

There do exist, however, circumstances in which this 4-vector sum is the same when the terms are evaluated at events that are simultaneous relative to any inertial system. This occurs, for example, if each of the $\kappa^{(a)}v^{(a)}(E_a)$ is independent of the event E_a along the world line of the particle (a), in which case $\kappa^{(a)}v^{(a)}(E_a) = \kappa^{(a)}v^{(a)}(E_a')$ for all E_a and E_a' along the particle's world line; thus,

$$\sum_{a=1}^{N} \kappa^{(a)}v^{(a)}(E_a) = \sum_{a=1}^{N} \kappa^{(a)}v^{(a)}(E_a') = \sum_{a=1}^{N} \kappa^{(a)}v^{(a)}, \tag{2.122}$$

a 4-vector that is an attribute of the particles and their motions and is defined independent of the events E_1, E_2, ... along the world lines of the particles.

Consider now a collision between a system of particles, the collision taking place in a finite region of space-time outside of which the world lines of all the

---

[*] This point is discussed also in A. Gamba, "Physical Quantities in Different Reference Systems According to Relativity," *American Journal of Physics*, 35: 83 (1967).

particles are straight. There exists a cone consisting of world points all in the absolute future relative to every world point in that finite interaction region of space-time and a similar cone in the absolute past. These cones can be shown on a space-time diagram (Figure 2.35) as the region interior to those null lines that are tangent to the boundary of the finite interaction region. That cone of the absolute past relative to all world points in the region of interaction consists of world points all of which occur *before the collision*, and the cone in the absolute future consists of world points all of which occur *after the collision*. Hence, we obtain the sum $\sum_{a=1}^{N} \kappa^{(a)} v_i^{(a)}$ by evaluating each $v^{(a)}$ at any event E_a before the collision in this absolute sense, the event lying on the world line of the particle. The sum is independent of the choice of events $E_1, E_2, \ldots$ as long as the events occur before the collision; in particular, they can be chosen to be events that are simultaneous relative to an inertial system S. Similarly, each $v_f^{(a)}$ is evaluated "after the collision" in the absolute sense, so the terms of $\sum_{a=1}^{N} \kappa^{(a)} v_f^{(a)}$ can be evaluated at events simultaneous relative to an inertial system S.

Before the collision, the 4-velocity of each particle is constant, so any linear combination $\sum_{a=1}^{N} \kappa^{(a)} v_i^{(a)}$ of the initial 4-vectors is constant. The $\kappa^{(a)}$ are arbitrary scalars. After the collision, the 4-velocity of each particle is again constant; therefore, any linear combination of the final 4-velocities $\sum_{a=1}^{N} \kappa^{(a)} v_f^{(a)}$ is constant.

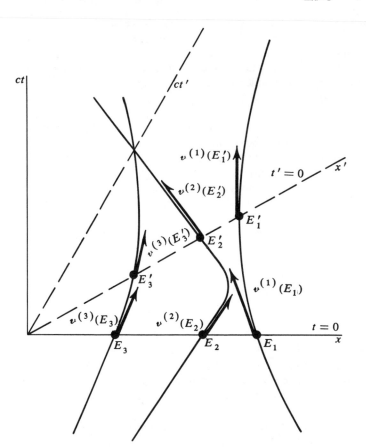

FIGURE 2.34 The sum $\sum_{a=1}^{N} \kappa^{(a)} v^{(a)}(E_a)$ evaluated at simultaneous events relative to S is not equal, in general, to the sum $\sum_{a=1}^{N} \kappa^{(a)} v^{(a)}(E'_a)$ evaluated at simultaneous events relative to S'.

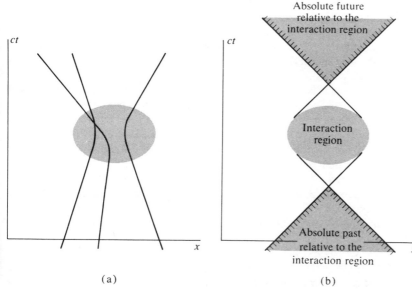

FIGURE 2.35 Before and after the collision defined in an absolute manner. (a) Boundary of the space-time region of interaction. (b) Absolute future and absolute past relative to the space-time region of interaction.

During the course of the collision, each 4-velocity $v^{(a)}$ may change as a result of the interactions and, in general, $v_i^{(a)}$ is not equal to $v_f^{(a)}$. Hence, the 4-vector $\sum_{a=1}^N \kappa^{(a)} v_i^{(a)}$ is not equal, in general, to the 4-vector $\sum_{a=1}^N \kappa^{(a)} v_f^{(a)}$ for arbitrary $\kappa^{(a)}$. Indeed, these two 4-vectors are equal for only one choice for $\kappa^{(a)}$ [aside from an arbitrary multiplicative constant that is independent of (a)], since, if there were other choices, we would have other standard nonrelativistic conservation laws in addition to those of momentum and energy. This choice is determined uniquely from the limiting case of the nonrelativistic conservation laws, (2.50) and (2.51), and the fact that the $\kappa^{(a)}$ are scalars. The $\kappa^{(a)}$ *must equal the rest masses $m^{(a)}$*. Hence, we are led to postulate the conservation law for elastic collisions:

$$\sum_{a=1}^N m^{(a)} v_i^{(a)} = \sum_{a=1}^N m^{(a)} v_f^{(a)}. \tag{2.123}$$

SUMMARY The nonrelativistic conservation laws of momentum and energy involve entities that transform like rotations and scalars under spatial rotations and therefore correspond to the spatial and time-like components of 4-vectors. There exists a cone of the absolute past relative to a finite interaction region in which any linear combination of the 4-velocities of a system of particles is constant. The same is true for a cone in the absolute future relative to the interaction region. The only linear combination of 4-velocities that is the same before and after the interaction contains the rest masses as coefficients.

2.2.4   *The general form of the conservation law of 4-momentum* *

We now consider the generalization of the conservation law applicable to inelastic collisions. The procedure for arriving at the form of this law is similar

---

* Feynman, Leighton, and Sands (vol. 1), Addison-Wesley, Sec. 16–5, p. 16–8.
Kittel, Knight, and Ruderman, McGraw-Hill, pp. 389–390.

to that used above: We examine the corresponding nonrelativistic laws, express them (within the approximation $v/c \ll 1$) in terms of entities that appear in relativistic mechanics, and then write these laws in covariant form. As before, we restrict our considerations to the values of dynamical variables outside the localized region of space-time in which the particles' world lines are curved (Figure 2.13).

The difference between an elastic and an inelastic collision according to non-relativistic mechanics is that the total kinetic energy is conserved in an elastic collision, whereas it is the total energy, and not the total kinetic energy alone, that is conserved in an inelastic collision. On the other hand, both the total momentum and the total mass are conserved during any collisions that can be described by nonrelativistic mechanics. Consider a collision, elastic or inelastic, in which N particles, each labeled by an integer a, $a = 1, 2, \ldots, N$, enter the space-time region of interaction and N' particles, each labeled by a primed integer a', $a' = 1', 2', \ldots, N'$, leave that region (Figure 2.36). The nonrelativistic conservation laws applicable to such a collision are

$$\sum_{a=1}^{N} m^{(a)}\mathbf{v}_i^{(a)} = \sum_{a'=1'}^{N'} m^{(a')}\mathbf{v}_f^{(a')}, \qquad \sum_{a=1}^{N} m^{(a)} = \sum_{a'=1'}^{N'} m^{(a')}. \qquad \text{N.R.} \qquad (2.124)$$

The procedure outlined above yields the following as the covariant form of these laws:

$$\sum_{a=1}^{N} m^{(a)}v_i^{(a)} = \sum_{a'=1'}^{N'} m^{(a')}v_f^{(a')}. \qquad (2.125)$$

The conservation law for elastic collisions, Equation (2.123), is a special case of this.

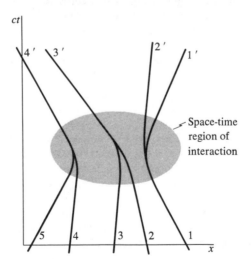

FIGURE 2.36 The particles labeled 1, 2, 3, 4, 5 enter the region of interaction and the particles labeled 1', 2', 3', 4' leave it.

The postulated conservation law (2.125) is consistent with the principle of relativity and reduces, for all $v^{(a)} \ll c$, to the correct nonrelativistic limit. These two conditions, however, do not ensure the validity of the postulate; that depends on the agreement of the postulated law with experiment. *The relativistic conservation law (2.125) is tested many times every day in laboratories in which high-energy physics is studied; to date, the validity of that law is unimpeachable.* (Some of the corroborating data are given in the examples and problems.)

It is interesting to note that the interactions between the particles of high-energy physics cannot be described within the framework of the concepts that we have considered thus far; it is necessary to use the concepts of quantum mechanics for that description. Nevertheless, the effects of the behavior during collisions between these particles is restricted by the relativistic conservation laws.

Each term such as $m^{(a)}v^{(a)}$ that appears in the conservation law (2.125) is the 4-momentum or energy-momentum vector $\not{p}^{(a)}$ associated with the appropriate particle. The sum of these vectors, all of which are evaluated either before or after the collision,

$$\mathscr{P}_i = \sum_{a=1}^{N} \not{p}_i^{(a)} \quad \text{or} \quad \mathscr{P}_f = \sum_{a'=1'}^{N'} \not{p}_f^{(a')} \tag{2.126}$$

is, respectively, the initial or final *total 4-momentum or total energy-momentum vector*. Relative to an inertial system S, the spatial 3-vector component of the total energy-momentum vector is the *total relativistic momentum* **P**,

$$\mathbf{P} = \sum_{a=1}^{N} \frac{m^{(a)}\mathbf{v}^{(a)}}{\sqrt{1 - [v^{(a)}/c]^2}}, \tag{2.127}$$

and the time-like component multiplied by c is the *total relativistic energy E*,

$$E = \sum_{a=1}^{N} \frac{m^{(a)}c^2}{\sqrt{1 - [v^{(a)}/c]^2}}. \tag{2.128}$$

The time-like component divided by c is the *total relativistic mass M*,

$$M = \sum_{a=1}^{N} \frac{m^{(a)}}{\sqrt{1 - [v^{(a)}/c]^2}}. \tag{2.129}$$

The conservation law states that the total 4-momentum $\mathscr{P}_i$ before the collision is equal to the total 4-momentum $\mathscr{P}_f$ after the collision. Thus, the components of $\mathscr{P}_i$ and $\mathscr{P}_f$, relative to a given inertial system, are equal, and the conservation law can be stated as a conservation law of relativistic momentum and a conservation law of relativistic energy (or of relativistic mass).

▶ *The conservation law of 4-momentum:*

$$\sum_{a=1}^{N} m^{(a)}v_i^{(a)} \equiv \mathscr{P}_i = \mathscr{P}_f \equiv \sum_{a'=1'}^{N'} m^{(a')}v_f^{(a')} \tag{2.130}$$

is equivalent to

▶ *the conservation law of relativistic momentum:*

$$\sum_{a=1}^{N} m_{vi}^{(a)}\mathbf{v}_i^{(a)} \equiv \mathbf{P}_i = \mathbf{P}_f \equiv \sum_{a'=1'}^{N'} m_{vf}^{(a')}\mathbf{v}_f^{(a')}, \tag{2.131}$$

or

$$\sum_{a=1}^{N} \frac{m^{(a)}}{\sqrt{1 - \{[v_i^{(a)}]^2/c^2\}}} \mathbf{v}_i^{(a)} = \sum_{a'=1'}^{N'} \frac{m^{(a')}}{\sqrt{1 - \{[v_f^{(a')}]^2/c^2\}}} \mathbf{v}_f^{(a')}, \tag{2.132}$$

and

▶ *the conservation law of relativistic energy:*

$$\sum_{a=1}^{N} m_{vi}^{(a)} c^2 \equiv E_i = E_f \equiv \sum_{a'=1'}^{N'} m_{vf}^{(a')} c^2 \tag{2.133}$$

or

$$\sum_{a=1}^{N} \frac{m^{(a)} c^2}{\sqrt{1 - \{[v_i^{(a)}]^2/c^2\}}} = \sum_{a'=1'}^{N'} \frac{m^{(a')} c^2}{\sqrt{1 - \{[v_f^{(a')}]^2/c^2\}}} \cdot \tag{2.134}$$

▶ *The conservation law of relativistic energy is equivalent to the conservation law of relativistic mass,*

$$\sum_{a=1}^{N} m_{vi}^{(a)} = \sum_{a=1}^{N} \frac{m^{(a)}}{\sqrt{1 - \{[v_i^{(a)}]^2/c^2\}}} \equiv M_i$$

$$= M_f \equiv \sum_{a'=1'}^{N'} \frac{m^{(a')}}{\sqrt{1 - \{[v_f^{(a')}]^2/c^2\}}} = \sum_{a'=1'}^{N'} m_{vf}^{(a')}. \tag{2.135}$$

We consider now some consequences of the energy-momentum conservation law in inelastic collisions.

The inelastic collision that is easiest to describe occurs when two identical particles collide and coalesce (Figure 2.37). Relative to some inertial system S, the motions are along a straight line and each particle approaches the point of collision with speed v. Let m be the mass of each particle before the collision and M that of the coalesced particle after the collision. If V is the speed of the particle M after the collision, then, according to the conservation law of energy-momentum,

$$\frac{m}{\sqrt{1 - (v^2/c^2)}} + \frac{m}{\sqrt{1 - (v^2/c^2)}} = \frac{M}{\sqrt{1 - (V^2/c^2)}} \tag{2.136}$$

and

$$\frac{mv}{\sqrt{1 - (v^2/c^2)}} + \frac{-mv}{\sqrt{1 - (v^2/c^2)}} = \frac{MV}{\sqrt{1 - (V^2/c^2)}}. \tag{2.137}$$

Therefore, from Equation (2.137), $V = 0$, and from Equation (2.136), the mass of the coalesced particle is

$$M = \frac{2m}{\sqrt{1 - (v^2/c^2)}} \cdot \tag{2.138}$$

The mass M of the composite particle is not equal to the sum of the masses of its constituent parts.

Let T_i be the total kinetic energy of the system before the collision:

$$T_i = 2 \left(\frac{mc^2}{\sqrt{1 - (v^2/c^2)}} - mc^2 \right) \cdot \tag{2.139}$$

The mass M can be written, therefore, as

$$M = 2m + \frac{T_i}{c^2}. \tag{2.140}$$

The kinetic energy T_i, which disappears in the collision as energy of motion, reappears after the collision associated with the inertial mass T_i/c^2. In a collision of two macroscopic objects, say two pieces of soft clay, the energy T_i appears in the form of internal energy that can be detected by an increase in the temperature of the composite object over the temperatures (which we assume

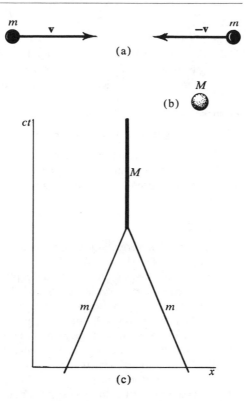

FIGURE 2.37 A simple inelastic collision. (a) The system before the collision. (b) The system after coalescence. (c) Space-time diagram of the collision.

were equal) of its component parts before the collision. Therefore, we must associate with the internal energy T_i an amount of inertial mass equal to T_i/c^2. (This point is discussed in greater detail in Section 2.2.5.)

SUMMARY The law of conservation of 4-momentum $\mathscr{P}_i = \mathscr{P}_f$ holds for inelastic as well as elastic collisions.

An object formed by the collision of its constituent particles has a mass that is T_i/c^2 greater than the rest masses of the constituent particles, where T_i is the initial kinetic energy of the system.

Example 2.9

Q. The explosion of 1 g of TNT (trinitrotoluene) results in an energy release of 4.2×10^3 J. Calculate the mass decrease per unit mass in a TNT explosion.

A. This amount of energy is equivalent to a mass of

$$\Delta m = \frac{4.2 \times 10^3 \text{ J}}{(3.00 \times 10^8 \text{ m/sec})^2} = 4.7 \times 10^{-14} \text{ kg}. \qquad (2.141)$$

Therefore, the mass decrease Δm per unit mass m in a TNT explosion is given by

$$\frac{\Delta m}{m} = \frac{4.7 \times 10^{-14} \text{ kg}}{1 \text{ g}} = 4.7 \times 10^{-11}. \qquad (2.142)$$

Problem 2.39

(a) How many 10-MeV electrons would have to be stopped by a 1-kg block to give that block an observable change in speed?
(b) How many 20-GeV electrons would it require?

Problem 2.40

The burning of 1 g of coal releases 2 to 3×10^4 J of energy.

(a) Calculate the mass decrease per unit mass in the process of burning coal.
(b) Experiments on combustion in chemistry demonstrate that the masses of the combustion products are equal to the masses of the materials that entered into the process. Explain.

Problem 2.41

(a) Calculate the energy, in joules, released by 1 ton of TNT. (The combustion of 1 g of TNT releases 4.2×10^3 J.)
(b) The energy that results from the explosion of a nuclear bomb can be measured in terms of the number of tons of TNT that produce the same amount of energy. Calculate the energy release in
 (i) the explosion of a 20-kiloton bomb,
 (ii) the explosion of a 50-megaton bomb.
(c) How much mass is converted to energy in the explosion of
 (i) a 20-kiloton nuclear bomb,
 (ii) a 50-megaton bomb?
(d) In some nuclear bomb explosions, about 10^{-4} of the mass of the bomb's core is converted to energy. Find the mass of the active material in such a 20-kiloton bomb.

Problem 2.42

The relation between the speed and the kinetic energy of fast electrons has been checked in an experiment* in which the speeds of the electrons were measured directly and their energies determined by measurement of the temperature rise they generated on being stopped in a metal disk. A galvanometer connected to a thermocouple on the disk was shown to deflect one division for each 0.80 J of internal energy increase in the disk. The charge was collected on a capacitor that discharged through a register that clicked once for each 7.6×10^{-8} C of charge that was discharged. The measurements are given in the following table in terms of the symbols defined as follows: electrons of measured speed v when stopped in the metal disk gave a galvanometer reading of g divisions when the register recorded r clicks.

v/c	g	r
0.960	$12\frac{1}{2}$	80
0.987	$36\frac{1}{2}$	80

---

* This experiment is described in W. Bertozzi, "Speed and Kinetic Energy of Relativistic Electrons," *American Journal of Physics*, *32*: 551 (1964), and can be seen in the film "The Ultimate Speed" by W. Bertozzi, produced by Educational Services Incorporated, Watertown, Mass.

(a) Plot curves showing the relativistic and nonrelativistic expressions for $(v/c)^2$ as a function of T.

(b) Determine the kinetic energies of the electrons from the table. Mark the corresponding points on the graph of (a).

(c) Does this result prove the validity of relativistic mechanics? Does it prove that nonrelativistic mechanics is not universally valid?

2.2.5 *The inertia of energy**

The result that the kinetic energy of a coalescing system contributes to the rest mass of the coalesced system is a special case of a general conclusion that we can derive from the energy-momentum conservation law and the 4-vector character of the energy-momentum vector. This derivation proceeds as follows: First we show that the total energy-momentum vector $\mathscr{P} = (E/c, \mathbf{P})$ of a system of particles corresponds to that of an object of relativistic mass E/c^2 moving with the velocity $c^2\mathbf{P}/E$. Then we show that the addition of any form of energy in the amount ΔE corresponds to a change in the relativistic mass of $\Delta E/c^2$. Finally, we show that the relativistic mass of a system is the measure of the inertia of the system that is called the (inertial) mass in newtonian mechanics. We conclude that *energy exhibits the property of inertia.*

We examine first some consequences of the 4-vector character of the total 4-momentum. Consider a system of N noninteracting particles of masses $m^{(1)}$, $m^{(2)}, \ldots$. The total energy-momentum vector of the system,

$$\mathscr{P} = \sum_{a=1}^{N} \not{p}^{(a)}, \tag{2.143}$$

is a time-like vector, since (Figure 2.38) the sum of any two time-like vectors is itself time-like. Since $\mathscr{P}$ is time-like, there exists an inertial system S_0 relative to which the spatial components of $\mathscr{P}$ are zero (Figure 2.39):

$$\mathscr{P} = \left(\frac{E_0}{c}, 0\right) \qquad \text{relative to } S_0, \tag{2.144}$$

where S_0 is the center-of-momentum system of the system of particles. Relative

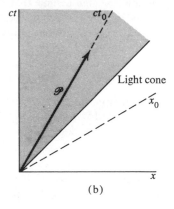

FIGURE 2.38   The sum $\mathscr{P} = \not{p}_1 + \not{p}_2$ of time-like 4-vectors pointing into the future is itself a time-like 4-vector pointing into the future. Since each $\not{p}_1$ lies within the future light cone, $\mathscr{P}$ must also.

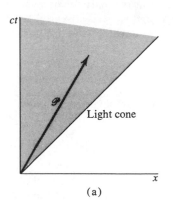

(a)

(b)

FIGURE 2.39   The inertial system S_0 relative to which a time-like vector has zero spatial coordinates. (a) $\mathscr{P}$ is a time-like 4-vector. (b) The spatial components of $\mathscr{P}$ are zero relative to S_0.

* Kacser, Prentice-Hall, Sec. 6.5, p. 168.
Kittel, Knight, and Ruderman, McGraw-Hill, pp. 404–405.
Resnick, John Wiley, Sec. 3.6, p. 131.
Taylor and Wheeler, W. H. Freeman, Sec. 13, p. 121.

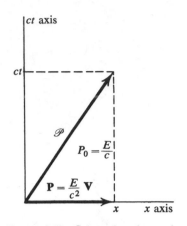

ct axis

FIGURE 2.40 $\mathscr{P}$ depends only on the system and its motion. Therefore, $\mathscr{P}$ must lie along the direction of the world line of the system and $\mathbf{V}/c = \mathbf{x}/ct = \mathbf{P}/P_0 = \mathbf{P}c/E$.

to this frame, the system is described by a 4-vector along the direction of the time axis. The components of $\mathscr{P}$ relative to any other inertial system S moving with velocity $-\mathbf{V}$ relative to S_0 are given by (see Problem 1.57)

$$\mathscr{P} = \left(\frac{E_0/c}{\sqrt{1 - (V^2/c^2)}}, \frac{(E_0/c^2)\mathbf{V}}{\sqrt{1 - (V^2/c^2)}} \right)$$

$$= \left(\frac{E}{c}, \frac{E}{c^2} \mathbf{V} \right), \tag{2.145}$$

where $E = E_0/[1 - (V^2/c^2)]^{1/2}$ is the total relativistic energy of the system. $\mathbf{V}$ is the velocity of the system relative to S, and the 4-vector $\mathscr{P}$ associated with the system describes a direction in space-time of a particle moving with velocity $\mathbf{V}$ (Figure 2.40).

We conclude that the total energy-momentum vector, because of its 4-vector nature, appears as the 4-momentum of a particle of relativistic mass E/c^2 traveling with velocity $\mathbf{V}$. Moreover, the relativistic mass E/c^2 and the velocity $\mathbf{V}$ are conserved if the particles in the system interact among themselves and then become noninteracting again.

Consider now the circumstances if any form of energy, of amount ΔE, is added to the system. The total 4-momentum becomes $\mathscr{P}_1$, another time-like 4-vector, which, relative to S, has the components

$$\mathscr{P}_1 = \left(\frac{E_1}{c}, \frac{E_1}{c^2} \mathbf{V}_1 \right), \tag{2.146}$$

where $E_1/c^2 = (E + \Delta E)/c^2$ corresponds to the total relativistic mass of the system of particles and $\mathbf{V}_1$ is the velocity of the inertial frame $S_{0'}$ in which $\mathscr{P}_1$ has zero spatial components. Therefore, $\mathbf{V}_1$ is the velocity of the system of particles relative to S. The energy ΔE that enters the system appears in the form of a contribution $\Delta E/c^2$ to the system's relativistic mass.

The next step in our demonstration that energy possesses inertia requires an analysis of an experiment that yields numbers that give the measure of the inertia of an object. Consider two particles of relativistic masses m_{v_1} and m_{v_2} that travel with velocities $\mathbf{v}_1$ and $\mathbf{v}_2$, respectively, before colliding. After the collision, we shall assume that there appear two particles of relativistic masses m_{v_3} and m_{v_4} traveling with velocities $\mathbf{v}_3$ and $\mathbf{v}_4$, respectively. Then the parameters m_{v_1}, m_{v_2}, m_{v_3}, and m_{v_4} satisfy the equations

$$m_{v_1} + m_{v_2} = m_{v_3} + m_{v_4}, \tag{2.147}$$

$$m_{v_1}\mathbf{v}_1 + m_{v_2}\mathbf{v}_2 = m_{v_3}\mathbf{v}_3 + m_{v_4}\mathbf{v}_4. \tag{2.148}$$

These equations describe, in the m's, a measurable property which is conserved [Equation (2.147)] that—since

$$m_{v_1} = \frac{|m_{v_3}\mathbf{v}_3 + m_{v_4}\mathbf{v}_4 - m_{v_2}\mathbf{v}_2|}{v_1} \tag{2.149}$$

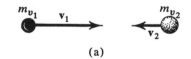

(a)

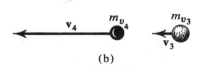

(b)

FIGURE 2.41 The equations $m_{v_1} + m_{v_2} = m_{v_3} + m_{v_4}$ and $m_{v_1}\mathbf{v}_1 + m_{v_2}\mathbf{v}_2 = m_{v_3}\mathbf{v}_3 + m_{v_4}\mathbf{v}_4$ provide a means of determining the measure of inertia m_v of a particle. (a) Before the collision. (b) After the collision.

according to Equation (2.148)—increases with a decrease in the magnitude v_1 of the velocity imparted to or delivered by the particle in the collision, other things being equal. Thus, for example, the parameter m_{v_1} provides *a measure of the inertia* of particle 1; Equations (2.147) and (2.148), valid for all collisions, define that measure (Figure 2.41).

The relativistic mass $m_v = m/[1 - (v^2/c^2)]^{1/2}$ is a measure of the inertia of a particle of rest mass m moving with speed v. For this reason, the concept of relativistic mass enters into many discussions on relativistic phenomena. On the other hand, the rest mass m is an invariant parameter associated with the particle that determines, with the speed v, the measure m_v of the inertia of the particle. Furthermore, m is a scalar, thus having the same value relative to every inertial system. For these reasons, the concept of rest mass is a more useful measure of inertia than is that of relativistic mass.

Consider now the system described above with 4-momentum $\mathscr{P}_1 = [E_1/c, (E_1/c^2)\mathbf{V}_1]$, the components taken relative to some inertial frame S. Suppose this system of particles collides with another system with 4-momentum $\mathscr{P}_2 = [E_2/c, (E_2/c^2)\mathbf{V}_2]$, the components measured relative to S. If two systems with 4-momentums $\mathscr{P}_3$ and $\mathscr{P}_4$ appear after the collision, the conservation law of energy-momentum gives

$$\frac{E_1}{c^2} + \frac{E_2}{c^2} = \frac{E_3}{c^2} + \frac{E_4}{c^2} \tag{2.150}$$

and

$$\frac{E_1}{c^2}\mathbf{V}_1 + \frac{E_2}{c^2}\mathbf{V}_2 = \frac{E_3}{c^2}\mathbf{V}_3 + \frac{E_4}{c^2}\mathbf{V}_4. \tag{2.151}$$

Comparison of Equations (2.147) and (2.148) with (2.150) and (2.151) shows that the measure of inertia M_{v_1} of the system of particles whose motion is described by Equation (2.146) is

$$M_{v_1} = \frac{E_1}{c^2} = \frac{E}{c^2} + \frac{\Delta E}{c^2}. \tag{2.152}$$

The energy ΔE added to the system of particles increases the inertia of the system, the increase in the relativistic mass of the system, a measure of its inertia, being $\Delta E/c^2$.*

This general conclusion is exemplified by the case studied earlier. Other examples will appear in Section 2.2.6.

SUMMARY   The total energy-momentum vector $\mathscr{P}$ of a system of particles, with components $(E/c, \mathbf{P})$ relative to S, corresponds to the 4-momentum of an object of relativistic mass $E/c^2 = M_v$ moving with the velocity $\mathbf{V} = \mathbf{P}/M_v$ relative to S. An addition of any form of energy ΔE to the system corresponds to a change of $\Delta M_v = \Delta E/c^2$ in the relativistic mass of the system. Furthermore, the relativistic mass of a system is the measure of the inertia of the system that corresponds to the inertial mass of newtonian mechanics. Thus, energy exhibits the property of inertia.

Problem 2.43

Calculate the mass decrease per century in a 100-W expenditure of power.

Problem 2.44

Approximately 1 J of energy can raise the temperature of a solid object of

---

* See the article by M. von Laue entitled "Inertia and Energy," in *Albert Einstein: Philosopher Scientist*, P. A. Schilpp (Ed.), Tudor Publishing, New York, 1949, reprinted in two volumes in paperback form as a Harper Torchbook, Harper and Row, New York, 1959.

mass 1 g by 1°C. Find the order of magnitude of the relative increase in mass of an object whose temperature is raised by 100°C.

2.2.6 The momentum associated with a flow of energy*

The fact that the amount of energy ΔE corresponds to the inertial mass $\Delta E/c^2$ suggests that it is necessary to associate some momentum with a flow of energy. The necessity for this can be shown from the energy-momentum conservation law and from the fact that the total energy-momentum vector of a system of noninteracting particles is a 4-vector.

Consider a system of particles with total 4-momentum $\mathscr{P}_i$ before the emission of any form of energy in the amount ΔE relative to some inertial frame S. After the emission the system has 4-momentum $\mathscr{P}_f$. The change in the 4-momentum,

$$\mathscr{P}_i - \mathscr{P}_f = \Delta\mathscr{P}, \tag{2.153}$$

is itself a 4-vector with a time-like component $\Delta E/c$ relative to S (Figure 2.42). In general, the spatial component $\Delta\mathbf{P}$ of $\Delta\mathscr{P}$ relative to S will not be zero, and this is the momentum associated with the energy flow. The velocity associated with the energy flow, as given by the argument of Figure 2.40, is

$$\mathbf{V} = \frac{c^2\,\Delta\mathbf{P}}{\Delta E}. \tag{2.154}$$

The speed V can be less than, equal to, or greater than c; we shall discuss each of these cases in turn.

If $V < c$, the 4-momentum $\Delta\mathscr{P}$ of the energy flow can be written in the form

$$\Delta\mathscr{P} = \left[\frac{\Delta Mc}{\sqrt{1-(V^2/c^2)}}, \frac{\Delta M\mathbf{V}}{\sqrt{1-(V^2/c^2)}}\right], \tag{2.155}$$

where

$$\Delta M = \frac{1}{c}\sqrt{\Delta\mathscr{P}\cdot\Delta\mathscr{P}} = \frac{1}{c}\sqrt{[(\Delta E)^2/c^2] - (\Delta\mathbf{P})^2}$$
$$= \frac{\Delta E}{c^2}\sqrt{1-(V^2/c^2)} \tag{2.156}$$

is the rest mass associated with the relativistic mass $\Delta E/c^2$ moving with speed V. Thus, *energy that is transmitted with a speed $V < c$ can be considered as a manifestation of the motion of an object of rest mass ΔM moving with the speed V.*

The particular case in which $V = 0$ is exemplified by the system being bombarded by or emitting an isotropic stream of particles. This case arises, for example, if an object is placed in an oven and heated uniformly on all sides.

In the case in which the speed of energy transmission is c, we cannot assign a rest mass.† There does not exist a reference inertial system in which the 4-momentum $\Delta\mathscr{P}$ of the energy flow has zero spatial components. However, it is necessary to associate a momentum with this energy flow; if the components

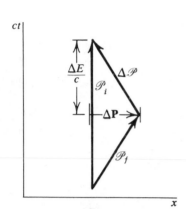

FIGURE 2.42 In general, the spatial component $\Delta\mathbf{P}$ of $(\mathscr{P}_i - \mathscr{P}_f)$ is not zero.

---

* Kacser, Prentice-Hall, Sec. 7.3, p. 190.
Kittel, Knight, and Ruderman, McGraw-Hill, pp. 397–400.
† However, energy that travels with the speed c does exhibit particle-like properties; in particular, a light beam of definite energy may display the behavior of an integral number of bundles, each such corpuscle having a well-defined energy relative to any inertial system. Each light corpuscle, containing a minimum nonzero amount of energy for that particular type of light, is called a *photon*. For the present, we shall use only the property that light carries energy and momentum, and we shall not use the corpuscular property. This property will be discussed briefly in Section 2.4.

of the 4-momentum $\Delta\mathscr{P}$ relative to S are $(\Delta E/c, \Delta\mathbf{P})$, then

$$c = V = \frac{c^2|\Delta\mathbf{P}|}{\Delta E} \qquad (2.157)$$

or

The momentum associated with the energy ΔE moving at speed c $\quad |\Delta\mathbf{P}| = \dfrac{\Delta E}{c}.$ $\qquad (2.158)$

The energy-momentum vector of energy transmitted with the speed c is a null vector:

$$\Delta\mathscr{P}\cdot\Delta\mathscr{P} = \left(\frac{\Delta E}{c}\right)^2 - \Delta\mathbf{P}\cdot\Delta\mathbf{P} = 0. \qquad (2.159)$$

These results show that a directed pulse of light, which contains energy, changes the momentum of any object upon which the light is incident. The corresponding pressure experienced by the object is called *radiation* or *light pressure.** If the light energy E is absorbed by the object, the momentum change experienced by the object is E/c; if the light energy E is completely reflected by the object, the momentum change is $2E/c$. Also, if the power in an absorbed light beam is W, the force experienced by the absorbing object is W/c.

Radiation pressure plays an important role in astronomical phenomena. For example, it is believed to be responsible *in part* for the fact that the tails of comets almost always are directed away from the sun (Figure 2.43).† Furthermore, radiation pressure also plays a role in stellar structure.‡

Consider now the case for which $V > c$, a circumstance that was considered earlier in Problems 1.100, 1.101, and 1.102. We also considered this case in Example 2.5, since it was shown there that an elastic collision takes place through a space-like energy-momentum transfer and

$$c < V = c^2 \frac{|\Delta\mathbf{P}|}{\Delta E} \quad \text{is equivalent to} \quad (\Delta E)^2 - c^2|\Delta\mathbf{P}|^2 < 0. \quad (2.160)$$

An energy-momentum transfer $(\Delta E/c, \Delta\mathbf{P})$ that travels faster than light is space-like, so

$$(\Delta E)^2 - c^2|\Delta\mathbf{P}|^2 = -\mu^2 c^4 \qquad (2.161)$$

determines a real invariant constant μ; μ corresponds to im, m being the (inertial) rest mass of a particle. The fact that the inertial mass m corresponding to the energy-momentum transfer is imaginary is not of consequence here, since, on the one hand, the inertial mass of a particle is defined by an experiment in which $v_{\text{particle}}/c \approx 0$ and, on the other hand, a space-like energy-momentum transfer appears to *all* observers to travel faster than c. Thus, although the energy-momentum transfer corresponds to an imaginary inertial mass, μ itself can be measured, and furthermore, is a real number and an invariant. Indeed, we can conceive that a special class of particles does exist that travels faster than c§

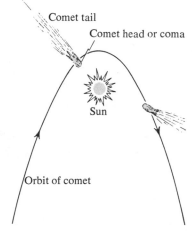

FIGURE 2.43 Almost all the tails of comets point away from the sun.

* See the discussion of the measurement of this pressure in the laboratory given in R. E. Pollock, "Resonant Detection of Light Pressure by a Torsion Pendulum in Air—An Experiment for Underclass Laboratories," *American Journal of Physics, 31*: 901 (1963).
† L. F. Biermann and R. Lüst, "The Tails of Comets," *Scientific American, 199*: 44, October 1958.
‡ See, for example, Chapter 16 of O. Struve, B. Lynds, and H. Pillans, *Elementary Astronomy*, Oxford Univ. Press, New York, 1959.
§ The possibility of the existence of particles that travel with speeds greater than c has been discussed in O. M. P. Bilaniuk, V. K. Deshpande, and E. C. G. Sudarshan, "'Meta' Relativity," *American Journal of Physics, 30*: 718 (1962), and particularly in G. Feinberg, "Possibility of Faster-Than-Light Particles," *Physical Review, 159*: 1089 (1967).

and that we can associate with the energy-momentum vector of each, not an inertial mass but rather a norm $-\mu^2 c^2$.

Let us examine briefly how a particle with a space-like energy-momentum would behave. Since

$$|E| = \frac{\mu c^2}{\sqrt{(v^2/c^2) - 1}}, \qquad |\mathbf{p}| = \frac{\mu v}{\sqrt{(v^2/c^2) - 1}}, \qquad (2.162)$$

we see that as $v \to c$, $|E| \to \infty$ and $|\mathbf{p}| \to \infty$. Indeed, the particle slows down as its energy increases. Alternatively, if the particle loses energy, it speeds up and, relative to one observer, it travels at an infinite speed when it has zero energy. Also, no matter how much energy we impart to the particle, as long as the energy remains finite we cannot slow down the particle to a speed equal to or less than c. Particles with these properties have not been observed to date, although in relativistic quantum mechanics, space-like energy-momentum transfers are described in terms of the exchange of "virtual" particles, particles that are neither real nor observable.

Consider now the space-time diagram of a space-like energy-momentum transfer from one system to another, as shown in Figure 2.44(a). Relative to the inertial system S, the energy ΔE is emitted by system 1 at A and absorbed by system 2 at B, where $\Delta E > 0$ (see Figure 2.40). However, relative to some other observer S' [Figure 2.44(b)], the event B occurs before the event A, so system 2 experiences its energy change at B before system 1 experiences the energy change at A. Since $(\Delta E/c, \Delta \mathbf{P})$ transforms as $(c\,\Delta t, \Delta \mathbf{r})$—for example,

$$c\,\Delta t' = \gamma(c\,\Delta t - \beta\,\Delta x), \qquad \Delta x' = \gamma(\Delta x - \beta c\,\Delta t),$$

$$\frac{\Delta E'}{c} = \gamma\left(\frac{\Delta E}{c} - \beta\,\Delta P_x\right), \qquad \Delta P_x' = \gamma\left(\Delta P_x - \frac{\beta\,\Delta E}{c}\right) \qquad (2.163)$$

—the fact that $\Delta t' = t_B' - t_A' < 0$ implies that $\Delta E' < 0$. That is, according to S', the energy change at B is negative, corresponding to an energy emission, and that at A is positive, corresponding to an energy absorption. Thus, although the two events A and B take place in a different order in time relative to S and S', each of the two observers notes an energy emission before the absorption takes place.

Finally, let us consider the use of a space-like energy-momentum transfer mechanism for sending signals that travel faster than light. By a signal, we

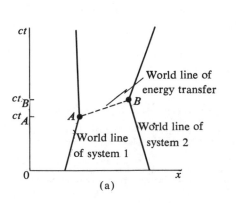
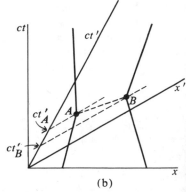

FIGURE 2.44   World lines of an energy transfer at speed $V > c$. (a) $t_B > t_A$. (b) $t_A' > t_B'$.

(a)

(b)

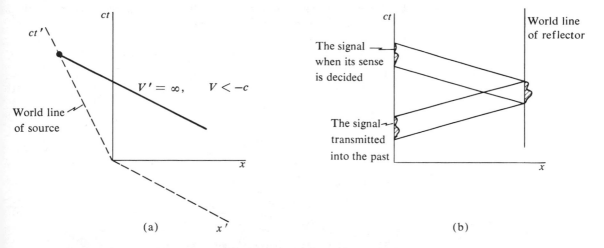

FIGURE 2.45 Two irreconcilable points of view. (a) A fast source of space-like Δ $_{\mathscr{P}}$ "emits" energy momentum into the past. (b) A signal carried by a space-like energy-momentum transfer mechanism into the past.

mean a statement, possibly in some code, whose sense is not predetermined before the instant at which it is sent. A beam of negative energy that travels into the past can be generated by the acceleration of the source to high speeds [Figure 2.45(a)]. A beam of this sort could be used to send a signal into the past [Figure 2.45(b)]; if such signals were possible, we could determine the sense of the signal long before we decide upon the signal. This contradiction shows that we cannot use a negative-energy energy-momentum transfer mechanism to send signals. This does not mean that such energy-momentum transfers cannot occur; it means only that we cannot control those energy-momentum transfers to the extent that we can use them for signaling [2]. This result does not accord with our everyday experiences with time-like energy-momentum transfer mechanisms, such as sound waves, but it is not in disagreement with behavior evident at the atomic level. At that level, phenomena cannot be controlled to the extent that our everyday experiences suggest.

SUMMARY   It is sometimes necessary to associate momentum with a flow of energy. The flow of energy associated with the change $\Delta\mathscr{P}$ in the 4-momentum of a system travels with the velocity $\mathbf{V} = c^2\,\Delta\mathbf{P}/\Delta E$ relative to an observer who describes $\Delta\mathscr{P}$ by the components $(\Delta E/c, \Delta\mathbf{P})$. A rest mass can be associated with the flow of energy if $V < c$. If $V = c$, the equivalent rest mass is zero, and the momentum $\Delta\mathbf{P}$ associated with the flow of energy ΔE satisfies the relation $|\Delta\mathbf{P}| = \Delta E/c$. Energy-momentum transfers that can be associated with speeds greater than c occur in elastic collisions, for example. The existence of particles that travel faster than c has been conjectured, but no such particles have been observed to date.

Example 2.10

   Q. Show that a pulse of energy transmitted at the speed of light can be considered as a particle of zero rest mass.

A. The energy-momentum and velocity of a particle of rest mass m, relative to an inertial system S, have the form

$$\not{p}_m = \left(\frac{m}{\sqrt{1 - (v_m^2/c^2)}} c, \; \frac{m}{\sqrt{1 - (v_m^2/c^2)}} \mathbf{v}_m \right) = \left(\frac{E_m}{c}, \mathbf{p}_m \right);$$

$$\mathbf{v}_m = \frac{c^2 \mathbf{p}_m}{E_m}. \tag{2.164}$$

Those of a pulse of energy traveling at speed c are given by

$$\not{p}_l = \left(\frac{E_l}{c}, \mathbf{p}_l \right) = \left(\frac{E_l}{c}, \frac{E_l}{c^2} \mathbf{v}_l \right); \qquad \mathbf{v}_l = \frac{c^2 \mathbf{p}_l}{E_l}, \qquad v_l = c. \tag{2.165}$$

Consider a particle of *fixed* relative energy E_m:

$$E_m = \frac{m}{\sqrt{1 - (v_m^2/c^2)}} c^2. \tag{2.166}$$

The speed v_m of this particle is given by

$$\frac{v_m}{c} = \sqrt{1 - (mc^2/E_m)^2}, \tag{2.167}$$

from which we see that v_m approaches c as m goes to zero. In the limit,

$$\lim_{\substack{m \to 0 \\ v_m \to c}} \frac{m}{\sqrt{1 - (v_m^2/c^2)}} = \frac{E_m}{c^2}, \tag{2.168}$$

the 4-momentum

$$\not{p}_{m=0} = \left(\frac{E_{m=0}}{c}, \frac{E_{m=0}}{c^2} \mathbf{v}_{m=0} \right) \tag{2.169}$$

is a null vector:

$$\not{p}_{m=0} \cdot \not{p}_{m=0} = \left(\frac{E_{m=0}}{c} \right)^2 - \left(\frac{E_{m=0}}{c^2} \right)^2 v_{m=0}^2 = 0. \tag{2.170}$$

Thus the properties of $\not{p}_{m=0}$ and $v_{m=0}$ are identical to those of $\not{p}_l$ and v_l, and a pulse of energy transmitted with speed c corresponds to a particle of zero rest mass and nonzero energy.

Example 2.11

Q. (a) The radiant energy emitted at the surface of the sun is 3.9×10^{26} W. Calculate the mass decrease of the sun per day.

(b) The energy travels with speed c. Calculate the force that this energy exerts on the earth on the assumption that all the radiant energy from the sun incident on the earth is absorbed by the earth.

(c) About 40% of the light incident on the earth is reflected away. (The ratio of light reflected from a sphere to light incident on it is called the *albedo* of the sphere.) Calculate the maximum force that the sun's radiant energy could exert on the earth.

A. (a) The mass equivalent of energy E is E/c^2, so the mass decrease per day is

$$\frac{3.9 \times 10^{26} \text{ J/sec}}{(3.0 \times 10^8 \text{ m/sec})^2} \times 1 \text{ day} \times \left(\frac{24 \text{ hr}}{\text{day}} \times \frac{3,600 \text{ sec}}{\text{hr}}\right) = 3.7 \times 10^{14} \text{ kg}.$$

$$(2.171)$$

(b) The energy flow of 3.9×10^{26} W is uniform through the surface of a sphere of area $4\pi(1.50 \times 10^{11} \text{ m})^2$ at the earth's distance from the sun. On this sphere, the earth appears as a circle of area $4\pi(6.4 \times 10^6 \text{ m})^2$. Therefore, the energy flow intercepted by the earth is

$$3.9 \times 10^{26} \text{ W} \times \frac{4\pi(6.4 \times 10^6 \text{ m})^2}{4\pi(1.50 \times 10^{11} \text{ m})^2} = 7.1 \times 10^{17} \text{ W}. \qquad (2.172)$$

The momentum per second incident on the earth is 7.1×10^7 W/c. The force F experienced by an object undergoing a momentum change of 7.1×10^{17} J/c·sec is

$$F = \frac{7.1 \times 10^{17} \text{ J/sec}}{3.0 \times 10^8 \text{ m/sec}} = 2.4 \times 10^9 \text{ N}. \qquad (2.173)$$

(c) Upon reflection, a light pulse of energy E undergoes a momentum change of magnitude $E/c - (-E/c) = 2E/c$. If only 40% of the energy is reflected directly back to the sun, the earth must undergo a momentum change of $0.4E/c$ in addition to E/c. Hence, the maximum force is

$$F_{\text{max}} = 1.4 \times 2.4 \times 10^9 \text{ N} = 3.4 \times 10^9 \text{ N}. \qquad (2.174)$$

Example 2.12

Immediately after Einstein propounded the theory of relativity, he asserted [3] that the amount of energy E is equivalent to the inertial mass E/c^2. Among his derivations of the inertia of energy, he presented one in which the relation $E = mc^2$ was based on the following two premises*:

(a) The center of mass of an isolated system is not displaced by an internal process.
(b) The momentum associated with a light pulse of energy E is E/c. (This result was known prior to the theory of relativity.)

This example derives premise (b) on the basis of (a) and the fact that the light energy E corresponds to the inertial mass $m = E/c^2$.

Q. Consider a closed box (Figure 2.46) with a light emitter at one end and a light absorber at the other, their separation distance being L. Suppose the mass is distributed so that the center of mass of the system is halfway between the emitter and the absorber. A light pulse of energy E emitted at one end and absorbed at the other corresponds to a transfer of mass E/c^2 and thus to a shift in the center of mass by a distance

$$\frac{-[-(E/c^2)(L/2)] + [(E/c^2)(L/2)]}{[M + (E/c^2)]} = \frac{(E/c^2)L}{M + (E/c^2)}, \qquad (2.175)$$

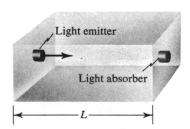

Light emitter

Light absorber

$\longmapsto$——————L——————$\longmapsto$

FIGURE 2.46   A closed system.

* Einstein's arguments for the inertia of energy from these two premises are discussed in E. Feenberg, "Inertia of Energy," *American Journal of Physics*, **28**: 565 (1960).

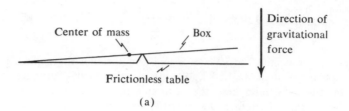

(a)

FIGURE 2.47 When the center of mass
is displaced, work can be performed by
the box while the center of mass is
returned to its original height. (a)
Before light emission. (b) After light
emission.

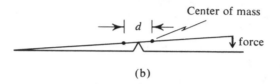

(b)

where $M + (E/c^2)$ is the total mass of the system. Work can be extracted
from such a shift in the internal position of the center of mass (Figure
2.47), resulting in the creation of energy. Therefore, the center of mass
cannot shift owing to an internal process alone. Show that the correct
result is obtained if a momentum E/c is associated with the light
energy E.

A. The center of mass of the system does not shift only if the material in the
box is being displaced as the light energy travels from one end of the
box to the other. This results if the light pulse carries a momentum,
say p, to be calculated. On being emitted, the light pulse transmits a
momentum $-p$ to the end of the box with the light emitter. Because of
this, that end of the box undergoes motion away from the light pulse,
and this movement is transmitted in the box's material to the other end
at the speed of sound—the speed, less than c, at which a disturbance
is transmitted. Before this movement reaches the other end, the light
pulse with its momentum p is absorbed there, imparting that momentum
to that end. This movement also is transmitted with the speed of sound
along the sides of the box. After the light is absorbed, the total momentum
of the box, $(-p) + p$, is zero and the center of mass is again at rest.
The material in the box possessed the momentum $-p$ for the time of
transit $t = L/c$ of the light pulse, corresponding to a motion of the mass
M of the material with speed $v = -p/M$ or, alternatively, a displacement
of $vt = -pL/Mc$. Therefore, the net displacement of the center of mass,

$$\frac{\sum m_i\, d_i}{\sum m_i} = \frac{M(-pL/Mc) + (E/c^2)L}{M + (E/c^2)}, \tag{2.176}$$

is zero only if

$$p = \frac{E}{c}. \tag{2.177}$$

Problem 2.45

The sun emits radiant energy at the rate of 3.9×10^{26} W. The *solar constant*
is defined as the energy per unit time incident on a unit area at the mean earth-
sun distance, the area being normal to the line to the sun.

(a) Calculate the solar constant.
(b) Calculate the maximum force that the sun's radiant energy can exert on your body. Express your answer as a percentage of your weight.

Problem 2.46

Calculate the relative change in the mass of the sun per century due to the emission of radiant energy.

Problem 2.47

Calculate the ratio of the force of the light pressure exerted by the sun on the earth to the gravitational force exerted by the sun on the earth.

Problem 2.48

(a) Light energy of W watts is incident normally on a flat surface that reflects the fraction ρ. Calculate the force, in newtons, experienced by the surface.
(b) Calculate the force experienced by the surface if the light is incident and reflected at an angle θ with the normal to the surface (Figure 2.48).

Problem 2.49

(a) Could you lift the mass equivalent of the light incident on the earth per day?
(b) Could you lift the mass equivalent of the radiation emitted from the sun per second?

Problem 2.50

How much mass is converted to energy in providing light from a 100-W bulb burning for 5 hr?

Problem 2.51

How much does 1 g of equivalent energy cost at the rate of 5¢ per kW·hr?

Problem 2.52

If the box of Figure 2.46 rests on a sufficiently rough table in the arrangement shown in Figure 2.47, the center of mass does shift and the process pictured there enables the box to do work. Explain.

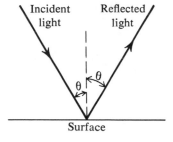

FIGURE 2.48 Relative directions of incident and reflected light.

Problem 2.53

Show that some dust particles in space are pushed out of the solar system by the radiation pressure from the sun. What can you say about the size of these particles?

Problem 2.54

Present-day rockets can launch a spaceship into space outside the effective range of the earth's gravitational force; for space voyages of distances of the order of a light-year, it will be necessary to accelerate spaceships to speeds (relative to the earth) near that of light. One suggestion* for the propulsion of such a spaceship is a *photon rocket*, a rocket in which the thrust is provided by an exhaust that moves at the speed of light. This problem involves a simple analysis of relativistic rocket thrust.† The relativistic equation of motion for the case in which the rocket thrust is active over a period of time can be developed in a manner similar to that for nonrelativistic motions (see Problem 2.7) and will not be discussed here.‡

(a) Consider a rocket of mass M in a force-free region of space. The rocket ejects a quantity of exhaust at the speed v' relative to the initial rest frame of the rocket; the final (rest) mass of the rocket is αM and its final speed is V. Show that

$$m_{v'}c^2 + \alpha M_V c^2 = Mc^2,$$
$$m_{v'}v' = \alpha M_V V.$$

Deduce the relation

$$\frac{V}{c} = \sqrt{1 - \alpha^2[1 + (V/v')]^2}.$$

(b) Consider another such rocket (a photon rocket) that emits light energy of amount E'. Show that

$$E' + \alpha_p M_V c^2 = Mc^2,$$

$$\frac{E'}{c} = \alpha_p M_V V$$

and deduce that

$$\frac{V}{c} = \sqrt{1 - \alpha_p^2[1 + (V/c)]^2}.$$

(c) Show that the ratio of the initial to final rest mass of the photon rocket of (b) is given by

$$\frac{M_{\text{initial}}}{M_{\text{final}}} = \left(\frac{c + V}{c - V}\right)^{1/2}.$$

(d) Show that $\alpha_p > \alpha$ for a given value of V. Discuss the consequences of this for interstellar space travel.

(e) Consider a photon rocket in free space, initially at rest relative to the earth, which is accelerated to a speed for which the time dilatation factor is 5%, deaccelerated to rest, and then returned to earth under the same circumstances. Calculate the fraction of the original mass of the rocket that returns to earth.§ What if the rocket ejects an exhaust at a speed $v' < c$?

---

* A discussion of the problems involved is given in G. Giannini, "Electrical Propulsion in Space," *Scientific American*, *204*: 57, March 1961.
† See J. R. Pierce, "Relativity and Space Travel," *Proceedings of the Institute of Radio Engineers*, *47*: 1053 (1959), reprinted in *Special Relativity Theory, Selected Reprints*, American Institute of Physics, New York, 1963.
‡ Details can be found in K. B. Pomeranz, "The Relativistic Rocket," *American Journal of Physics*, *34*: 565 (1966).
§ A discussion of the fuel requirements for space travel is given in the articles by E. Purcell, "Radio-astronomy and Communication Through Space," and S. von Hoerner, "The General Limits of Space Travel," in *Interstellar Communication*, A. G. W. Cameron (Ed.), W. A. Benjamin, New York, 1963.

Problem 2.55

Draw a graph of E versus v for a particle that travels faster than c and for which μ equals the mass of the electron.

Problem 2.56

Draw graphs of the apex angle θ of the Čerenkov radiation (Problem A1.3) emitted by a charged particle that travels faster than c against

(a) the speed v of the particle,
(b) the ratio $E/\mu c^2$ of the particle.

Problem 2.57

Let $\mathbf{v}$, with $v > c$, be the velocity of a particle relative to S and $\mathbf{v}'$ be the velocity relative to S'. Let $\mathbf{V}$ be the velocity of S' relative to S. Use the result of Problem 1.78 to show that

$$\frac{v'^2}{c^2} = 1 + \left(\frac{v^2}{c^2} - 1\right) \frac{1 - (V^2/c^2)}{[1 - (\mathbf{v}\cdot\mathbf{V}/c^2)]^2},$$

and hence that $v' > c$ if $v > c$.

2.3 The Covariant Equation of Motion for a Particle Experiencing a Given External Force

Newton's equation of motion

$$m \frac{d^2\mathbf{r}}{dt^2} = \mathbf{F} \qquad \text{N.R.} \qquad (2.178)$$

describes the behavior of a particle experiencing the force $\mathbf{F}$ in circumstances in which every pertinent relative speed involved is much less than c. The force function $\mathbf{F}$ describes the effects of the environment on the particle's motion, and in general, this environment is itself dependent on that motion. For example, if the particle of mass m is one of a pair of interacting particles, the environment of m is the other member of the pair whose motion is determined by that of m (Figure 2.49). In some circumstances, however, the motion of m does not change the environment to any appreciable extent, so $\mathbf{F}$ is independent of that motion. In this case, $\mathbf{F}$ is said to be a (given) *external force*.

This section is concerned with the generalization of the nonrelativistic equation of motion applicable to circumstances in which relative speeds of the order of c are manifest. We must find for this purpose a form for the equation of motion that is compatible with both the principle of relativity and the principle of the invariance of the speed of light, or in other words, a covariant generalization of Newton's equation of motion.

The covariant equation of motion will contain, on one side, a dynamical variable, corresponding to $d^2\mathbf{r}/dt^2$, that is sufficient to determine the motion in terms of an appropriate set of initial conditions. On the other side of the equation there will appear the dynamical variable corresponding to $\mathbf{F}$ that describes the effects of the environment on the motion. The environment may depend on the motion, as occurs in the nonrelativistic case, but the environment at one instant cannot depend on the simultaneous nature of the motion in a

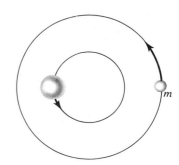

FIGURE 2.49 A simple example of a case in which the motion of m alters the environment of m.

relativistic theory. This and other similar problems can be avoided if we restrict our considerations to the effects of a given external environment, as we will in this section. Thus, we restrict our discussions below to finding the co-variant equation of motion; we will leave out any inquiry into the mutual effects of interactions on a system of particles.

The simplest covariant equation of motion, that for a free particle, can be deduced from the equation $v(t_0) = v_0$, where t_0 is the proper time of the particle and v_0 is the initial velocity. This equation can be written also in a covariant form independent of the initial conditions such as the 4-vector v_0 that takes on different values for different motions. The constant 4-vector v_0 does not change along the world line relative to S, so $dv_0/dt = 0$. Consider the corresponding derivative

$$\frac{dv(t_0)}{dt_0} = \frac{1}{\sqrt{1 - (v^2/c^2)}} \frac{dv(t_0)}{dt}. \tag{2.179}$$

This derivative is a 4-vector, since the product of the 4-vector Δv and the scalar $1/\Delta t_0$ is itself a 4-vector. Hence, a covariant equation that describes the motion of a particle that experiences no force is

$$\frac{dv(t_0)}{dt_0} = 0. \tag{2.180}$$

This equation can be written in a form more typical of equations of motion by multiplication with the mass m of the particle:

$$0 = m \frac{dv(t_0)}{dt_0} = \frac{dmv(t_0)}{dt_0} = \frac{dp}{dt_0}. \tag{2.181}$$

The spatial component of this 4-vector equation relative to an inertial system is equivalent to

$$\frac{d\mathbf{p}}{dt} = 0, \tag{2.182}$$

an equation of motion similar in form to the nonrelativistic equation

$$0 = \frac{d(m\mathbf{v})}{dt} = m \frac{d\mathbf{v}}{dt}. \qquad \text{N.R.} \tag{2.183}$$

Problem 2.58

Show, from the results of Problem 2.8, that the covariant equation

$$\frac{dv(t_0)}{dt_0} = 0$$

ensures that, relative to any inertial system S,

$$\frac{d^2\mathbf{r}}{dt^2} = 0.$$

Problem 2.59

Show that the covariant equation

$$\frac{dv(t_0)}{dt_0} = 0$$

is equivalent, relative to an inertial system S, to the scalar equation

$$\frac{d\gamma}{dt} = 0, \qquad \left[\gamma = \frac{1}{\sqrt{1 - (v^2/c^2)}}\right]$$

and the vector equation

$$\frac{d(\gamma \mathbf{v})}{dt} = 0.$$

State explicitly what the velocity $\mathbf{v}$ means.

2.3.1   The covariant equation of motion*

We wish to find a covariant equation that describes the motion of a particle experiencing a given external force. The procedure we use is analogous to that with which we obtained the covariant equations for conservation laws—we begin with the 3-vector form of the equation of motion relative to one inertial system and then generalize this equation to a covariant form. In the present case, however, we do not know at the start the 3-vector form of the equation of motion relative to any inertial system, but we do know the form, namely Newton's equation, valid for circumstances in which *all* the relative speeds involved are sufficiently small. Therefore, we start with Newton's equation of motion.

This equation describes very accurately the motion of a particle in those cases in which all the pertinent relative speeds, including the speed of the particle relative to the reference inertial system, are sufficiently smaller than c. It is always possible to choose inertial systems relative to which the particle's speed is sufficiently small. Indeed, the motion of the particle itself distinguishes, at each event on the world line of the particle, one particular inertial system in which the velocity of the particle relative to that inertial system is undoubtedly sufficiently smaller than c. This is the inertial system in which the particle is instantaneously at rest at that event. It is important to note that *this special inertial system is defined by the motion under consideration* and is not defined in an absolute manner that would violate the principle of relativity.

Consider the motion of a particle at any instant of time t relative to an (arbitrary) inertial system S. Let $\mathbf{v}(t)$ denote the velocity relative to S of the particle at that time. There exists another inertial system $S_0(t)$ that moves relative to S with a constant velocity $\mathbf{V}$ equal to the velocity $\mathbf{v}(t)$ of the particle at time t relative to S (Figure 2.50). The instantaneous velocity $\mathbf{v}_0(t)$ of the particle relative to $S_0(t)$ at the S time t is zero, since

$$\mathbf{v}_0(t) = \frac{\mathbf{v}(t) - \mathbf{V}}{1 - [\mathbf{v}(t) \cdot \mathbf{V}/c^2]} = 0. \qquad (2.184)$$

The inertial system in which the particle is instantaneously at rest at S time t is called the (instantaneous) *rest* (or proper) *inertial system* of the particle at the S time t (Figure 2.51). Note that, for an accelerated particle, the rest system $S_0(t)$ at one instant t of S time may not coincide with the rest system $S_0(t')$ at another time t' (Figure 2.52).

---

* Kacser, Prentice-Hall, Sec. A.6, p. 124; Sec. 6.1, p. 154; Sec. 6.2, p. 156.

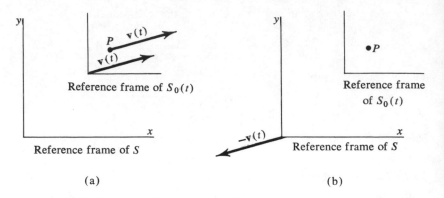

FIGURE 2.50   Motions of the reference frames and of the particle P at the S time t. (a) Motions from the point of view of S. (b) Motions from the point of view of $S_0(t)$.

(a)   (b)

Consider the motion of a particle that is described by Newton's equation of motion, with a given force function, relative to an inertial system in which the speed of the particle is sufficiently smaller than c. At the S time t, Newton's equation of motion gives a valid description of the motion relative to $S_0(t)$:

$$m\frac{d\mathbf{v}_0}{dt_0} = \mathbf{F}_0 \qquad \text{N.R.} \tag{2.185}$$

where the subscript zero indicates that the entity so labeled is defined relative to the instantaneous rest system $S_0(t)$, that inertial system in which $\mathbf{v}_0(t) = 0$. A given external force, $\mathbf{F}_0$ does not depend on the motion of the particle; it depends only on the position of the particle relative to its environment and on intrinsic properties of the particle, such as its electric charge.

We know the form of the equation of motion, Equation (2.185), relative to one inertial system, so our next problem is to determine the form of the equation of motion relative to any other inertial system. According to the principle of relativity, the equation of motion must be expressible in covariant form, so we look for a covariant equation that reduces to Equation (2.185) in the instantaneous rest system of the particle.

A time interval measured relative to $S_0(t)$ is equal to the proper time interval Δt_0 associated with the motion of the particle, where Δt_0 is a scalar quantity. Thus, the acceleration measured in the $S_0(t)$ system at S time t, $d\mathbf{v}_0/dt_0$, is the derivative with respect to the scalar t_0 of the spatial part of the velocity 4-vector

FIGURE 2.51   Coordinate axes of $S_0(t)$ at S time t. (a) World line of the particle and the axes of the $S_0(t)$ inertial system relative to S. (b) World line of the particle relative to $S_0(t)$.

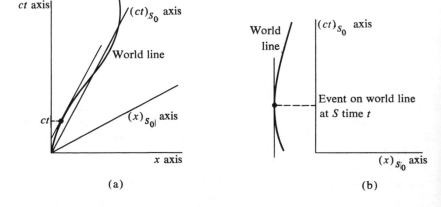

(a)   (b)

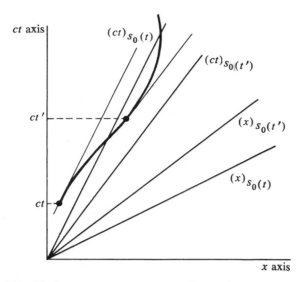

FIGURE 2.52   The instantaneous rest system $S_0(t)$ at S time t may not coincide with the instantaneous rest system $S_0(t')$ at S time t'.

$$v = \left(c \frac{dt}{dt_0}, \frac{d\mathbf{r}}{dt_0} \right).$$

Note: Even though v has the components $(c, 0, 0, 0)$ relative to $S_0(t)$, $\Delta \mathbf{v}_0$ will not be equal to zero unless the particle experiences no acceleration. These statements are compatible, since the instantaneous rest system, $S_0(t)$, depends on the time. The time-like component, relative to $S_0(t)$, of dv/dt_0 is zero:

$$\begin{aligned}
\frac{d}{dt_0}\left(c \frac{dt}{dt_0} \right)\bigg|_{v=0} &= \frac{1}{\sqrt{1 - (v^2/c^2)}} \frac{d}{dt}\left(c \frac{1}{\sqrt{1 - (v^2/c^2)}} \right)\bigg|_{v=0} \\
&= \frac{(v/c)(dv/dt)}{[1 - (v^2/c^2)]^2}\bigg|_{v=0} = 0.
\end{aligned}$$
(2.186)

Therefore, the left-hand side of the equation of motion (2.185) is given by the nonvanishing components, which are spatial, of the 4-vector $m \, dv/dt_0$.

We can define a 4-vector by specifying its components relative to one inertial reference system. Therefore, we define a force 4-vector, f, to be that 4-vector with components $(0, \mathbf{F}_0)$ relative to the instantaneous rest frame of the particle. The components of the force 4-vector relative to any other inertial system can be obtained by the appropriate Lorentz transformation; for example, if $S_0(t)$ and hence the particle are moving with the velocity $\mathbf{v} = v\hat{x}$ relative to S at S time t, then the components (f_t, f_x, f_y, f_z) of f relative to S are given by

$$f_t = \frac{F_{0x} v/c}{\sqrt{1 - (v^2/c^2)}}, \qquad f_x = \frac{F_{0x}}{\sqrt{1 - (v^2/c^2)}}, \qquad f_y = F_{0y}, \qquad f_z = F_{0z}. \quad (2.187)$$

The 4-force f is called the *Minkowski force*, because it was Minkowski who first expressed the equation of motion in covariant form.

The motion of a macroscopic object is described by the equation of motion (2.185) in those cases in which the speed of the particle relative to its environment is much smaller than c. Under these circumstances, the equation of motion (2.185) relative to the instantaneous rest frame of the particle is given by the nonvanishing components of the covariant equation

$$m \frac{dv}{dt_0} = f. \tag{2.188}$$

This suggests, but does not prove, that the covariant form of the equation of motion is the 4-vector equation (2.188). It is conceivable that the covariant form of the equation of motion is an equation between, for example, two tensors of the second order (see Problem 1.125) with the components of $m_0 \, d\mathbf{v}_0/dt_0$ and $\mathbf{F}_0$ determining three of the components of each of these tensors. It is also conceivable that the equation of motion takes on a form different from Equation (2.188) in those cases in which the speed of the particle relative to its environment is comparable to c. However, experiments support the suggestion that the covariant form of the equation of motion for macroscopic objects is a relation between 4-vectors; the covariant equation,

$$\text{Covariant equation of motion:} \qquad m \frac{dv}{dt_0} = f, \tag{2.189}$$

is the form of the equation of motion for macroscopic objects that is valid relative to every inertial system.

Under some circumstances, this equation is suitable also for the description of the motion of a single constituent particle of matter, and not only for the conglomerations of constituent particles that form the objects of our everyday experience. However, as you will see later in your studies of these constituent particles, it is necessary in general to use other concepts, those of quantum mechanics, to describe the behavior on an individual basis of the constituent particles of matter.

S U M M A R Y   Newton's equation of motion $\mathbf{F} = m\mathbf{a}$ can be generalized to a covariant equation that describes the motions of objects in circumstances in which pertinent relative speeds of the order of c are involved. The covariant equation relates the 4-vector or Minkowski force f, which is determined by the environment, to the rate of change of the 4-velocity with respect to the proper time of the particle: $m(dv/dt_0) = f$.

Example 2.13

Consider the motion of a particle of mass m described relative to an inertial system S by

$$x = \frac{c}{\omega} \cos \omega t \tag{2.190}$$

during the time $0 \le t < \pi/2\omega$* (Figure 2.53). The velocity of the particle relative to S is

$$v = -c \sin \omega t, \tag{2.191}$$

so the 4-velocity v has components relative to S given by

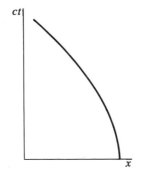

FIGURE 2.53   The world line described by $x = (1/\alpha) \cos (\alpha ct)$.

---

* The speed of the motion described by Equation (2.190) is c at $t = \pi/2\omega$. However, at that instant, the energy E and the momentum p of the particle are infinite, so the relation $E^2 - p^2 c^2 = m^2 c^4$ is satisfied even though $E/p \to c$.

$$v = \left(\frac{c}{\cos \omega t}, \; -c \tan \omega t\right).$$ (2.192)

The Minkowski force can be calculated from the covariant equation of motion:

$$\mathcal{f} = m\frac{dv}{dt_0} = m\frac{1}{\sqrt{1 - (v^2/c^2)}}\frac{d}{dt}\left(\frac{c}{\cos \omega t}, \; -c \tan \omega t\right)$$

$$= \frac{m}{\cos \omega t}\left(\frac{\omega c \sin \omega t}{\cos^2 \omega t}, \; \frac{-\omega c}{\cos^2 \omega t}\right) = \left(\frac{m\omega c \sin \omega t}{\cos^3 \omega t}, \; \frac{-m\omega c}{\cos^3 \omega t}\right).$$ (2.193)

The nonvanishing component F_{0x} of the Minkowski force relative to the instantaneous rest frame of the particle is given by

$$F_{0x} = \frac{\sqrt{1 - (v^2/c^2)}f_t}{v/c} \quad \text{or} \quad F_{0x} = \sqrt{1 - (v^2/c^2)}f_x$$ (2.194)

to be

$$F_{0x} = -\frac{m\omega c}{\cos^2 \omega t} = -\frac{mc^3}{\omega x^2}.$$ (2.195)

Note that F_{0x} is referred to the instantaneous rest system, whereas x is referred to the system S.

Problem 2.60

The history of a moving particle, relative to an inertial system S, is represented on a space-time diagram by the hyperbola

$$x^2 - c^2t^2 = \alpha^2, \quad y = z = 0.$$

(a) Show that the form of the equation for the world line is the same relative to all inertial systems related to S by the velocity $\mathbf{V} = \beta c\hat{x}$.
(b) Find $\mathbf{v} = v(t)\hat{x}$ relative to S.
(c) Find dt_0 in terms of t and dt.
(d) Show that $\alpha^2 \, d^2x/dt_0^2 = c^2x$.

Problem 2.61

Show that $\mathcal{f}\cdot v = 0$ by proving that this is true in the rest system S_0 of the particle.

Problem 2.62

At time t relative to S, a particle is moving with velocity $\mathbf{v} = v\hat{y}$ relative to S. Find the components of the Minkowski force $\mathcal{f}$ relative to S in terms of the components $(0, \mathbf{F}_0)$ relative to the instantaneous rest system S_0 of the particle.

Problem 2.63

Show that $\mathcal{f}\cdot\mathcal{f} = -\mathbf{F}_0\cdot\mathbf{F}_0 = -F_0^2$.

Problem 2.64

Define a 4-vector $\mathcal{N}$ by the equation $\mathcal{N} = [d(\mu v)/dt_0]$, where μ is a scalar. Show that $\mathcal{N}\cdot v = 0$ if μ is an invariant—that is, if $d\mu/dt_0 = 0$. *Hint:* Differentiate the equation $v\cdot v = c^2$.

2.3.2 *Motion along a straight line under a given external force**

We now examine some of the consequences of the relativistic equation of motion (2.189). We start with the simplest cases involving the points to be discussed and proceed to more complicated cases.

Consider a particle moving along the x axis of an inertial system S under a force that always lies along the x_0 direction in the instantaneous rest system of the particle:

$$\mathbf{F}_0 = F\hat{x}_0. \tag{2.196}$$

Relative to S, the nonvanishing components of the Minkowski force are

$$f_0 = \gamma F \frac{v}{c} \quad \text{and} \quad f_1 = \gamma F, \tag{2.197}$$

where v is the instantaneous velocity (in the one-dimensional vector notation in which the vector is denoted by its component along the x direction) of the particle relative to S and $\gamma = [1 - (v^2/c^2)]^{-\frac{1}{2}}$. The covariant equation of motion, $m\, dv/dt_0 = f$, has the following component equations relative to S:

$$m\gamma \frac{d}{dt}(\gamma c) = f_0 = \frac{\gamma F v}{c} \tag{2.198}$$

and

$$m\gamma \frac{d}{dt}(\gamma v) = f_1 = \gamma F. \tag{2.199}$$

The factor γ appears on each side of the equations, which can then be simplified to the forms

$$m \frac{d}{dt}(c^2\gamma) = Fv \tag{2.200}$$

and

$$m \frac{d}{dt}(\gamma v) = F. \tag{2.201}$$

Equation (2.201) describes how the velocity v changes under the action of the external force F and reduces, in the limit of low speeds ($|v|/c \ll 1$), to the non-relativistic form of the equation of motion,

$$m \frac{dv}{dt} = F. \qquad \text{N.R.} \tag{2.202}$$

The relativistic form of this equation, (2.201), cannot be written in this form, but it can be expressed in a form similar to a nonrelativistic equation equivalent to (2.202), namely,

$$\frac{d}{dt}(mv) = F \quad \text{or} \quad \frac{dp_0}{dt} = F, \qquad \text{N.R.} \tag{2.203}$$

where $p_0 = mv$ is the nonrelativistic momentum. Since m is an invariant, Equation (2.201) is equal to

$$\frac{d}{dt}(m\gamma v) = \frac{d}{dt}(m_v v) = F \quad \text{or} \quad \frac{dp}{dt} = F. \tag{2.204}$$

---

* Feynman, Leighton, and Sands (vol. 1), Addison-Wesley, Sec. 15–8, p. 15–9; Sec. 15–9, p. 15–10. Kittel, Knight, and Ruderman, McGraw-Hill, pp. 408–410. Resnick, John Wiley, Sec. 3.5, p. 119.

The *physical force* relative to S, $\mathbf{F} = F\hat{x}$, is equal to the time derivative of the relativistic momentum as measured in the inertial system S.

The relativistic equation of motion (2.204) is similar to the nonrelativistic form (2.203); the difference lies in the mass coefficient m_v. The relativistic mass m_v increases with increasing speed $|v|$, whereas the (rest) mass m is a scalar. The effect of the dependence of the relativistic mass on the speed can be seen from the following example: Consider a region of the motion in which the physical force F is constant. The change in the velocity per unit time under the action of the force decreases as v increases, since m_v increases with v and the change in $m_v v$ in the time interval Δt is given by

$$\Delta(m_v v) = F \,\Delta t. \tag{2.205}$$

We see again that the relativistic mass is a measure of the inertia of the particle and that the relativistic mass,

$$m_v = \frac{m}{\sqrt{1 - (v^2/c^2)}} = m + \frac{T}{c^2}, \tag{2.206}$$

is increased above its rest value m by T/c^2, a contribution from the relativistic kinetic energy T. Recall that it is this same measure of inertia that plays the role of the inertial mass coefficients in collisions.

Let us now consider the time-like component relative to S of the equation of motion, (2.200):

$$\frac{d}{dt}(m\gamma c^2) = \frac{d}{dt}(m_v c^2) = Fv. \tag{2.207}$$

In order to obtain an interpretation for this equation, we consider first the familiar case of the nonrelativistic limit, $|v|/c \ll 1$. In this circumstance, we can expand $mc^2\gamma$ as a power series in v/c, $mc^2\gamma = mc^2[1 + v^2/(2c^2) + \cdots]$, and obtain, from Equation (2.207),

$$\frac{d}{dt}(\tfrac{1}{2}mv^2) = Fv. \qquad \text{N.R.} \tag{2.208}$$

This is the nonrelativistic relation between the rate of change of the kinetic energy $\tfrac{1}{2}mv^2$ and the work done by the force per unit time, $Fv = F\,dx/dt$, since $F\,dx$ is the work done by the force $F\hat{x}$ while the particle undergoes the displacement $(dx)\hat{x}$ in the time dt.

A comparison of Equations (2.207) and (2.208) suggests that, in relativistic mechanics, we interpret $d(m_v c^2)/dt$ as the time rate of change in the kinetic energy. This suggestion is consistent with our earlier definition of $T = m_v c^2 - mc^2$ as the relativistic kinetic energy, since

$$\int_{v=0}^{v=v} \frac{d(m_v c^2)}{dt}\,dt = \int_{v=0}^{v=v} d(m_v c^2) = m_v c^2 - mc^2. \tag{2.209}$$

We see from Equation (2.207) that the time rate of change of the relativistic kinetic energy is equal to the work done by the physical force $F\hat{x}$ per unit time. Also, we have that

$$T_2 - T_1 = \int_1^2 F\,dx. \tag{2.210}$$

SUMMARY   In the case in which the spatial component of the Minkowski force is along the direction of motion, the covariant equation of motion reduces

to
$$\frac{dp}{dt} = F \quad \text{and} \quad \frac{dT}{dt} = Fv;$$

in this case, F is the spatial component of the Minkowski force in the instantaneous rest system and is called the physical force.

Example 2.14

The Minkowski force experienced by a particle undergoing the motion

$$x = \frac{c}{\omega} \cos \omega t \tag{2.211}$$

was calculated in Example 2.13 to have the components $(0, -mc^3/\omega x^2)$ relative to the instantaneous rest system. Therefore, relative to the system S, the physical force is $-mc^3/\omega x^2$, and the equation of motion is

$$\frac{dp}{dt} = -\frac{mc^3}{\omega x^2}. \tag{2.212}$$

Note that the solution of the force equation $dp/dt = -k/x^2$ is $x = (k/mc^2)$ $\times \cos(mc^3t/k)$ only for the initial conditions $x_0 = k/mc^2$, $v_0 = 0$.

Example 2.15

The increase in the inertia of a particle with increasing speed can also be seen from the transformation equation for collinear velocities, Equation (1.119):

$$v_S = \frac{V + v_{S'}}{1 + (Vv_S/c^2)}, \tag{2.213}$$

where $v_S \hat{x}$ is the velocity of a particle with respect to an inertial system S, $v_{S'} \hat{x}'$ is its velocity with respect to another inertial system S', and $V\hat{x}$ is the velocity of S' with respect to S. Let S' be the instantaneous rest frame of the particle at some event on its world line, so that $V = v(t)$, the velocity relative to S at that event. Let $v_{S'} = a_0 \Delta t_0$ be the change in the velocity with respect to the rest frame in the proper time interval Δt_0, where $a_0 = F_0/m_0$ is the acceleration experienced by the particle relative to S' at that event. Then $v_S - v = (dv/dt) \Delta t$ is the corresponding change in the velocity as measured in S and is given, to first order in Δt_0, by

$$\frac{dv}{dt} \Delta t = v_S - v = \frac{v + a_0 \Delta t_0}{1 + (va_0 \Delta t_0/c^2)} - v$$

$$= (v + a_0 \Delta t_0)\left(1 - \frac{va_0 \Delta t_0}{c^2}\right) - v \tag{2.214}$$

$$= a_0\left(1 - \frac{v^2}{c^2}\right) \Delta t_0 = a_0\left(1 - \frac{v^2}{c^2}\right)^{3/2} \Delta t.$$

Hence,

$$a = \frac{dv}{dt} = a_0\left(1 - \frac{v^2}{c^2}\right)^{3/2} \tag{2.215}$$

or

$$a_0 = \frac{1}{[1 - (v^2/c^2)]^{3/2}} \frac{dv}{dt} = \frac{d}{dt}\left(\frac{1}{\sqrt{1 - (v^2/c^2)}} v\right), \tag{2.216}$$

from which we see that an acceleration a_0 in the instantaneous rest frame results in a smaller acceleration in another inertial system as a consequence of the properties of space and time.

Example 2.16

Q. A particle moves from rest at the origin at time $t = 0$ relative to S under a constant physical force $\mathbf{F} = mg\hat{x}$.

    (a) Calculate the velocity and the position of the particle at time t.
    (b) Show that, for $|v|/c \ll 1$, the answers to (a) reduce to the appropriate nonrelativistic form.

A. (a) The equation of motion is

$$m \frac{d}{dt}\left[\frac{v}{\sqrt{1 - (v^2/c^2)}}\right] = mg. \tag{2.217}$$

Since $v = 0$ at $t = 0$, one integration yields

$$\frac{v}{\sqrt{1 - (v^2/c^2)}} = gt, \tag{2.218}$$

which can be squared and then solved for (the positive) v to give

$$v = \frac{gt}{\sqrt{1 + [(gt)^2/c^2]}}. \tag{2.219}$$

Note that $|v| < c$ for all values of t and that $|v| \to c$ as $t \to \infty$. This behavior results from the fact that the relativistic mass $m_v = \gamma m$ increases with increasing speed $|v|$. The position x at time t is given by the differential equation

$$\frac{dx}{dt} = \frac{gt}{\sqrt{1 + [(gt)^2/c^2]}} \tag{2.220}$$

and the initial condition $x = 0$ at $t = 0$ to be

$$x = \int_0^t \frac{gt'\, dt'}{\sqrt{1 + [(gt')^2/c^2]}}$$
$$= \frac{c^2}{g}[\sqrt{1 + (g^2 t^2/c^2)} - 1]. \tag{2.221}$$

The graphs of v and x as functions of t are shown in Figure 2.54.

    (b) Formulas (2.219) and (2.221) can be expanded in powers of gt/c for $|gt/c| \ll 1$ as

$$v = gt\left[1 - \frac{1}{2}\left(\frac{gt}{c}\right)^2 + \cdots\right] \tag{2.222}$$

and

$$x = \frac{c^2}{g}\left[1 + \frac{1}{2}\left(\frac{g^2 t^2}{c^2}\right) + \cdots - 1\right]$$
$$= \tfrac{1}{2}gt^2 + \cdots. \tag{2.223}$$

For $|gt/c| \ll 1$, $v \approx gt$ and $|v|/c \ll 1$. Therefore, for $|v|/c \ll 1$,

$$v = gt \quad \text{and} \quad x = \tfrac{1}{2}gt^2, \tag{2.224}$$

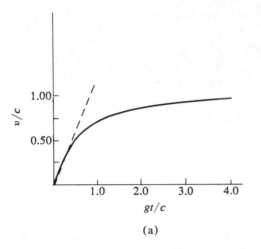

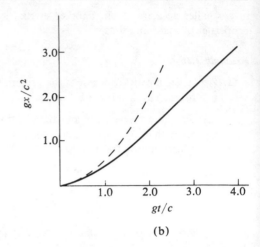

FIGURE 2.54 The motion of a particle experiencing a constant physical force along its direction of motion. (a) $v/c = (gt/c)/\sqrt{1 - (gt/c)^2}$. The dashed line is the nonrelativistic result, $v/c = gt/c$. (b) $gx/c^2 = \sqrt{1 + (gt/c)^2} - 1$. The dashed curve is the nonrelativistic result, $gx/c^2 = \frac{1}{2}(gt/c)^2$.

the nonrelativistic formulas for a particle undergoing a constant acceleration g.

Problem 2.65

Consider a space traveler who experiences a constant acceleration of magnitude $g = 9.8$ m/sec^2 in his rest frame.

(a) Find the speed of the space traveler 1 year after he starts from rest.
(b) Find the time dilatation factor at the end of that year.
(c) Find the proper time of the space traveler that has elapsed in that interval.
(d) How far has the space traveler moved in that time?

Problem 2.66

A particle moves from rest relative to S under the constant physical force $\mathbf{F} = mg\hat{x}$, where $g = 9.8$ m/sec^2.

(a) Calculate the times t relative to S at which the velocity deviates from the nonrelativistic prediction by
    (i) 1%,
    (ii) 10%.
(b) Calculate the times t relative to S at which the position deviates from the nonrelativistic prediction by
    (i) 1%,
    (ii) 10%.

Problem 2.67

Consider a stellar object with the same radius as the earth. Assume that the acceleration due to gravity g' at the surface of the object is given by Newton's law of gravitation. Estimate the minimum mass of such a stellar object for which it is necessary to use relativistic mechanics to describe free-fall motions similar to those encountered in everyday life.

Problem 2.68

A particle moves from rest relative to S under the constant physical force $\mathbf{F} = mg\hat{x}$.

(a) Show that, for t sufficiently large that $gt/c \gg 1$, the velocity v is given by

$$\frac{v}{c} = 1 - \frac{c^2}{2(gt)^2}.$$

(b) Show that the relativistic momentum p is given by the product (force) × (time).
(c) Compare the behavior of v and p for large t, and also compare the relativistic behavior of each with its behavior as predicted by newtonian nonrelativistic mechanics.

Problem 2.69

Explain why the relativistic value of T is greater than the nonrelativistic expression $\frac{1}{2}mv^2$. Why does $T \to \infty$ as $|v| \to c$?

Problem 2.70

A particle of mass m undergoes straight-line motion under the constant relative force $mg\hat{x}$.

(a) Show that, for $gt/c \ll 1$, $T = \frac{1}{2}m(gt)^2$, the nonrelativistic result.
(b) Show that, for $gt/c \gg 1$, $T = (mg)(ct)$. Interpret the two factors, mg and ct.

Problem 2.71

A particle in straight-line motion undergoes a constant acceleration $g\hat{x}$ relative to S (for a time less than that required for its speed to reach the value c). The particle was at rest relative to S at time $t = 0$. Calculate the components, relative to S, of

(a) the physical force,
(b) the Minkowski force.

Problem 2.72

Consider a particle undergoing motion along the x axis of an inertial system S described by the equation of motion

$$m\frac{d^2x}{dt_0^2} = f,$$

where the Minkowski force f has components $(T, X, 0, 0)$ relative to S. Show in the following manner that $f \cdot v = 0$:

(a) From the definition of Δt_0 show that

$$\left(\frac{dt}{dt_0}\right)^2 - \frac{1}{c^2}\left(\frac{dx}{dt_0}\right)^2 = 1$$

and hence that

$$\frac{dt}{dt_0}\frac{d^2t}{dt_0^2} = \frac{1}{c^2}\frac{dx}{dt_0}\frac{d^2x}{dt_0^2}.$$

(b) Show that

$$T\frac{dt}{dt_0} = \frac{1}{c}X\frac{dx}{dt_0},$$

so that $cT = Xv$.

(c) Show from the results of (b) that $f \cdot v = 0$.

Problem 2.73

A particle of mass m undergoes motion along the x axis of an inertial system S under the physical force $\mathbf{F} = -m\omega^2\mathbf{x}$, where ω is a positive constant. The maximum amplitude of the motion is a.

(a) Show that the equation of motion can be written in the form

$$v\frac{d}{dx}\left[\frac{v}{\sqrt{1 - (v^2/c^2)}}\right] = -\omega^2x.$$

(b) Show that

$$\frac{c^2}{\sqrt{1 - (v^2/c^2)}} = c^2 + \frac{\omega^2}{2}(a^2 - x^2).$$

(c) Let $\gamma(x) = 1/\sqrt{1 - (v^2/c^2)}$. Show that the period T of the motion is given by

$$T = \frac{4}{c}\int_0^a \frac{\gamma(x)}{\sqrt{\gamma^2(x) - 1}}\,dx.$$

(d) Show that, for $\omega a/c \ll 1$, $T = 2\pi/\omega$, the nonrelativistic form.

(e) Show that, if $\omega a/c$ is small,

$$T \approx \frac{2\pi}{\omega}\left(1 + \frac{3}{16}\frac{\omega^2a^2}{c^2}\right).$$

Problem 2.74

Consider an object whose behavior is described by the equation

$$\frac{d(\mu v)}{dt_0} = \mathcal{N},$$

where $\mu(t_0)$ is the relativistic mass of the object at proper time t_0 and $\mathcal{N}$ is the generalized Minkowski force given by $\mathcal{N} = bv$, with b a constant. Note that $\mathcal{N} \cdot v \neq 0$.

(a) Show that the object does not experience any acceleration.

(b) Show that $\mu = bt_0 + \mu_0$, where $\mu_0 = \mu(0)$.

(c) Find an expression for μ in terms of the time t relative to any inertial system S.

Problem 2.75

An object of variable relativistic mass $\mu(t_0)$ undergoes straight-line motion along the x axis of S described by the equation of motion

$$\frac{d(\mu v)}{dt_0} = \mathcal{N},$$

with the components of $\mathcal{N}$ relative to S being $(T, X, 0, 0)$.

(a) Show that

$$\frac{d\mu}{dt_0} = \frac{\mathcal{N} \cdot v}{c^2}.$$

(b) Let $(T_0, X_0, 0, 0)$ be the components of $\mathcal{N}$ in the instantaneous rest system of the object. Show that

$$\frac{d\mu}{dt_0} = \frac{1}{c} T_0.$$

(c) Find $d\mu/dt_0$ in terms of the instantaneous velocity of the object relative to S and the components of $\mathcal{N}$ relative to S.

2.3.3 Relativistic effects of a force in an arbitrary direction*

We now consider the case, more general than that studied above, in which the force and the motion do not lie along the same line at all times. Under these circumstances, the equation of motion at time t takes on a simple form relative to S if we choose the direction of the x axis to be along the instantaneous direction of the velocity $\mathbf{v}(t)$ at the time t. The more general form of the equation is obtained in Example 2.17 and the problems.

Let the components of the Minkowski force f, relative to the instantaneous rest system, be

$$f = (0, F_{0x}, F_{0y}, F_{0z}). \tag{2.225}$$

Then, relative to S, the components of f, as given by a Lorentz transformation, are

$$f = (f_t, \mathbf{f}) = \left(\gamma F_{0x} \frac{v}{c}, \gamma F_{0x}, F_{0y}, F_{0z} \right). \tag{2.226}$$

Relative to S, the force law at time t takes the form

$$m\gamma \frac{d}{dt}(\gamma c) = \gamma \frac{d}{dt}(m_v c) = f_t, \qquad m\gamma \frac{d}{dt}(\gamma \mathbf{v}) = \gamma \frac{d}{dt}(m_v \mathbf{v}) = \mathbf{f}, \tag{2.227}$$

or

$$\frac{d}{dt}(m_v c) = F_{0x} \frac{v}{c} \tag{2.228}$$

and

$$\frac{d}{dt}(m_v \mathbf{v}) = \sqrt{1 - (v^2/c^2)}\, \mathbf{f}. \tag{2.229}$$

The 3-vector $\sqrt{1 - (v^2/c^2)}\, \mathbf{f}$ is the force that would be obtained by a measurement in S of the time rate of change of the relativistic momentum $\mathbf{p} = m_v \mathbf{v}$; this is the *physical force* $\mathbf{F}$,

$$\mathbf{F} = \sqrt{1 - (v^2/c^2)}\, \mathbf{f}, \tag{2.230}$$

with components

$$\mathbf{F} = (F_{0x}, \sqrt{1 - (v^2/c^2)} F_{0y}, \sqrt{1 - (v^2/c^2)}\, F_{0z}). \tag{2.231}$$

---

* Kacser, Prentice-Hall, Sec. 6.4, p. 161; Sec. 6.6, p. 174; Sec. 7.2, p. 182.
Kittel, Knight, and Ruderman, McGraw-Hill, pp. 395–396, 400–401, 410–413.
Resnick, John Wiley, Sec. 3.5, p. 119.

The equation of motion (2.229) can be written in terms of the physical force as

$$\frac{d}{dt}(m_v\mathbf{v}) = \frac{d\mathbf{p}}{dt} = \mathbf{F}, \qquad (2.232)$$

similar to one form of the nonrelativistic equation of motion.

The time-like component of the covariant equation of motion $m\, dv/dt_0 = \slashed{f}$ is

$$m\gamma \frac{d(\gamma c)}{dt} = f_t, \qquad (2.233)$$

where $(f_t, \mathbf{f})$ are the components of the Minkowski force relative to S. Relative to S_0, $v = (c, 0)$ and $\slashed{f} = (0, \mathbf{F}_0)$, so $v \cdot \slashed{f} = 0$. This equation can be expressed in terms of the components of v and $\slashed{f}$ relative to S as

$$f_t = \frac{\mathbf{f} \cdot \mathbf{v}}{c}. \qquad (2.234)$$

Therefore,

$$m\gamma \frac{d(\gamma c)}{dt} = \frac{\mathbf{f} \cdot \mathbf{v}}{c}. \qquad (2.235)$$

Since the physical force $\mathbf{F}$ is given by $\mathbf{f} = \gamma\mathbf{F}$ (Equation 2.230), the time-like component of the covariant equation of motion can be written as

$$\frac{d(m_v c^2)}{dt} = \mathbf{F} \cdot \mathbf{v}. \qquad (2.236)$$

Hence, the relativistic kinetic energy is given by

$$T = m_v c^2 - mc^2 = mc^2 \left[\left(1 - \frac{v_x^2 + v_y^2 + v_z^2}{c^2} \right)^{-\frac{1}{2}} - 1 \right], \qquad (2.237)$$

since the change in the relativistic kinetic energy between the motion with velocity $\mathbf{v}_1$ and that with $\mathbf{v}_2$ is equal to the work done by the physical force along the path of the motion:

$$\Delta T = T_2 - T_1 = m_{v_2} c^2 - m_{v_1} c^2$$

$$= \int_{v=v_1}^{v=v_2} \frac{d(m_v c^2)}{dt}\, dt = \int_{\substack{\text{path from motion with} \\ \text{velocity } \mathbf{v}_1 \text{ to motion with} \\ \text{velocity } \mathbf{v}_2}} \mathbf{F} \cdot \mathbf{v}\, dt$$

$$= \int_{\substack{\text{path from motion with} \\ \text{velocity } \mathbf{v}_1 \text{ to motion with} \\ \text{velocity } \mathbf{v}_2}} \mathbf{F} \cdot d\mathbf{r}. \qquad (2.238)$$

Moreover, this change in the relativistic kinetic energy is a function only of the initial and final speeds of the particle; it does not depend on the manner in which this change in speed is brought about.

The covariant form of the equation of motion can be written in terms of the energy-momentum vector as

$$\frac{d\slashed{p}}{dt_0} = \slashed{f}; \qquad (2.239)$$

the spatial components of this equation $d\mathbf{p}/dt = \mathbf{F}$ describe the motion, and the time component $d(m_v c^2)/dt = \mathbf{F} \cdot \mathbf{v}$ describes the rate at which the relativistic energy changes.

SUMMARY   The covariant equation of motion

$$m \frac{dv}{dt_0} = f$$

is equivalent, relative to an inertial system S, to the equations

$$\frac{d\mathbf{p}}{dt} = \mathbf{F} \quad \text{and} \quad \frac{dT}{dt} = \mathbf{F} \cdot \mathbf{v},$$

where $\mathbf{F} = \gamma \mathbf{f}$ is the physical force.

Example 2.17

Q. Consider a particle experiencing the Minkowski force f with components $(f_t, \mathbf{f})$ relative to an inertial system S.

(a) Find the physical force $\mathbf{F}$ experienced by the particle when it is moving with velocity $\mathbf{v} = v_x \hat{x} + v_y \hat{y} + v_z \hat{z}$.

(b) The particle undergoes an acceleration, relative to S, in the x direction at an instant when its velocity is $\mathbf{v} = v_x \hat{x} + v_y \hat{y} + v_z \hat{z}$. Show that the y and z components of the physical force are not zero, and express them in terms of F_x.

A. (a) The covariant equation of motion $m \, dv/dt_0 = f$ has spatial components relative to S given by

$$\gamma \frac{d}{dt} (m\gamma \mathbf{v}) = \mathbf{f} \quad \text{or} \quad \frac{d\mathbf{p}}{dt} = \sqrt{1 - (v^2/c^2)} \, \mathbf{f}, \qquad (2.240)$$

with $v^2 = v_x^2 + v_y^2 + v_z^2$. Therefore, the physical force $\mathbf{F}$ equal to the time rate of change of the relativistic momentum $d\mathbf{p}/dt$ is given by

$$\mathbf{F} = \sqrt{1 - (v^2/c^2)} \, \mathbf{f}. \qquad (2.241)$$

This equation is the same as Equation (2.230) even though, since $\mathbf{v} \neq v\hat{x}$, the components of $\mathbf{F}$ are not given by (2.231) (see Problem 2.78).

(b) Since the acceleration $d\mathbf{v}/dt$ of the particle is in the x direction,

$$\frac{dv_y}{dt} = \frac{dv_z}{dt} = 0$$

and

$$\frac{dm_v}{dt} = \frac{d}{dt} \left[\frac{m}{\{1 - [(v_x^2 + v_y^2 + v_z^2)/c^2]\}^{1/2}} \right]$$

$$= \frac{m_v v_x}{c^2 - v^2} \frac{dv_x}{dt}. \qquad (2.242)$$

Therefore, the components of the physical force $\mathbf{F}$ are given by the equation of motion $d\mathbf{p}/dt = \mathbf{F}$ to be

$$F_x = \frac{d(m_v v_x)}{dt} = \frac{dm_v}{dt} v_x + m_v \frac{dv_x}{dt} = \left(\frac{m_v v_x^2}{c^2 - v^2} + m_v \right) \frac{dv_x}{dt}$$

$$= m_v \frac{(c^2 - v_y^2 - v_z^2)}{c^2 - v^2} \frac{dv_x}{dt}, \qquad (2.243)$$

and $\qquad F_y = \dfrac{d(m_v v_y)}{dt} = \dfrac{dm_v}{dt} v_y = \dfrac{m_v v_x v_y}{c^2 - v^2} \dfrac{dv_x}{dt},$

$$F_z = \frac{m_v v_x v_z}{c^2 - v^2} \frac{dv_x}{dt}. \qquad (2.244)$$

Hence, the components F_y and F_z are not zero and are given by

$$F_y = \frac{v_x v_y}{c^2 - v_y^2 - v_z^2} F_x \quad \text{and} \quad F_z = \frac{v_x v_z}{c^2 - v_y^2 - v_z^2} F_x. \quad (2.245)$$

The y and z components of the physical force do not vanish, even though the acceleration is only in the x direction, for the following reason: The acceleration of the particle in the x direction, which results in a change in v_x, produces a change in the relativistic mass m_v and, therefore, a change in the y and z components $m_v v_y$ and $m_v v_z$ of the momentum, even though v_y and v_z remain constant. The force components, $F_y = (dm_v/dt)v_y$ and $F_z = (dm_v/dt)v_z$, are required to produce the resulting change in the corresponding components of the momentum.

Problem 2.76

A particle of charge q experiences a magnetic force that is given by $\mathbf{F} = q\mathbf{v} \times \mathbf{B}$, where $\mathbf{B}$ is a constant vector.

(a) Show that the relativistic mass m_v and the magnitude p of the relativistic momentum do not change under the action of the force.

(b) Show that the particle moves along a helix, by proving each of the following statements: The component of the velocity $\mathbf{v}_\parallel$ in the direction of $\mathbf{B}$ remains constant, as does the magnitude of the component $\mathbf{v}_\perp$ of the velocity perpendicular to $\mathbf{B}$. Relative to a point that moves with the velocity $\mathbf{v}_\parallel$ of magnitude $\mathbf{v} \cdot \mathbf{B}/B$, the particle moves in a circle of radius $\rho = m_v v_\perp / qB$, where $\mathbf{v}_\perp = \mathbf{v} - \mathbf{v}_\parallel$.

Problem 2.77

A particle experiences a Minkowski force f at one event on its world line. The components of f relative to the particle's instantaneous rest frame are $(0, \mathbf{F}_0)$, those relative to an inertial system S are $(f_t, \mathbf{f})$, and those relative to another inertial system S' are $(f_{t'}, \mathbf{f}')$. The velocity of S' relative to S is $\mathbf{V}$.

(a) Show that

$$\mathbf{f} = \mathbf{F}_0 + \mathbf{v}\frac{\mathbf{v}\cdot\mathbf{F}_0}{v^2}(\gamma - 1),$$

where $\gamma = [1 - (v^2/c^2)]^{-\frac{1}{2}}$, and derive from this that

$$f_t = \frac{\mathbf{f}\cdot\mathbf{v}}{c}.$$

(b) Let $\gamma_V = [1 - (V^2/c^2)]^{-\frac{1}{2}}$. Show that

$$\mathbf{f}' = \mathbf{f} + \mathbf{V}\frac{\mathbf{V}\cdot\mathbf{f}}{V^2}(\gamma_V - 1) - \mathbf{V}\gamma_V\frac{\mathbf{v}\cdot\mathbf{f}}{c^2}.$$

(c) Let $\gamma' = [1 - (v'^2/c^2)]^{-\frac{1}{2}}$. Show that the transformation law for the physical force is given by

$$\gamma' \mathbf{F}' = \gamma \left\{ \mathbf{F} + \mathbf{V} \left[(\gamma_V - 1) \frac{\mathbf{V} \cdot \mathbf{F}}{V^2} - \gamma_V \frac{\mathbf{v} \cdot \mathbf{F}}{c^2} \right] \right\}.$$

(d) Show that the result of (c) reduces to Equation (2.231) in the case where $\mathbf{v} = \mathbf{V} = v\hat{x}$.

Problem 2.78

A particle experiences a Minkowski force f with components $(0, \mathbf{F}_0)$ relative to its instantaneous rest frame at an instant when the particle is moving with velocity $\mathbf{v}$ relative to an inertial frame S.

(a) Show that the physical force with respect to S is given by

$$\mathbf{F} = \sqrt{1 - (v^2/c^2)}\, \mathbf{F}_0 + \frac{\mathbf{v}}{v^2} \mathbf{v} \cdot \mathbf{F}_0 [1 - \sqrt{1 - (v^2/c^2)}].$$

(b) Show that the result of (a) reduces to Equation (2.231) for the special case $\mathbf{v} = v\hat{x}$.

Problem 2.79

A particle of mass m moving with constant relativistic momentum $p_0 \hat{x}$ with respect to an inertial system S at S times $t < 0$ experiences a constant physical force $\mathbf{F} = F\hat{y}$ for $t \geq 0$.

(a) Show that at S time $t \geq 0$

$$p_x = p_0, \qquad p_y = Ft, \qquad p_z = 0.$$

(b) Show that at S time $t \geq 0$

$$E^2 = E_0^2 + (Fct)^2,$$

where

$$E_0^2 = m^2 c^4 + p_0^2 c^2.$$

(c) Find v_x, v_y, v_z at S time $t \geq 0$.
(d) Explain why v_x changes with t.
(e) Find $x(t)$, $y(t)$, and $z(t)$.

Problem 2.80

Two particles collide elastically. Let the interaction between the particles be described by the Minkowski force $f_{i(j)}$, the Minkowski force on the ith particle due to the presence of the jth, and the corresponding physical force $\mathbf{F}_{i(j)}$.

(a) Show that the total momentum is conserved if the physical forces satisfy Newton's third law

$$\mathbf{F}_{1(2)} = -\mathbf{F}_{2(1)}.$$

(b) Let t_{i0} be the proper time of the ith particle. Show that the equations of motion

$$\frac{dp_1}{dt_{10}} = f_{1(2)}, \qquad \frac{dp_2}{dt_{20}} = f_{2(1)}$$

do not necessarily lead to the conservation law of the total 4-momentum if $f_{1(2)} = -f_{2(1)}$ were true. Explain.

2.4 Nuclei and Fundamental Particles

Relative speeds comparable to c of macroscopic objects have not been obtained in the laboratory, in part because of the tremendous energies that would be required for their achievement. Indeed, the usefulness in the laboratory of the relativistic laws of motion is limited to applications involving nuclei and fundamental particles.* Therefore, at this point we introduce some of the pertinent topics of interest in the fields of nuclear physics and fundamental particle physics. This study will provide illustrative examples of relativistic effects, and, through checks with experiments, it should give you some confidence in the laws of relativity theory.

The relativistic conservation law of 4-momentum provides a valid relation between the observed initial and final states of free motion of these particles. Consequently, in circumstances involving relative speeds all much smaller than c, the conservation laws of nonrelativistic mechanics are valid. Conservation laws, however, do not provide a complete description of the motions; in fact, the details of the behavior of nuclei and fundamental particles cannot be described adequately by the concepts we have considered up to this point. Indeed, at the present time, a comprehensive theory that describes this detailed behavior is not existent. There is a theory, quantum mechanics,† that correctly describes the behavior of atoms, behavior that lies intermediate to the level of our everyday experiences and the level of particle behavior. Thus, the principles of quantum mechanics are more relevant to the description of particle behavior than are the principles of newtonian or classical, as opposed to quantum mechanical, relativistic mechanics. Some of the consequences of the principles of quantum mechanics will be described in the following outline of nuclear and particle processes. However, the method of presentation of these and some other facts in this section differs from that in the preceding parts of the book; the following contains some statements for which little corroborative evidence is presented. The justification for statements about such matters will be left as topics for further study.

The particles are introduced below in an order that corresponds to the chronological order in which they were discovered or began to play an important role in physics: the electron, the photon, the nucleus, the nucleon, the neutrino, antiparticles, the muon and the π-meson, strange particles, and resonances. You can supplement the description given below by reading some of the books and articles listed as references in this section.‡

It is worthwhile to digress at this point, before entering into a description of the fundamental particles, to consider sources of highly energetic particles and methods by which these particles can be "observed." We will restrict our

---

* In this context, the word "fundamental" means only that a particle so described plays an important part in the basic structure of matter; the study of fundamental particles collectively is expected to provide the foundation for an understanding of that structure. The word does not imply that a fundamental particle is necessarily an ultimate constituent of matter nor that one fundamental particle may not be considered as a composite structure of others.

† An introduction to the ideas of quantum mechanics, written for the general reader, is given in B. Hoffmann, *The Strange Story of the Quantum*, Dover, New York, 1959.

‡ Additional general references for this section are K. W. Ford, *The World of Elementary Particles*, Blaisdell, Waltham, Mass., 1963; C. E. Swartz, *The Fundamental Particles*, Addison-Wesley, Reading, Mass., 1965; D. H. Frisch and A. M. Thorndike, *Elementary Particles*, Momentum Book No. 1, D. Van Nostrand, Princeton, N.J., 1964; C. S. Cook, *Structure of Atomic Nuclei*, Momentum Book No. 8, D. Van Nostrand, Princeton, N.J., 1964; and other references listed in C. E. Swartz, "Resource Letter SAP-1 on Subatomic Particles," *American Journal of Physics*, **34**: 1079 (1966).

discussion to the simplest examples and will reference more extensive accounts for further information.

The early work on high-energy particles was accomplished by use of natural sources of these particles. For example, in their early studies of nuclei, Rutherford and his colleages used α particles, the nuclei of helium atoms, which are emitted by certain naturally radioactive substances. However, these particles are not so energetic as those from another natural source; particles showing extreme relativistic behavior occur naturally in *cosmic rays*.*

Cosmic rays are the flux of energy in the forms, first, of high-energy particles that are incident upon the earth's atmosphere from outer space and, second, of particles that the incident primary flux produces in the earth's atmosphere. The high-energy particles from outer space are called primaries, and their flux consists mainly of protons together with a few α particles and even smaller numbers of other nuclei.†

Cosmic-ray particles have been observed with much higher energies than those to which it has been possible to accelerate particles in the laboratory. For this reason, the study of cosmic rays has provided much information on high-energy processes. The means by which these particles obtain such high energies are not understood at present, but this problem is under extensive study.‡

Today it is not necessary to rely on such natural sources as cosmic rays or radioactive materials for fast-moving particles. Various types of machines have been developed that generate and accelerate beams of charged particles, such as electrons, protons, and α particles, to high energies.§

The detection of particles in much of the early research in nuclear and particle physics was performed with the aid of simple apparatus such as scintillation foils, which when struck by an α particle emit a small flash of light that can be seen in a darkened room by an observer whose eyes are adjusted to the darkness. Another important piece of instrumentation of that period was the cloud chamber (Figure 2.55), invented by the English physicist Charles T. R. Wilson (1869–1959). This chamber is an instrument that can be used to show the paths of charged particles that traverse its interior. It consists of a cylinder, filled with air saturated with a vapor, whose volume can be changed by the movement of a piston. An energetic particle will ionize the atoms of the gas in the chamber along the path of the charged particle. A sensing device can be set up that is triggered by an incoming charged particle and that, in turn, causes the piston to move and expand the air in the cylinder. The gas becomes supersaturated, and the ions created along the path of the charged particle act as nuclei for the condensation of liquid droplets. These droplets produce a visible track along the path of each charged particle; such tracks can be photographed

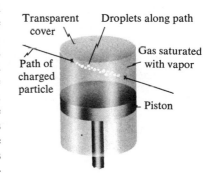

FIGURE 2.55   Wilson cloud chamber.

---

* See B. Rossi, *Cosmic Rays*, McGraw-Hill, New York, 1964, and the references listed in J. R. Winckler and D. J. Hofmann, "Resource Letter CR-1 on Cosmic Rays," *American Journal of Physics*, *35*: 2 (1967).

† Some properties of the primary flux of cosmic rays are described in L. Scarsi, "Cosmic Radiation," *American Journal of Physics*, *28*: 213 (1960).

‡ Some measurements that are designed to provide information on the origin of cosmic rays are described in B. Rossi, "High-Energy Cosmic Rays," *Scientific American*, *201*: 134, November 1959.

§ See, for example, R. R. Wilson, "Particle Accelerators," *Scientific American*, *198*: 64, March 1958; R. R. Wilson and R. Littauer, *Accelerators*, Anchor Books, Garden City, N.Y., 1960; and Chapter 9 of D. L. Livesey, *Atomic and Nuclear Physics*, Blaisdell, Waltham, Mass., 1966. A huge accelerator that began operation in 1966 is described in E. L. Ginzton and W. Kirk, "The Two-Mile Electron Accelerator," *Scientific American*, *205*: 49, November 1961.

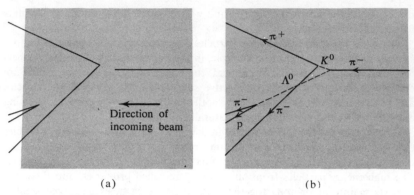

(a)   (b)

FIGURE 2.56 These fundamental particles were identified by their dynamical behavior in a bubble chamber picture. (a) The lines show the tracks of fundamental particles as seen in a bubble chamber picture. (Only particles with electric charges leave tracks.) (b) Interpretation: An incoming charged particle π^- is responsible for the creation of two electrically neutral particles, K° and Λ°, which in turn are responsible for the creation of other charged particles, namely $\pi^+ - \pi^-$ and $\pi^- - $ p, respectively.

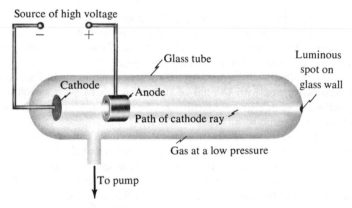

FIGURE 2.57 Cathode ray tube. The anode is at a higher electric potential than is the cathode; thus, electric charge, emitted at the cathode, streams toward the anode and on to the end of the tube. Electric and magnetic forces can be exerted on the cathode ray beam, and they show that it is a beam of negative charges; in fact, cathode rays are a beam of electrons.

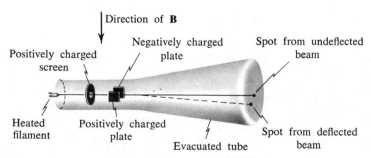

FIGURE 2.58 Schematic diagram of Thomson's e/m apparatus.

and examined at leisure. More refined types of chambers* of the same basic character are used in today's research in nuclear and particle physics (Figure 2.56).

2.4.1 The electron

The electron was discovered as a result of experiments, started about a century ago, that were performed to investigate the conduction of electricity in gases.† In these experiments, studies were made of the transmission, under the action of a high voltage, of electric charge through closed tubes containing gases at various pressures (Figure 2.57). The relation between science and technology is illustrated in the progress of these studies, since advances in these investigations depended, for example, on the refinement of techniques for evacuating the gas from the closed tubes.‡

The studies culminated in 1897 with the measurement of the charge-to-mass ratio of the particle that we now call the electron, by the English physicist J. J. Thomson (1856–1940). The electric-charge carrier emitted at the negative cathode in an evacuated tube was deflected by electric and magnetic forces in turn (Figure 2.58), and from a measurement of the deflection the ratio e/m_e was calculated (see Problem 2.81).

Thomson's experiments led to a crude determination of the mass and charge of the electron, although they did not demonstrate conclusively that it carried an elementary unit of charge or even that an elementary charge unit existed. These facts were shown by the oil-drop experiment of the American physicist Robert A. Millikan (1868–1953). The essence of this experiment was the balancing of the gravitational force on a drop of oil with an electric force exerted on the charged drop (Figure 2.59). The drops were charged with a small number of electrons so that the discreteness of the number of charges was evident from Millikan's measurements (see Problem 2.82). The magnitude of the charge of the electron is the elementary charge unit e:

$$e = 1.602 \times 10^{-19} \, \text{C}. \qquad (2.246)$$

The charge of an electron is $-e$.

By convention, the electron is denoted by e or sometimes by e⁻ to indicate that it carries a negative electric charge.§ This particle plays an important role in chemical phenomena and in the physics of solids.¶ Indeed, in principle,

---

* See H. Yagoda, "The Tracks of Nuclear Particles," *Scientific American*, *194*: 41, May, 1956; D. A. Glaser, "The Bubble Chamber," *Scientific American*, *192*: 46, February 1955; and G. K. O'Neill, "The Spark Chamber," *Scientific American*, *207*: 36, August 1962.
† The research that led to the discovery of the electron is described in D. L. Anderson, *The Discovery of the Electron*, Momentum Book No. 3, D. Van Nostrand, Princeton, N.J., 1964, and pp. 349–366 of E. Whittaker, *A History of the Theories of Aether and Electricity—The Classical Theories*, Thomas Nelson, New York, 1951.
‡ One application of modern technology to studies that developed directly from these investigations of the conduction of electricity in gases is described in L. Spitzer, Jr., "The Stellarator," *Scientific American*, *199*: 28, October 1958.
§ The letter *e* denotes the positive elementary charge, but in practice, no confusion arises as a result of this double usage. In printed matter, the symbol for the elementary charge is italicized, whereas the symbol for electron, like that for any particle, appears in roman type.
¶ See any college-level introductory text on chemistry or D. K. Sebera, *Electronic Structure and Chemical Bonding*, Blaisdell, Waltham, Mass., 1964. The part that electrons play in the physics of the solid state of matter is described in A. T. Stewart, *Perpetual Motion*, Anchor Books, New York, 1965, and in A. Holden, *The Nature of Solids*, Columbia Univ. Press, New York, 1965.

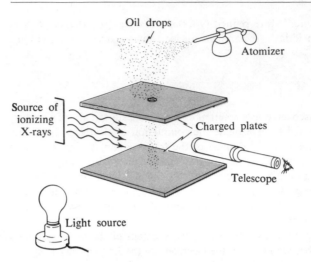

FIGURE 2.59   A schematic picture of the Millikan oil-drop experiment.

nearly all the properties of ordinary matter can be deduced mathematically in terms of the motions of the negatively charged electrons around or past positively charged nuclei. We shall not consider here the part played by the electron in the study of the properties of ordinary matter; our concern is with the nature of the electron itself and its function in the study of nuclei and of other fundamental particles.

How can we describe or identify a minute particle such as the electron? We certainly cannot label it, say by writing "electron" on it, for this writing would require attaching to the electron a multitude of particles of ink, particles that are themselves much more massive than the electron. For similar reasons, we cannot identify an electron by its color or shape. We can describe an electron only by presenting those dynamical variables that determine its behavior and do not change throughout the history of the electron. For example, the (rest) mass m_e describes the inertial behavior of the electron, and the electric charge $q_e = -e$ gives a measure of the electric forces that an electron experiences.

The parameters m_e and q_e are not sufficient, however, to completely describe the behavior of electrons. It is necessary to assign to an electron an intrinsic angular momentum called *spin*. The concept of angular momentum suggests rotational motion, and that of intrinsic angular momentum suggests the spinning of an extended object about an internal axis (Figure 2.60). However, within the framework of the concepts that we have considered thus far, it is impossible for an object to possess such an intrinsic angular momentum unless that object has a composite structure. The composite structure is necessary in order that every component term $\mathbf{r} \times \mathbf{p}$ relative to some internal point is not zero (Figure 2.61). Thus, the concept of an intrinsic angular momentum is not consistent, within the framework of the concepts we have considered, with the idea of a particle without structure. However, within the framework of quantum mechanics, it is possible to attribute to a particle without structure an intrinsic angular momentum. The possible values of the magnitude of this intrinsic angular momentum are

$$0, \frac{1}{2}\left(\frac{h}{2\pi}\right), \left(\frac{h}{2\pi}\right), \frac{3}{2}\left(\frac{h}{2\pi}\right), \ldots, \quad h = (6.6256 \pm 0.0005) \times 10^{-34} \text{ J sec}, \quad (2.247)$$

according to the results of quantum mechanics. The constant h is called Planck's constant in honor of the German physicist Max Planck (1858–1947) who introduced it. It is conventional to express the intrinsic angular momentum or spin of a particle in units of $(h/2\pi)$. Within this convention, the possible values of the magnitude of the spin S of a particle are 0, 1/2, 1, 3/2, 2, ...

The magnitude of the spin S_e of an electron is 1/2. This spin plays a role in the behavior of the electron in two important ways. In the first place, the spin of the electron contributes, together with the orbital angular momentum associated with the motion of the electron as a whole, to the total angular momentum of the electron. The conservation law of angular momentum is applicable, under the appropriate circumstances, only to the total angular momentum of the electron and not just to that associated with the electron's motion.

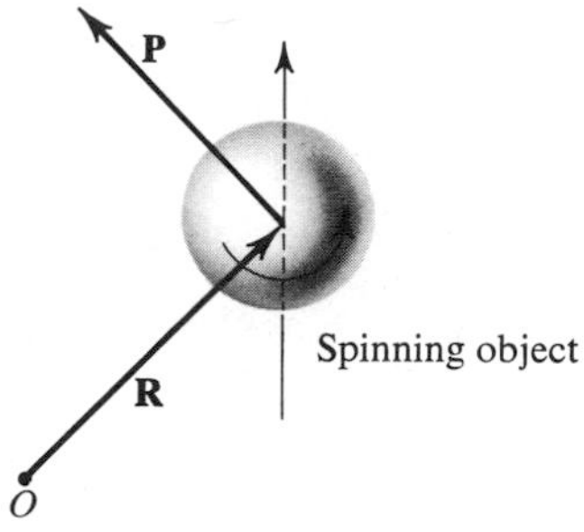

Second, it is possible to distinguish between two electrons by the directions of their intrinsic angular momentums. The fact that two electrons may be distinguishable from one another owing to their spins has important consequences for atomic and molecular structure.*

A fundamental particle is described by those parameters that determine the behavior of that particle. For many purposes, the electron can be described as a particle with the following properties:†

$$q_e = -e, \qquad m_e = (9.1091 \pm 0.0004) \times 10^{-31} \text{ kg}, \qquad S_e = \tfrac{1}{2}. \quad (2.248)$$

It may be possible to distinguish two electrons from each other by the direction of their spins, but other than this, the description of an electron requires so few parameters that it would appear that we cannot distinguish one electron from any other. Could we not add a little bit of mass or a small amount of charge to an electron? The answer is no. All negatively charged electrons are identical, and they can exist only with the properties, the given values of mass, charge, and spin, just described. This property of the discreteness of the fundamental particles still awaits explanation.

SUMMARY   A fundamental particle such as the electron is specified by those dynamical parameters that describe the behavior of the particle. These parameters include the mass and the electric charge of the particle and an intrinsic spin or angular momentum. It may not be possible to associate the rotation of a constituent element of a fundamental particle with this intrinsic spin.

Problem 2.81

This problem provides an analysis of Thomson's e/m experiment, which is illustrated in Figure 2.58.

(a) Consider an electron moving freely with the nonrelativistic speed v until it enters a region of constant electric force directed perpendicularly to the

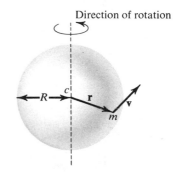

FIGURE 2.60   A spinning object possesses an intrinsic angular momentum in addition to the orbital angular momentum $\mathbf{R} \times \mathbf{P}$ due to the center-of-mass motion of the object as a whole.

FIGURE 2.61   The contribution $m\mathbf{r} \times \mathbf{v} = \mathbf{r} \times \mathbf{p}$ of the constituent particle of mass m to the angular momentum about c is not zero only if the radius R of the object (and hence every $\mathbf{r}$) is not zero.

---

* See G. Gamow, "The Exclusion Principle," *Scientific American*, *201*: 74, July 1959.
† One other parameter, called the lepton number, is required for a complete description of the behavior of electrons. However, we shall consider here only briefly the properties associated with this parameter. It is necessary to assign, in addition, another physical quantity, called the intrinsic magnetic moment. However, the intrinsic magnetic moment can be calculated in terms of the charge, mass, and spin.

initial velocity of the electron. Let E denote the magnitude of the electric force per unit charge. Show that the angle θ, through which the electron is deviated from its original path, is—assuming that θ is small compared with unity—$\theta = (eE/m)(l/2v^2)$. The distance along which the electron experiences the electric force is l.

(b) An electron with the same initial speed experiences the force from a field of constant magnetic induction **B** over the same distance. Show (using the same assumption) that the angle through which the electron is deflected is $\theta' = (evB/m)(l/2v^2)$.

(c) Thomson adjusted the magnetic induction so that $\theta = \theta'$; his measurements gave the following results:

$$l = 5 \text{ cm},$$

$$E = 1.5 \times 10^4 \text{ N/C},$$

$$\theta = \frac{8}{110} \text{ rad},$$

$$B = 5.5 \times 10^{-4} \text{ T}.$$

Calculate the ratio e/m and the initial speed v from these results.

Problem 2.82

This problem provides an analysis of the Millikan oil-drop experiment illustrated in Figure 2.59. Consider a spherical oil drop of radius a and density ρ.

(a) Show that the gravitational force experienced by the drop is $(4\pi/3)a^3\rho g$.

(b) Show that the buoyancy of the air of density ρ_a experienced by the drop reduces the effective gravitational force experienced by the drop to $(4\pi/3)a^3(\rho - \rho_a)g$.

(c) A spherical drop of radius a moving slowly with the speed v through air experiences a force of air resistance given by $F_a = 6\pi\eta av$, where η is a constant called the viscosity. Show that the oil drop achieves a constant speed of $v = 2\pi a^3(\rho - \rho_a)g/(9\pi\eta a)$.

(d) The oil drop attains a charge q from the ionizing effect of the X-rays. The plates are charged so that a constant electric force per unit charge of magnitude E is exerted upward on the drop. Show that the drop attains a terminal speed $v_e = [|q|E - (4\pi/3)a^3(\rho - \rho_a)g]/(6\pi\eta a)$.

(e) One set of Millikan's measurements gave the values

$$\rho = 8.96 \times 10^2 \text{ kg/m}^3, \qquad \rho_a = 1.2 \text{ kg/m}^3,$$
$$\eta = 1.84 \times 10^{-5} \text{ kg·m/sec}, \qquad E = 4.97 \times 10^5 \text{ N/C},$$
$$v = 4.43 \times 10^{-2} \text{ cm/sec}, \qquad v_e = 3.48 \times 10^{-2} \text{ cm/sec}.$$

Calculate the charge on the oil drop.

(g) Results typical of those obtained by Millikan for the charge on one oil drop at various times are, in units of 10^{-10} Fr (1 Fr = 1 Franklin = 3.336×10^{-10} C):

19.7,	19.7,	24.6,	24.6,	24.6,	24.6,	24.6,	24.6,	29.6,	29.6,
29.6,	34.5,	34.5,	34.5,	39.4,	39.5,	39.5,	44.4,	44.4,	44.4,
49.4,	49.4,	49.4,	53.9,	53.9,	53.9,	59.1,	59.1,	59.1,	59.1,
63.7,	63.7,	68.7,	68.7,	68.7,	78.3,	83.2.			

Mark these results on a scale from 0 to 85, and from this calculate a value for e in coulombs.

Problem 2.83

An electron is accelerated from rest through a potential charge of V volts.

(a) What is the kinetic energy of the electron, in electron volts?
(b) What is the kinetic energy of the electron, in joules?
(c) Plot the final speed v of the electron as a function of V. Mark on this graph that point where the nonrelativistic formula for kinetic energy gives the correct result to within 1% and also the point where the energy associated with the rest mass of the electron is 1% of the total energy.

Problem 2.84

What is the speed of an electron that is traveling with a kinetic energy equal to its rest energy?

Problem 2.85

In 1909, Bucherer measured the ratio q_e/m_v for fast electrons and obtained results that are equivalent to those shown in the table.

Show that these results are consistent with the relativity theory but not with newtonian mechanics. (Assume that q_e does not change with the speed of the electron.)

v/c	$-q_e/m_v$, C/kg
0.3173	1.662×10^{11}
0.3787	1.629
0.4281	1.591
0.5154	1.511
0.6870	1.284

Problem 2.86

The energy of one primary cosmic-ray particle has been measured to be 1.0×10^{20} eV.

(a) How many joules is this equivalent to?
(b) Find $(c - v)/c$ for an electron of this energy and calculate the Lorentz contraction factor $\sqrt{1 - (v^2/c^2)}$.

2.4.2 The photon*

Light possesses one of the attributes—namely, momentum—that we usually associate with a particle; the energy E traveling in one direction with the speed c has a momentum of magnitude E/c associated with its motion. However, the fact that we can assign a momentum to an energy-transport process does not endow this energy with corpuscular properties. In particular, light exhibits wave-like properties that are incompatible, in the framework of our usual ideas, with particle-like behavior. Within the framework of the concepts of quantum mechanics, on the other hand, the boundary between a wave-like nature and a particle-like nature is not distinct. In fact, light shows wave-like properties in some processes and particle-like properties in others. Also, as you will see later in your studies, particles exhibit wave-like properties in some circumstances.

---

* Kacser, Prentice-Hall, Sec. 7.4, p. 198; Sec. 7.5 p. 200.

The corpuscular theory of light was believed by scientists prior to the turn of the nineteenth century, because this theory was consistent with their knowledge of the properties of light, such as the fact that light propagates in straight lines. Shortly after this, however, wave-like properties of light were demonstrated conclusively and the wave character of light became accepted by all scientists; the corpuscular theory was discarded because of the belief in the incompatibility of a wave-like nature and a particle-like nature. Furthermore, those properties, such as that of straight-line propagation, that had supported the corpuscular theory were found to be explainable on the basis of a wave-like nature because of the smallness of the wavelengths ($\sim 10^{-6}$ m) of light.

The belief that light is purely wave-like in character has now been demolished. In a paper published in 1905,* Einstein considered the processes of emission and absorption of light. He noted that the experiments that supported the wave theory involved time-averaged light intensities, so these experiments were not necessarily in conflict with a corpuscular theory. His investigations led him to the conclusion that light exhibited corpuscular-like properties in some circumstances. Einstein inferred that light travels, is emitted, and is absorbed, in discrete packets. The energy E of the ultimate bundle associated with light of frequency ν is given by

$$E = h\nu, \qquad (2.249)$$

where h is Planck's constant. (The appearance of Planck's constant is characteristic of equations describing quantum behavior.) The basic corpuscle of light energy is called a photon. Thus, a photon is the ultimate unit of light, and a light beam of frequency ν consists of an integral number n of photons. The energy of such a beam is $nh\nu$, and if the light is traveling in one direction, the beam possesses a momentum of magnitude $nh\nu/c$.

The energy E of a photon is the fourth component of the 4-momentum $\mathscr{P}$, so the value of E depends on the reference frame relative to which the energy is measured. The permissible values of E range over all positive numbers. The same photon will possess different energies (or correspond to different frequencies) relative to different inertial systems. A monochromatic light beam is distinguished only by its frequency (which depends on the reference frame), its intensity (which depends on the number of photons), and its direction of polarization; hence, aside from the two basic choices of polarization direction, all photons are equivalent.

The photon is a particle with zero electric charge, zero rest mass, and an intrinsic spin of unity (in units of $h/2\pi$). The photon is denoted by the symbol γ.

$$\text{Photon } \gamma: \qquad m_\gamma = 0, \qquad q_\gamma = 0, \qquad S_\gamma = 1. \qquad (2.250)$$

The frequency $\nu = E/h$ associated with a photon extends over the range from zero to infinity, depending on the inertial reference frame in which the frequency is measured. Visible light consists of photons with frequencies of a restricted range in which light can be detected by the human eye, corresponding to wavelengths from 4,000 to 7,000 Å. Relative to another reference frame, such

---

* This was the same year in which Einstein, an unknown 26-year-old patent clerk, published his papers on the special theory of relativity and Brownian motion. A translation of part of the paper of interest here is reproduced with explanatory comments in Chapter 17 of M. H. Shamos, *Great Experiments in Physics*, Holt, Rinehart & Winston, New York, 1959. A translation of the complete paper is given in A. B. Arons and M. B. Peppard, "Einstein's Proposal of the Photon Concept—a Translation of the Annalen der Physik Paper of 1905," *American Journal of Physics, 33*: 367 (1965).

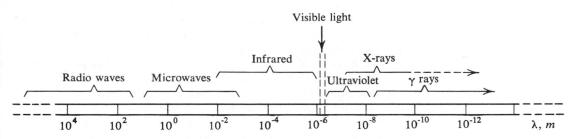

FIGURE 2.62   The electromagnetic spectrum.

light may appear in other forms, as shown in Figure 2.62. All these types of
radiation are called *electromagnetic radiation*, since they are emitted and
absorbed only by electrically charged particles.

The existence of electromagnetic waves was inferred theoretically, before
special relativity was known, by the Scottish physicist James Clerk Maxwell
(1831–1879), who formulated a classical theory of the interaction between
electric charged particles. This theory satisfies the principles of special relativity,
and indeed, it was an examination of the basis for the successes of this theory
that led Einstein to the theory of special relativity.

Maxwell's classical theory of the interactions between electric charged particles
admits solutions in which the field of the force of interaction travels through
space as waves with the speed c. The fact that radiation that travels with the
speed c is emitted by oscillations of electric charges was demonstrated experimentally in 1888 by Heinrich Hertz. Einstein's investigations showed that these
waves occur in bundles of energy, the elementary unit of energy of a wave of
frequency ν being $h\nu$.

Photons are emitted by a charged particle in processes in which energy-
momentum is exchanged by the charged particle with some other charged system.
For example, a fast electron that is slowed down in a collision with a nucleus
can emit photons (Figure 2.63). Photons that are emitted in this decelerating
process are called *bremsstrahlung*, the German word for "braking radiation."

Bremsstrahlung forms the main component of X-rays, a form of electromagnetic radiation that was discovered in 1895 by the German physicist
Wilhelm Konrad Röntgen (1845–1932). Röntgen's experiment is illustrated
in Figure 2.64. The radiation that caused the barium salts to fluoresce was in-
visible and, as Röntgen showed, could travel through opaque material. The
radiation was unlike anything known to that date and, since the symbol x is
often used to designate an unknown, Röntgen called them X-rays. Röntgen

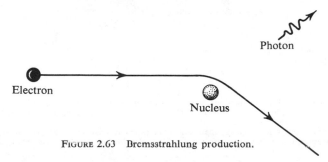

FIGURE 2.63   Bremsstrahlung production.

showed, by passing X-rays through a series of slits, that they traveled in straight lines, and since they were not deviated by electric forces, that they did not possess an electric charge. Shortly thereafter, it was shown that X-rays exhibited wave-like properties, with wavelengths about 10^{-10} m for X-rays produced by voltages of the order of 10^4 V. The waves were also found to be transverse. Hence, the X-rays were identified as a form of electromagnetic radiation.

Conclusive proof of the corpuscular nature of electromagnetic radiation was obtained in experiments performed in 1922 by the American physicist Arthur Holly Compton* (1892–1962). Compton measured the wavelengths of X-rays scattered by electrons in matter [Figure 2.65(a)] and found that one component of the scattered wave showed a change in wavelength that depended only on the angle of scatter θ. If X-rays exhibited wave-like properties only, the electrons would oscillate with the same frequency as the incident radiation and would reradiate the absorbed energy in all directions in the form of electromagnetic radiation of that frequency. The change in the frequency, or the shift in wavelength, could be explained only on the basis that the X-rays consisted of photons of incident energy $E = h\nu$ and momentum of magnitude $P = h\nu/c$ that underwent an elastic collision with the electrons [Figure 2.65(b)], and emerged at the angle θ with final energy $E' = h\nu'$ and momentum of magnitude $P' = h\nu'/c$ (Figure 2.66). The final energy, calculated from the conservation law of energy-momentum, is given by Equation (2.89):

$$\frac{1}{E'} - \frac{1}{E} = \frac{1}{m_e c^2}(1 - \cos\theta). \qquad (2.251)$$

The change $\lambda' - \lambda$ in the wavelength is determined by the equation $E = h\nu$ for the frequency of the photon and the relation $\lambda\nu = c$:

$$\lambda' - \lambda = \frac{h}{m_e c}(1 - \cos\theta). \qquad (2.252)$$

Equation (2.252) determined the correct wavelength change of the scattered X-rays in Compton's experiment. This change in the wavelength of the X-rays with scattering angle is known as the *Compton effect.*†

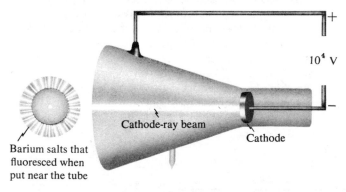

FIGURE 2.64 The cathode-ray tube experiment by which Röntgen discovered X-rays.

10^4 V

Cathode-ray beam

Cathode

Barium salts that fluoresced when put near the tube

---

* Compton described his investigations, and the attitudes of other American physicists toward his corpuscular interpretation of the results, in A. H. Compton, "The Scattering of X Rays as Particles," *American Journal of Physics, 29*: 817 (1961).
† Extracts from Compton's original papers, and some explanatory remarks, are given in Appendix 5 of M. H. Shamos, *Great Experiments in Physics*, Holt, Rinehart & Winston, New York, 1959.

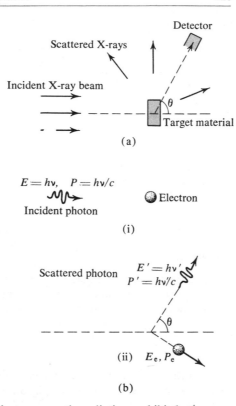

(a)

$E = h\nu, \quad P = h\nu/c$

Incident photon

Electron

(i)

Scattered photon

$E' = h\nu'$
$P' = h\nu'/c$

θ

(ii) E_e, P_e

(b)

FIGURE 2.65 The Compton effect. (a) Schematic diagram of the Compton experiment. (b) Schematic diagram of a Compton scattering: (i) before the collision; (ii) after the collision.

SUMMARY   Light and other electromagnetic radiations exhibit both wave-like and particle-like behaviors. These radiations consist of bundles of energy, the basic unit of radiation of frequency ν being $E = h\nu$. Particle-like behavior is demonstrated by the Compton effect.

Problem 2.87

(a) Calculate the energy of a photon of yellow-green light, $\lambda = 0.55 \, \mu$.
(b) The human eye can observe flashes of light in which only $\sim 5 \times 10^{-17}$ J of light energy enter the eye. About 10% of the photons reach the retina. Estimate the number of photons that enter the eye from such a flash and the number of photons that enter the retina.*

Problem 2.88

(a) A photon of frequency ν is observed in the inertial reference frame S. Find the frequency ν' of the photon relative to another inertial system S' moving with the speed V in the direction of motion of the photon.
(b) A photon of wavelength 0.55 μ (yellow-green light) is observed in the frame S. Calculate typical values of V in order that this photon be observed in the various parts of the electromagnetic spectrum (Figure 2.62) relative to S'.

---

* See, for example, Chapter 3 of H. T. Epstein, *Elementary Biophysics—Selected Topics*, Addison-Wesley, Reading, Mass., 1963.

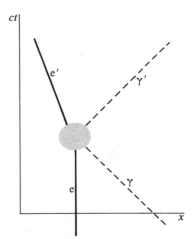

FIGURE 2.66 Space-time diagram for the Compton effect: a photon γ and an electron e collide and move off as the photon γ' and the electron e'.

Problem 2.89

Compton used X-rays of wavelength $\lambda = 0.711$ Å in his 1922 experiments. Calculate* the wavelengths of the X-rays scattered at

(a) 45°,
(b) 90°,
(c) 135°.

Problem 2.90

Compton observed that one component of the scattered X-rays had the same wavelength as the incident beam. Show that a 0.711-Å photon scattered from an electron bound to the atom, in which case the mass that undergoes recoil is the atomic mass $\sim 10^5 m_e$, does not undergo any appreciable change in wavelength.

Problem 2.91

Derive the Compton formula (2.252) from the energy-momentum conservation law applied to the photon and the electron. *Hint:* Use, with appropriate simplifications, the procedure employed in deriving Equation (2.89).

Problem 2.92

(a) Plot, as a function of θ, the wavelength of the scattered X-rays when a monochromatic X-ray beam of wavelength 0.213 Å is incident on a target.
(b) How would you modify this graph so that it applies to another wavelength for the incident beam?

Problem 2.93

(a) Show that 4-momentum is not conserved if a free electron emits a photon as shown in Figure 2.67.
(b) Assume that a fast electron with kinetic energy $T \gg m_e c^2$ decelerates in a collision with a heavy nucleus ($M \gg m_e$). Show that the electron can emit a photon whose energy may range from 0 to T and that energy momentum can be conserved. *Hint:* The nucleus can absorb the momentum of the electron without absorbing any appreciable amount of energy.
(c) Duane and Hunt showed experimentally in 1915 that the minimum wavelength λ_{min} of the X-rays emitted by electrons that had been accelerated through a potential difference V was given by

$$\lambda_{min} = \frac{1.24 \times 10^{-6}}{V} \text{ V} \cdot \text{m}.$$

Derive this relation.
(d) Calculate the minimum potential difference through which electrons must be accelerated in order to produce X-ray wavelengths of
  (i) 0.711 Å,
  (ii) 0.213 Å.

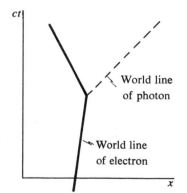

FIGURE 2.67 Space-time diagram of the emission of a photon by an electron.

ct

World line of photon

World line of electron

x

---

* A comparison of these theoretical values and the results of Compton's experiments is shown on p. 356 of M. H. Shamos, *op. cit.*

Problem 2.94

In his 1905 paper on photons, Einstein applied the theory to the photo-electric effect. This effect is the ejection of electrons from the surface of metals by the action of light (Figure 2.68). Suppose that work in the amount ϕ (called the work function) is required to extract an electron from the surface of a piece of a given metal. Suppose further that an electron in the metal absorbs a photon of frequency ν.

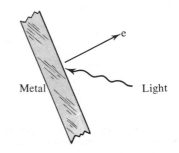

FIGURE 2.68   The photoelectric effect.

(a) Show that the maximum kinetic energy of the electron is given by $T_{max} = h\nu - \phi$.
(b) The threshold frequency ν_0 associated with a given metal is the minimum frequency of light that can result in the emission of photoelectrons. Find ν_0 in terms of ϕ.
(c) The work function for sodium is 2.46 eV. Find the threshold frequency for the production of photoelectrons from sodium.

2.4.3   The atomic nucleus*

The hypothesis of the atomic structure of matter provides a cornerstone for chemistry and the physics of matter. For example, the elementary kinetic theory of gases pictures a gas as consisting of small hard spheres flying, between elastic collisions with one another, through otherwise empty space. The word atom comes from the Greek word $\alpha\tau o\mu o\varsigma$ meaning indivisible—a property that describes the behavior of the particles relevant to stoichiometry in chemistry or the kinetic theory of gases in physics. Nevertheless, the structure of the atom is a legitimate field of investigation. The discovery of the basic property of this structure was published by Ernest Rutherford in 1911. To see how this discovery came about,† we shall discuss briefly how one can obtain information on an object the size of the atom, about 10^{-10} m across.

The structures of the objects of everyday experience are revealed to us by visible light, via wavelengths greater than 10^{-7} m. These wavelengths correspond to several thousand times the size of an atom and so we cannot use visible light, an extremely coarse probe for the atom (see Problem 2.97), to examine the structure of the atom. However, we can examine its structure by using a different type of beam, one of a more sensitive probe, and observing the consequent scattering. In the early decades of this century Rutherford and his colleagues had the use of such a probe at hand from α-particle sources. Due to previous investigations of Rutherford and others, it was known that α particles were doubly positive-charged helium atoms, and were thus expected to be of com-parable size, or smaller, than the atoms of heavier elements. Therefore α-particle sources provided the "light" with which Rutherford could "illuminate" atoms.

Just as we see an object by the visible light it scatters, we can examine the structure of atoms by the distribution of the α particles that atoms deflect (Figure 2.69). The incident beam should be sufficiently dilute that the α particles

* Kacser, Prentice-Hall, Sec. 7.1, p. 180.
† The story of this exciting era is given in A. Romer, *The Restless Atom*, Anchor Books, Garden City, N.Y., 1960, and in a chapter entitled "The Age of Rutherford," in E. Whittaker, *A History of the Theories of Aether and Electricity—The Modern Theories 1900–1926*, Thomas Nelson and Sons, N.Y., 1953.

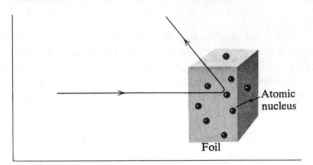

FIGURE 2.69 An exaggerated diagram of a typical event in the scattering of an α particle in the foil.

do not interact with one another and thus interfere with the scattering distribution from the atoms. The beam should also be uniform and parallel so that the relative number of α particles scattered in any direction provides a measure of the interactions that they experience and not a measure of nonuniformities in the beam. Finally, the α particles in the incident beam should be mono-energetic, for the deflection of a projectile depends on its incident energy as well as on the force it experiences.

The intensity of the incident beam is measured by

$$\text{Intensity} = \frac{\text{number of particles crossing a given plane}}{(\text{cross-sectional area of the beam})(\text{time of observation})}. \quad (2.253)$$

The number of particles scattered per unit time into a given solid angle $d\Omega$ is proportional to the incident intensity, and the proportionality constant is called the differential scattering cross section $d\sigma$:

$$d\sigma = \frac{\text{number of particles scattered into } d\Omega \text{ per unit time}}{\text{incident intensity}}. \quad (2.254)$$

The differential cross section $d\sigma$ has units of area, and the appropriate unit for our present study is the barn, given by

$$1 \text{ barn} = 1 \text{ b} = 10^{-28} \text{ m}^2. \quad (2.255)$$

The number of particles scattered into a solid angle $d\Omega$ is proportional to $d\Omega$ for sufficiently small solid angles. Thus the ratio $d\sigma/d\Omega$ is an appropriate quantity with which to measure scattering distributions.

For projectiles of a given energy, the variation of $d\sigma/d\Omega$ with angle provides a measure of the interaction between the projectile and the target that results in the scattering. For scattering by a fixed hard sphere, most projectiles that hit the sphere are scattered through relatively large angles, and in fact [see Problem 2.95(d)] $d\sigma/d\Omega$ is a constant for all angles (Figure 2.70). For scattering by an inverse-square force [see Problem 2.95(e)], called *Rutherford scattering*, there is some scattering in the backward directions, but most of the scattering is through small angles (Figure 2.71). The scattering by a square-well force [Problem 2.95(f)] is intermediate between these two cases; there is more scatter-

FIGURE 2.70 Differential scattering cross-section coefficient for scattering by a fixed smooth, hard sphere of radius a.

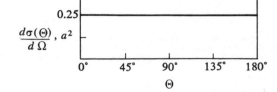

$\frac{d\sigma(\Theta)}{d\Omega}, a^2$

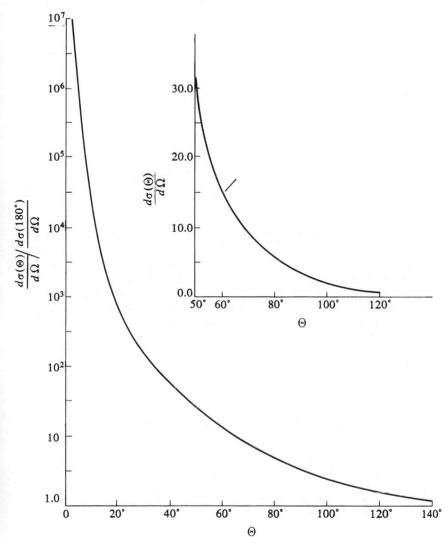

FIGURE 2.71 Rutherford scattering cross-section coefficient. Note the logarithmic vertical scale on the large graph.

ing through small angles than through large angles (Figure 2.72). Indeed, for a sufficiently high energy there is no scattering at all in the backward directions.

The measurements in Rutherford's laboratory of the scattering of α particles by atoms (Figure 2.73) were performed by Hans Geiger (1882–1947) and Ernest Marsden (1889–). Their results are shown in Figure 2.74 together with a plot of the differential cross section for Rutherford scattering by a fixed inverse-square force. The coincidence of their results with that theoretical curve shows that the scattering experienced by the charged α particles is due to a fixed coulomb (inverse-square) force to the closest distance of approach, about 3×10^{-15} m (as compared to the 10^{-10} m size of the atom). For example, deviations from that theoretical curve would have appeared at angles around 40° had the force been different at 6×10^{-15} m, as shown in Figures 2.75 and

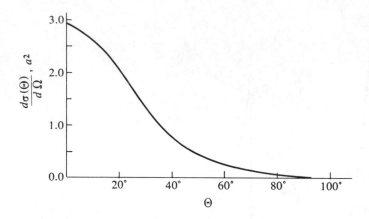

FIGURE 2.72 Differential scattering cross-section coefficient for scattering by a fixed square well of depth $V_0 = E$ and radius a.

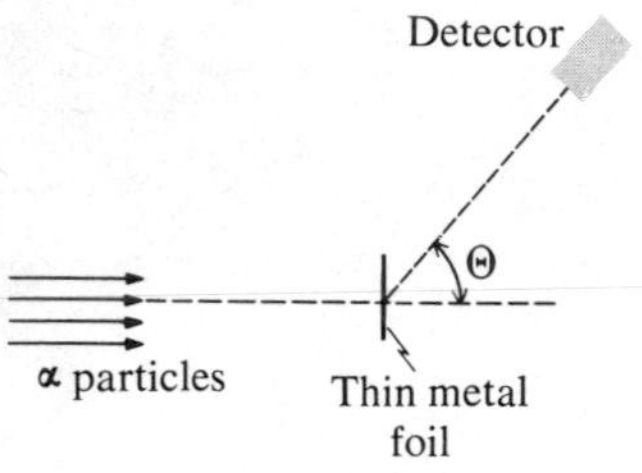

2.76. The scattering through these large angles also shows that the force center must have been massive compared to the α particle or the force center would have been knocked aside in the collision, and the α particle not deviated very much. Rutherford concluded from an analysis of the results of Geiger and Marsden that the atom contains a central core, the *nucleus*, which has a diameter of a few femtometers, and contains almost all the mass of the atom. The Rutherford model of the hydrogen atom is shown in Figure 2.77.

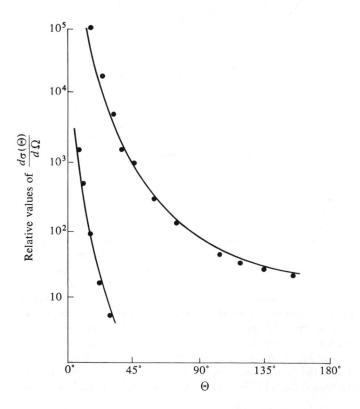

FIGURE 2.73 The scattering experiment of Geiger, Marsden, and Rutherford.

FIGURE 2.74 The circled dots are measurements of the number of scintillations counted at the given angle by Geiger and Marsden. The curves are plots of $A/\sin^4 \frac{1}{2}\Theta$.

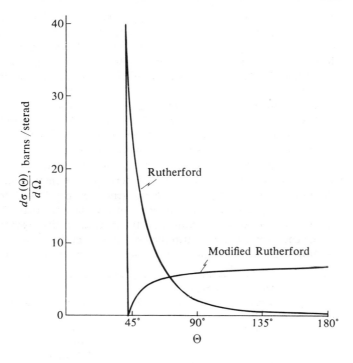

FIGURE 2.75 $d\sigma(\Theta)/d\Omega$ for scattering by a coulomb force to 6×10^{-15} m and a hard-sphere force at 6×10^{-15} m. The curve coincides with the Rutherford cross section for $\Theta < 38.9°$.

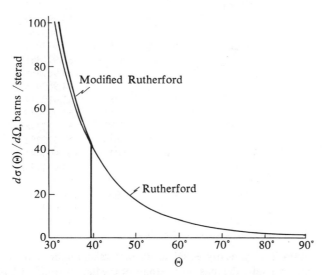

FIGURE 2.76 $d\sigma(\Theta)/d\Omega$ for scattering by a pure coulomb force to 6×10^{-15} m and no force acting over distances less than that. The curve differs very little from the Rutherford cross section for $\Theta < 30°$.

The nucleus, the core of the atom, is very small in size compared to the atom, even though it contains most of the atomic mass. (Nuclei range in size from 1 to 7×10^{-15} m, and atoms, from 1 to 3×10^{-10} m.) In this respect, an atom is similar to the solar system, with the sun corresponding to the nucleus and the planets in their orbits to the electrons about the nucleus. The size of the atom itself is determined by the volume in space swept out by the electrons.

The chemical properties of an atom are determined by the position of the

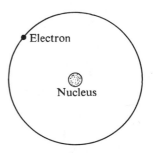

FIGURE 2.77 The Rutherford model of the hydrogen atom.

species of the atom in the periodic table. The various species of atoms can be listed in order according to their positions in the periodic table; such an ordering determines the *atomic number* Z of the atom. A brilliant young associate of Rutherford's, Henry Gwyn-Jeffries Moseley (1887–1915), showed, through an analysis of the characteristic X-rays emitted by atoms that had been bombarded by fast electrons, that the atomic number determines the electric charge on the nucleus of the atom. The electric charge of the nucleus of the atom of atomic number Z is Z positive fundamental charges.* Thus, a neutral atom consists of the nucleus of charge Ze and Z electrons, each of charge $-e$.

The masses of many nuclei have been determined by their inertial effects; ionized atoms are deflected by forces that act on the net electric charge of the atom, and the amount of deflection depends on the inertial mass experiencing the force.† Nuclear masses are given generally in terms of the mass of the corresponding neutral atom: The mass of the nucleus equals that of the atom less the mass of the Z electrons plus the mass equivalent (generally negligible) of the energy necessary to strip the electrons from the nucleus. Typical values of atomic masses are shown in Table 2.1.‡ The mass of any given neutral

TABLE 2.1 *Selected Atomic Masses*

Atomic element	Chemical symbol	Z	M, u	A = integer nearest M
Hydrogen	H	1	1.007825	1
			2.014102	2
			3.016050	3
Oxygen	O	8	15.994915	16
			16.999133	17
			17.999160	18
Sulfur	S	16	31.972074	32
			32.971461	33
			33.967865	34
			34.969031	35
Iron	Fe	26	53.939617	54
			55.934936	56
			56.935398	57
			57.933282	58
Silver	Ag	47	106.905094	107
			108.904756	109
Gold	Au	79	196.96654	197
Uranium	U	92	234.04090	234
			235.04391	235
			238.05077	238

atom is very nearly equal to an integral number A of atomic mass units or, alternatively, the mass of A atoms of the lightest form of hydrogen. The integer A is called the *mass number* of the given atomic species.

---

* See Chapter 1 of R. D. Evans, *The Atomic Nucleus*, McGraw-Hill, New York, 1955.
† Some methods of measuring atomic masses are described briefly in H. E. Duckworth, "Weighing Atoms," *American Journal of Physics*, 25: 503 (1957), and in more detail in Chapter 2 of R. D. Evans, *op. cit.*
‡ A recent atomic mass table is given in J. H. E. Mattauch, W. Thiele, and A. H. Wapstra, "1964 Atomic Mass Table," *Nuclear Physics*, 67: 1 (1965). The mass M of a nucleus is given there in terms of the mass excess, $M - A$, all in (unified) atomic mass units u.

A species of atom can be denoted by the chemical name, in abbreviation, with the mass number in the right superscript position.* The atomic number can be written in the left subscript position, even though giving both the atomic number and the chemical symbol is redundant.

A species of *atom* described by given values of Z and A is called a *nuclide*. Nuclides with the same atomic number Z are *isotopes* of one another, and nuclides with the same mass number A are *isobars*. For example, three isotopes of hydrogen have been observed, $_1H^1$, $_1H^2$, and $_1H^3$. The nuclei of these isotopes are simpler in structure than are other nuclei, so the study of these nuclei has been important to the science of nuclear physics. For this reason, these nuclei are given special names; the proton p is the nucleus of the $_1H^1$ nuclide, the deuteron d that of $_1H^2$, and the triton t that of $_1H^3$. One isotope of helium, $_2He^3$, has the same mass number as $_1H^3$, so $_1H^3$ and $_2He^3$ are isobars.

It is necessary to associate a spin, or intrinsic angular momentum, to each species of nuclei, as was the case for the electron. For example, the spin, in units of $h/2\pi$, of $_1H^2$ is 1, that of $_8O^{16}$ is 0, and that of $_{92}U^{235}$ is 7/2.

Since nuclei show greater effects of structure than do electrons, they can be described in part by parameters that will give information on their sizes and some properties of their shapes. The size of a given nucleus can be determined by scattering experiments involving different projectiles. The size of a nucleus measured by a scattering experiment depends upon the type of interaction that is effective in the scattering. For example, the scattering of electrons† by nuclei determines the distribution of electric charge in the nucleus and, hence, an effective radius for the charge of the nucleus. These experiments have shown that, for all but the lightest nuclei, the effective radius of the charge distribution is given by

$$R = r_0 \times A^{\frac{1}{3}}, \qquad r_0 = 1.2 \text{ fm}. \qquad (2.256)$$

Other types of scattering experiments result in a similar form, $r_0 \times A^{\frac{1}{3}}$, for the nuclear radius, the only difference lying in the measured value of r_0.

The volume of a nucleus is proportional to R^3; hence, scattering measurements show that the volume is proportional to A, the mass of the nucleus. Thus, measurements show that all nuclei except the lightest have approximately the same density. This result indicates that the interaction responsible for holding together the constituents of the nucleus differs in some respects from the coulomb electrostatic and the newtonian gravitational forces. Consider a system of particles bound together by an inverse-square force, as the latter forces are. The force holding any one of these particles to the system is the vector sum of the component forces exerted by all the other particles in the system; thus, the force of attraction that each constituent particle experiences increases as the number of particles of the system increases [Figure 2.78(a)]. Therefore, such forces would be expected to compress the heavier nuclei more and would result in a density that increases with an increasing number of particles. On the other hand, the interactions responsible for binding the constituent particles in a nucleus result in equal densities for all nuclei. This suggests that each particle

---

* The Commission for Symbols, Units and Nomenclature has recommended the format $^{16}_8O$. However, "because of typographical difficulties," the American Institute of Physics uses the notation $_8O^{16}$.

† Measurements of the size of the nucleus by electron scattering experiments are described in R. Hofstadter, "The Atomic Nucleus," *Scientific American*, *195*: 55, July 1956.

FIGURE 2.78 A comparison of the
effects of long- and short-range forces.
(a) A particle P experiencing an inverse-
square force undergoes a greater force
when a larger number of particles are
present. This results in an increased
density with an increased number of
particles. (b) A particle P experiencing
forces due to its nearest neighbors only
does not undergo a greater force when
more particles are added to the system.
Thus, the density does not depend on
the number of particles present.

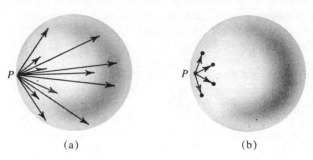

(a)   (b)

in the nucleus experiences interactions only from its nearest neighbors [Figure
2.78(b)], so that the nuclear interactions probably act only over a very short
range. This result is verified by other experimental evidence that we shall examine
shortly.

A study of chemical reactions provides information on molecular structure
and molecular interactions. Similarly, a study of reactions involving nuclei
yields information on nuclear structure and the interactions that nuclei experience. The first nuclear reactions studied were those that occur spontaneously
in the more massive nuclei. These emit particles in a process called *radio-
activity*.* The radiations emitted were found to be of three different types,
which Rutherford named α, β, and γ rays, respectively. Rutherford showed that
α particles were doubly ionized helium atoms or, in other words, helium nuclei;
β rays are electrons (and their positively charged counterparts called positrons
described in Section 2.4.6) that are emitted in the decay of certain radioactive
nuclei and that are created at the instant of decay: γ rays are high-energy radiations that travel with the speed of light and are identical in all respects, except
frequency or energy, with photons of visible light.

In 1919, Rutherford produced the first artificial transmutations, a change of
an atom from one chemical species to another induced by artificial means
(Figure 2.79). In this classic experiment, Rutherford bombarded nitrogen with α
particles to produce the reaction

$$_2\text{He}^4 + {}_7\text{N}^{14} \rightarrow {}_8\text{O}^{17} + {}_1\text{H}^1. \qquad (2.257)$$

This reaction can be written in an abbreviated form, which is standard† today:

$$\text{N}^{14}(\alpha, p)\text{O}^{17}. \qquad (2.258)$$

The particle emitted in this experiment, the nucleus of the lightest species of
hydrogen, was called the *proton* by Rutherford, after the Greek word for first.
Because the mass of a nucleus is approximately an integral multiple of the proton's
mass, it was believed that protons might be the basic mass unit of nuclei. The
proton p is a fundamental particle that is described in part by the following
dynamical variables:

$$m_\text{p} = (1.67252 \pm 0.00008) \times 10^{-27}\ \text{kg}, \qquad q_\text{p} = e, \qquad S_\text{p} = \tfrac{1}{2}. \qquad (2.259)$$

---

* Intense radioactive sources were prepared first by the remarkable Polish–French chemist and
physicist Marie Curie (1867–1934) and her husband, the French physicist Pierre Curie (1859–1906).
Their life has been described by their daughter Ève Curie, *Madame Curie*, Doubleday, Garden City,
N.Y., 1937.
† The general form is

Initial (incoming outgoing) final
nuclide (particle(s), particle(s)) nuclide.

The development of artificial transmutations opened the way to modern studies of nuclear structure and nuclear interactions; changes are induced in the structure of nuclei, and the resulting effects are investigated.* This procedure is followed in laboratories all over the world today, and these studies are rapidly increasing our understanding of nuclear structure and nuclear reactions.

Nuclear reactions can be described only within the framework of quantum mechanics, but the initial and final systems in these reactions are related by conservation laws, some of which we have already encountered. Nuclear reactions satisfy the *conservation law of electric charge*: the total electric charge after the reaction equals the total electric charge before. For example, in the β decay of $_{83}Bi^{210}$, the emission of an electron increases the atomic number Z by 1, so a nucleus of $_{84}Po^A$ results, where A is the mass number of the daughter nucleus. Thus, the conservation law of electric charge determines the atomic number Z of the daughter nucleus, although not its mass number A. Another conservation law that applies to nuclear reactions is the *conservation law of mass number*: the total mass number before the reaction equals the total mass number after the reaction. For example, the mass number of $_{83}Bi^{210}$ is not changed in its β decay, the emission of an electron of mass number 0, and so the daughter nucleus in this reaction is that of $_{84}Po^{210}$.

Conservation of 4-momentum also applies to nuclear reactions. This is illustrated by the fact that, in the spontaneous β decay of a nucleus at rest, the time-like component of the 4-momentum, the mass, of the parent nucleus is equal to or greater than the mass of the daughter nucleus plus that of an electron (see Section 2.4.5). For example, the mass of the neutral atom $_{83}Bi^{210}$ is 209.984121 u, while that of $_{84}Po^{210}$, which includes one more electron corresponding to the particle emitted from $_{83}Bi^{210}$, is 209.982876 u. The excess mass, 0.001245 u, appears in the form of energy of the amount 1.16 MeV.

SUMMARY   The nucleus of an atom can be described by its atomic number Z, which determines the electric charge on the nucleus, its mass number A, and its spin. The radius of the nucleus is $r_0 \times A^{\frac{1}{3}}$, where $r_0 \sim 1$ fm. The interactions between the constituents of nuclei have a very short range.

Some nuclear reactions, such as radioactive decay, are natural or spontaneous, while others are induced by artificial means. The total electric charge and the total mass number are conserved in all nuclear reactions.

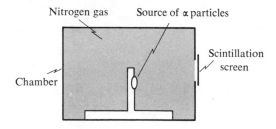

Nitrogen gas   Source of α particles

Scintillation screen

Chamber

FIGURE 2.79   Rutherford's experiment showing artificial transmutation. Scintillations were observed on the screen even when there was sufficient matter between the α particle source and the screen to stop the α particles. Rutherford interpreted the scintillations as being due to fast protons knocked out of nitrogen nuclei in the process $N^{14}(\alpha, p)O^{17}$.

Example 2.18

The radioactive decay of a nuclide obeys the following laws:

---

* See, for example, C. S. Cook, *Structure of Atomic Nuclei*, Momentum Book No. 8, D. Van Nostrand, Princeton, N.J., 1964.

(a) The probability of decay per unit time, λ, is the same for all atoms of the species.

(b) The probability of decay per unit time does not depend on the age of the atom.

Consider a system that contains $N(t)$ atoms of a particular species at time t. The probability of an atom decaying in a short time interval Δt is given by $-\Delta N(t)$, the number that decay in that time interval, divided by the number $N(t)$ present. Alternatively, it is equal to the probability of decay per unit time, λ, multiplied by the time interval Δt. Hence,

$$\frac{-\Delta N(t)}{N(t)} = \lambda \, \Delta t \tag{2.260}$$

or

$$\frac{dN(t)}{dt} = -\lambda N(t). \tag{2.261}$$

Therefore, if N_0 atoms are present at time $t = 0$, at time t the number of atoms that have not decayed is

$$N(t) = N_0 e^{-\lambda t}. \tag{2.262}$$

The *half-life* $T_{\frac{1}{2}}$ is that time interval in which the probability of a given atom decaying is equal to $1/2$:

$$\frac{1}{2} = \frac{N(T_{\frac{1}{2}})}{N_0} = e^{-\lambda T_{\frac{1}{2}}}, \tag{2.263}$$

so that

$$T_{\frac{1}{2}} = \frac{\ln 2}{\lambda} = \frac{0.693}{\lambda}. \tag{2.264}$$

The *mean life* τ is the average lifetime per atom. The sum of the lifetimes of all the atoms is given by

$$L = \sum_{\text{all } t} t \times (\text{number of atoms that decay in the time interval}$$
$$\text{between } t \text{ and } t + dt)$$

$$= \int_0^\infty t\,[\lambda N(t)\,dt] = \int_0^\infty t\lambda N_0 e^{-\lambda t}\,dt \tag{2.265}$$

$$= \frac{N_0}{\lambda}.$$

Therefore, the mean life τ is given by

$$\tau = \frac{L}{N_0} = \frac{1}{\lambda}. \tag{2.266}$$

The mean life and the half-life are related by

$$T_{\frac{1}{2}} = 0.693\tau \quad \text{or} \quad \tau = 1.44 T_{\frac{1}{2}}. \tag{2.267}$$

Example 2.19

Q. Two particles have 4-momenta $\not{p}_1$ and $\not{p}_2$, respectively, with the components $(E_1/c, \mathbf{p}_1)$ and $(E_2/c, \mathbf{p}_2)$ relative to an inertial system S. Calculate the energy, the magnitude of the momentum, and the speed of particle 2 relative to the rest frame of particle 1.

A. Let the components, relative to the rest system of particle 1, of $\not{p}_2$ be $(E_{21}/c, \mathbf{p}_{21})$. The corresponding components of $\not{p}_1$ are $(m_1 c, 0)$. We note that the scalar product of $(m_1 c, 0)$ with any 4-vector involves only the time-like component of that 4-vector. Hence,

$$\not{p}_1 \cdot \not{p}_2 = \left(\frac{E_{21}}{c}\right)(m_1 c) = m_1 E_{21}, \tag{2.268}$$

and

$$E_{21} = \frac{\not{p}_1 \cdot \not{p}_2}{m_1} = \frac{(E_1 E_2/c^2) - \mathbf{p}_1 \cdot \mathbf{p}_2}{m_1}. \tag{2.269}$$

The magnitude of the momentum $\mathbf{p}_{21}$ is given by the equation

$$m_2^2 c^2 = \left(\frac{E_{21}}{c}\right)^2 - |\mathbf{p}_{21}|^2 \tag{2.270}$$

to be

$$|\mathbf{p}_{21}| = \sqrt{(E_{21}/c)^2 - m_2^2 c^2} = \frac{\sqrt{(\not{p}_1 \cdot \not{p}_2)^2 - m_1^2 m_2^2 c^4}}{m_1 c}. \tag{2.271}$$

The speed v_{21} of particle 2 relative to particle 1 is given by

$$v_{21} = \frac{c^2 |\mathbf{p}_{21}|}{E_{21}} = \frac{c \sqrt{(\not{p}_1 \cdot \not{p}_2)^2 - m_1^2 m_2^2 c^4}}{\not{p}_1 \cdot \not{p}_2}. \tag{2.272}$$

Problem 2.95

(a) Show that the scattering of a uniform parallel beam by a central force results in a distribution that is symmetrical about the line, through the force center, parallel to the direction of the incident beam.

(b) Show that the solid angle (Figure 2.80) into which are scattered particles with impact parameters between S and $S + dS$ (Figure 2.81) is given by

$$d\Omega = |2\pi \sin \Theta \, d\Theta|.$$

(c) Show that

$$\frac{d\sigma}{d\Omega} = -\frac{S \, dS}{\sin \Theta \, d\Theta}.$$

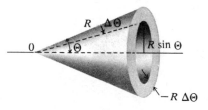

FIGURE 2.80 Solid angle $\Delta\Omega$.

Hint: Equate the number of incident particles with impact parameters between S and $S + dS$ to the number scattered through angles between Θ and $\Theta + d\Theta$.

(d) Show that, for scattering by a hard sphere of radius a (see Problem 2.3),

$$\frac{d\sigma}{d\Omega} = \frac{a^2}{4}.$$

(e) Show that the distribution for scattering of particles of incident energy E by a fixed inverse-square force $\mathbf{F} = k\hat{r}/r^2$ (see Problem 2.5) is given by

$$\frac{d\sigma}{d\Omega} = \frac{1}{4}\left(\frac{k}{2E}\right)^2 \frac{1}{\sin^4(\Theta/2)}.$$

(f) Show that the distribution for scattering of particles of energy $E = V_0$ incident on the fixed square well of Problem 2.4 is given by

$$\frac{d\sigma}{d\Omega} = \frac{a^2[\sqrt{2}\cos(\Theta/2) - 1][\sqrt{2} - \cos(\Theta/2)]}{2\cos(\Theta/2)[3 - 2^{3/2}\cos(\Theta/2)]^2}.$$

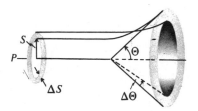

FIGURE 2.81 Definition of ΔS and $\Delta\Theta$. Note that $\Delta\Theta$ is negative.

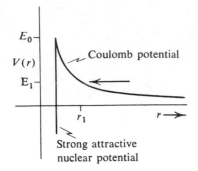

FIGURE 2.82 An α particle with energy E_1 encountering the potential barrier shown can penetrate only to the distance r_1 from the center. An α particle must have a minimum energy of E_0 to escape from the central region. Both these results are a consequence of newtonian mechanics.

Problem 2.96

(a) The uranium nucleus $_{92}U^{238}$ emits α particles with an energy of 4.2 MeV. Assume that the potential energy of interaction between the α particle and the residual nucleus is a repulsive coulomb potential up to the separation distance r_1 and a strong attractive potential for $r < r_1$, similar to that shown in Figure 2.82. Find the minimum value, according to newtonian mechanics, of r_1.

(b) Experiments show that α particles incident on the residual nucleus experience pure coulomb scattering up to separation distances of 9×10^{-15} m. Find the value, according to newtonian mechanics, of the minimum energy E_0 with which an α particle can escape from $_{92}U^{238}$ on the basis of this information.

(c) Mark the values of E_0, E_1, and r_1 on Figure 2.82 and explain why the results of (a) and (b) show that newtonian mechanics does not describe correctly the interaction of the α particle and the residual nucleus.

Problem 2.97

We see an object by light waves in a manner analogous to the way in which a canoeist detects the presence of an island by a comparison of the wave patterns on the open water and on the water on the lee side of the island. Find the sizes of the islands that, for water waves of 4-ft distance from crest to crest, correspond to an atom of diameter 10^{-8} cm and a nucleus of diameter 10^{-12} cm for light waves of 5×10^{-7} m distance from "crest" to "crest." Would you expect that a canoeist could detect the islands from such wave patterns?

Problem 2.98

(a) Estimate the density of nuclear matter.

(b) Estimate the weight, in tons, of 1 cm³ of nuclear matter.

(c) Calculate the ratio of the mass of the electrons to that of the nucleus for the nuclides listed in Table 2.1.

Problem 2.99

(a) Calculate the energy that is released in the reaction $Li^7(p, \alpha)He^4$.
    Unified atomic mass units:

$$_3Li^7 - 7.016004; \quad _1H^1 - 1.007825; \quad _2He^4 - 4.002603.$$

(b) Calculate the energy that is released in the reaction $Li^6(d, \alpha)He^4$.
    Unified atomic mass units:

$$_3Li^6 - 6.015125; \quad _1H^2 - 2.014102.$$

(c) The energies calculated above were determined from mass measurements and from the relation $E = Mc^2$. Compare these with the measured energies:*

$$Li^7(p, \alpha)He^4 - 17.28 \pm 0.03 \text{ MeV,}$$

$$Li^6(d, \alpha)He^4 - 22.20 \pm 0.04 \text{ MeV.}$$

---

* N. M. Smith, Jr., *Physical Review, 56*: 548 (1939).

Can these results be explained within the framework of newtonian mechanics? State the reasons for your answer.

(d) Calculate a value for c from values obtained by mass and energy measurements in (a), (b), and (c). Note that you can measure the speed of light in experiments in which light plays no role.

Problem 2.100

(a) The nucleus of an atom of total mass M is in an excited state of energy ΔE above its ground state, the state of lowest energy. The nucleus decays from rest by γ emission to its ground state. Show that the frequency of the emitted photon is given by

$$\nu = \frac{\Delta E}{h}\left(1 - \frac{1}{2}\frac{\Delta E}{Mc^2}\right),$$

if the recoil speed V of the nucleus satisfies the relation $V/c \ll 1$.

(b) Calculate the frequency of the photon emitted in the decay of an excited state of $_6C^{12}$ that lies 4.43 MeV above the ground state of $_6C^{12}$.

$$M_{_6C^{12}} = 12.00 \text{ u}.$$

Problem 2.101

Nuclide 1 undergoes α particle decay into nuclide 2.

(a) Show that $Z_1 = Z_2 + 2$ and $A_1 + 4 = A_2$.

(b) Show that the decay cannot occur spontaneously unless $M_1 \geq M_2 + M_{_2He^4}$. Explain why the last term is the mass of the neutral $_2He^4$ atom and not that of the α particle.

Problem 2.102

(a) Show that the following α particle decays satisfy the conservation laws of electric charge and mass number, and are consistent with that of 4-momentum.

$$_{94}Pu^{240} \xrightarrow[\alpha \text{ decay}]{} {}_{92}U^{236} \xrightarrow[\alpha \text{ decay}]{} {}_{90}Th^{232} \xrightarrow[\alpha \text{ decay}]{} {}_{88}Ra^{228}$$

Unified atomic mass units:

$_2He^4$ — 4.002603; $_{94}Pu^{240}$ — 240.05388; $_{92}U^{236}$ — 236.04564;

$_{90}Th^{232}$ — 232.03812; $_{88}Ra^{228}$ — 228.03114.

(b) Calculate the energy, in megaelectron volts, released in the α decays of (a).

Problem 2.103

There is reason to believe that when the earth came into existence two isotopes of uranium, U^{235} and U^{238}, were formed in approximately equal quantities. However, today, U^{235} constitutes only 0.7% of natural uranium whereas U^{238} constitutes the rest. The half-life of U^{235} is 0.71×10^9 yr and that of U^{238} is 4.50×10^9 yr. Calculate the age of the earth.

Problem 2.104

The half-life of $_{92}U^{238}$ is 4.5×10^9 yr, and that of $_{92}U^{234}$ is 2.5×10^5 yr.

(a) Calculate the mean lives of these nuclides.
(b) The solar system is approximately 5×10^9 years old. If a mass M of each of these nuclides was present at the birth of the solar system, what percentage of that original mass has not decayed?

Problem 2.105

A radioactive nuclide A decays with a mean life of $1/\lambda_A$ into a nuclide B that is also radioactive, and B decays with a mean life of $1/\lambda_B$ into a nuclide C.

(a) Show that, if $N_A = N_A(t)$ is the number of atoms of A present at time t,

$$\frac{dN_B}{dt} = N_A\lambda_A - N_B\lambda_B.$$

(b) Suppose that $N_A(0) = N_A^0$ and $N_B(0) = 0$. Show that

$$N_A = N_A^0 e^{-\lambda_A t}.$$

Show that

$$N_B = N_A^0(xe^{-\lambda_A t} + ye^{-\lambda_B t})$$

only if

$$-x\lambda_A - \lambda_A + x\lambda_B = 0$$

and

$$x + y = 0.$$

Hence, show that

$$N_B = N_A^0 \frac{\lambda_A}{\lambda_B - \lambda_A}(e^{-\lambda_A t} - e^{-\lambda_B t}).$$

(c) The *activity* of a radioactive nuclide is equal to the number of decays of that nuclide per unit time. Show that, under the conditions of (b), the activity of nuclide B is $N_B\lambda_B$. Explain why the activity is not equal to dN_B/dt.
(d) Nuclide $_{83}Bi^{210}$ decays by β emission with a half-life of 5.0 days into $_{84}Po^{210}$; $_{84}Po^{210}$ decays by α emission with a half-life of 138 days into $_{82}Pb^{206}$. A pure sample of $_{83}Bi^{210}$ exists at time $t = 0$. Calculate the following:
  (i) the time at which only 1% of the atoms present are those of $_{83}Bi^{210}$;
  (ii) the time at which one-half of the atoms present are $_{84}Po^{210}$;
  (iii) the time at which 99% of the atoms are those of $_{82}Pb^{206}$;
  (iv) the activity of $_{84}Po^{210}$ and $_{83}Bi^{210}$ at the end of two days, seven days, 100 days, and 500 days. *Hint:* Use the fact that $T_B \gg T_A$.

Problem 2.106

(a) A 20-GeV electron is incident on a proton at rest. Calculate the energy and momentum of this system and the speed of the center of momentum of this system.
(b) Calculate the energy available in the center-of-momentum reference frame.

Problem 2.107

Two electrons, each of energy 500 MeV, move toward each other from opposite directions and collide.

(a) Calculate the speed of one of the electrons relative to the other.

(b) Calculate the energy required by an electron, in collision with a stationary electron, in order that the energy available in the center-of-momentum reference frame is 1,000 MeV.

Problem 2.108

A proton of energy E is incident on a stationary proton.

(a) Calculate, and plot on a graph, the energy available in the center-of-momentum reference frame for $E > 1$ GeV.
(b) Mark the points on the graph for the following energies:
　(i) 3 GeV (produced by the Brookhaven Cosmotron),
　(ii) 6 GeV (produced by the Berkeley Bevatron),
　(iii) 30 GeV (produced by accelerators at Brookhaven and at CERN, the European Center for Nuclear Research).

Problem 2.109

The first stage of the 2-mi accelerator at Stanford is designed to produce a beam of 20-GeV electrons. (Ultimately, it will yield electrons with energies over 40 GeV.)

(a) Calculate the Lorentz contraction factor for the 20-GeV electrons. How long is the accelerator relative to the rest frame of the 20-GeV electrons?
(b) Assume that these electrons experience a constant relative force down the length of the accelerator. Calculate the force each electron experiences.
(c) Use newtonian mechanics to calculate the distance through which an electron would have to experience the force of (b) in order to reach the speed c, if the newtonian description were valid.
(d) If newtonian mechanics described the behavior of fast electrons, through what distance would the force of (b) have to act on the electrons in order that they gain energy in the amount of 20 GeV?

Problem 2.110

The 1.0×10^{20}-eV cosmic-ray particle whose effects were recently observed was probably a proton.

(a) Calculate the ratio $(c - v)/c$ for such a particle.
(b) Calculate the Lorentz contraction factor for such a proton. *Hint:* Show that $\sqrt{1 - (v^2/c^2)} = \sqrt{2[(c - v)/c]}$ for such a fast particle.
(c) Calculate the time, relative to the proton's rest frame, required for our Galaxy to move past the proton. (The width of our Galaxy is about 3×10^4 parsec, where 1 parsec $= 3 \times 10^{13}$ km.)

2.4.4　The neutron and the nucleon

The facts that the mass of a nucleus is approximately an integral multiple A of the mass of the proton and that the charge of a nucleus is a smaller integral multiple Z of the charge of the proton suggest that a nucleus of mass number A consists of A particles, Z protons, and $A - Z$ neutral particles each with the mass of the proton. This neutral counterpart to the proton was discovered in 1932 by the English physicist James Chadwick (1891–) in a study of nuclear reactions induced in boron by α particles,* although the existence of this particle had been conjectured by Rutherford before. This neutral particle is called the *neutron*.†

The neutron n is similar to the proton but has zero charge. The neutron has a spin of $\frac{1}{2}(h/2\pi)$, as does the proton, and is described by the following properties:

$$m_n = (1.67482 \pm 0.00008) \times 10^{-27} \text{ kg}, \qquad q_n = 0, \qquad S_n = \tfrac{1}{2}. \qquad (2.273)$$

Unlike the proton, however, the *free* neutron, a neutron outside the nucleus, is not stable but undergoes β decay into a proton with a half-life of 12 min.

The discovery of the neutron immediately provided a firm foundation for the explanation of nuclear structure. A nucleus of mass number A and atomic number Z consists of Z protons and $N = A - Z$ neutrons. The neutron number is N, and nuclides with equal neutron numbers are called isotones. The energy B that is equivalent to the difference between the mass of Z protons and N neutrons and the mass of the bare nucleus M',

$$\frac{B}{c^2} = ZM_p + NM_n - M' = \Delta M, \qquad (2.274)$$

represents the energy that would be required to decompose the nucleus into Z protons and N neutrons. This energy is called the *binding energy* of the nucleus. The relation $B = \Delta M c^2$ between the binding energy B and the mass difference ΔM has been verified in many experiments and provides strong corroboration for the relativistic equation $E = Mc^2$.

The binding energy per constituent particle, proton or neutron, for some of the stable nuclei is shown in Figure 2.83. It can be seen from there that the binding energy per particle is fairly constant over the range of the more massive nuclei. We would not expect this result if each particle experienced a force due to all the other particles in the system (see Figure 2.78), so the constancy of the binding energy per particle suggests again that the interparticle forces have a very short range and are exerted between near neighbors only.

The binding energy per particle is not constant, however, and this fact can be used to explain the energy release in two types of nuclear processes that have important and well-known practical applications. In one of these, a heavy unstable nucleus splits into two approximately equal-sized parts with the release of a large amount of energy in a process called *fission*.‡ Energy is released, since the total binding energy of the unstable nucleus is less than that of the two product nuclei, as can be seen from Figure 2.83. Fission provides the energy in nuclear reactors; it is the process involved in the explosion of small nuclear weapons, and is the trigger for the large weapons.

In the opposite process, the lighter nuclei combine to form a larger nucleus with the release of a large amount of energy. This process is called *fusion*.

---

* Chadwick's discovery is described in Section 9.3a of D. L. Livesey, *Atomic and Nuclear Physics*, Blaisdell, Waltham, Mass., 1966. An extract from Chadwick's account of his discovery, with explanatory comments, is given in *Great Experiments in Physics*, M. H. Shamos (Ed.), Holt, Rinehart & Winston, New York, 1959.

† Speculations on the existence of, and the unsuccessful searches for, the neutron prior to Chadwick's discovery are described in N. Feather, "A History of Neutrons and Nuclei, Part 2," *Contemporary Physics*, *1*: 257 (1960).

‡ Personal recollections of the discovery of the fission process are given in O. Hahn, "The Discovery of Fission," *Scientific American*, *198*: 76, February 1958. See also E. B. Sparberg, "A Study of the Discovery of Fission," *American Journal of Physics*, *32*: 2 (1964).

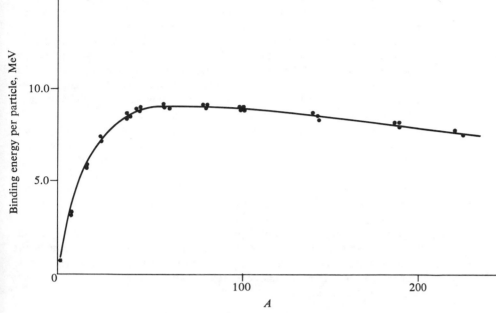

FIGURE 2.83   Typical values of the binding energy per particle.

The energy release results from the fact that the total binding energy of the light component nuclei is less than that of the composite nucleus, as can be seen from the extreme left-hand side of Figure 2.83. The fusion process is the main source of stellar energy.* The large nuclear weapons obtain their energy from the fusion process. Also, the hope exists that this process can be used to provide electrical power from the heavy hydrogen, deuterium, present in the oceans.†

Nuclear reactions can be used also to produce elements that are not sufficiently stable to be found naturally.‡

Neutrons are useful as probes for examining the structure of nuclei, since, because they are neutral, they do not interact with the electric charge of the nuclei but experience only the nuclear part of the forces. On the other hand, their lack of electric charge does lead to difficulties in their use, since, for example, they cannot be observed directly in cloud chambers or in bubble chambers, nor can they be accelerated by electric forces.§ Neutrons are useful also in studies of crystal structure.¶

The neutron and the proton have more features in common than mass values and the fact that they are the constituents of nuclei. Indeed, the neutron and the proton can be considered as different aspects of one and the same particle.

---

* See Chapter 9 of F. Hoyle, *Astronomy*, Doubleday, Garden City, N.Y., 1962.
† See R. F. Post, "Fusion Power," *Scientific American, 197*: 73, December 1957.
‡ See G. T. Seaborg and A. R. Fritsch, "The Synthetic Elements: III," *Scientific American, 208*: 68, April 1963, and G. T. Seaborg, "The Man-Made Chemical Elements Beyond Uranium," *Physics Today, 15*: 19, August 1962.
§ How some of these difficulties are overcome is described in L. Cranberg, "Fast-Neutron Spectroscopy," *Scientific American, 210*: 79, March 1964.
¶ For a description of this use and for other information on the neutron, see D. J. Hughes, *The Neutron Story*, Anchor Books, Garden City, N.Y., 1959.

This particle is called the *nucleon*; its two forms, the neutron and the proton, are distinguished only by their electric charge. The small difference in the masses of the neutron and proton is believed to represent the mass equivalent of the difference in the energies required to produce their electromagnetic properties.

That the neutron and proton are different aspects of the same particle is supported by a study of the interactions between neutrons and protons, between protons and protons, and between neutrons and neutrons. These studies show that that part of the interaction that is not electromagnetic in origin is the same for both neutrons and protons. These interactions are studied in scattering experiments, such as those described in Section 2.4.3. However, the interactions between nucleons cannot be described in terms of newtonian mechanics; in particular, the vector $m\mathbf{a}$ for a nucleon cannot be measured in a nucleon-nucleon collision. (Hence, it is not correct to speak of nucleon-nucleon forces, although the word "forces" is used often in such phrases to denote "interactions.") The concepts of quantum mechanics must be used to describe the nucleon-nucleon interactions and their effects.

The results of the interactions between two particles of similar masses, such as the proton and the neutron, are most conveniently described relative to the center of momentum system or, in nonrelativistic cases, to the center-of-mass reference frame. For example, Newton's equation for the relative position vector $\mathbf{r}(t)$ of two particles is

$$\mu \frac{d^2\mathbf{r}}{dt^2} = \mathbf{F}(\mathbf{r}), \qquad (2.275)$$

where $\mu = m_1 m_2/(m_1 + m_2)$ is the reduced mass and $\mathbf{F}$ is the force experienced by one particle due to the presence of the other. The differential cross section in the center-of-mass frame is determined by the force between the particles and the relative incident energy only. The cross section in the laboratory frame, on the other hand, appears distorted into the forward direction by the center-of-mass motion (see Problem 2.111).

The measured values of some of the cross sections for neutron-proton and

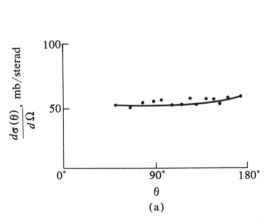

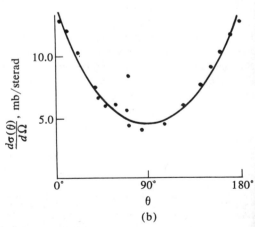

(a)   (b)

FIGURE 2.84 Center-of-mass differential scattering cross-section coefficient $d\sigma(\theta)/d\Omega$ for neutron-proton scattering. (a) Kinetic energy of neutron relative to the laboratory frame = 14.1 MeV. (b) Kinetic energy of the neutron relative to the laboratory frame = 90 MeV. [Plotted from data taken from W. N. Hess, *Reviews of Modern Physics*, *30*: 368 (1958).]

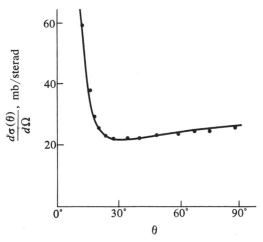

FIGURE 2.85 Center-of-mass $d\sigma(\theta)/d\Omega$ for proton-proton scattering for an incident proton energy of 19.8 MeV. [Plotted from data taken from W. N. Hess, *Reviews of Modern Physics, 30*: 368 (1958).]

proton-proton scattering experiments are shown in Figures 2.84 and 2.85. Now, although the results of these experiments cannot be explained on the basis of newtonian mechanics, we can use that mechanics to gain some insight into the interactions or forces involved. The rapid increase in the proton-proton scattering cross section (Figure 2.85) that appears at small angles can be interpreted as Rutherford scattering; as we might expect, this increase can be explained on the basis of the electric repulsion of the protons. Aside from this, though, we see that the differential cross sections are approximately constant and are similar to the case in which one hard sphere is incident upon another (Figure 2.86). Indeed, we can use the measured cross sections to estimate the radius of the corresponding hard sphere. For proton-proton scattering at 19.8 MeV, we have r^2 approximately equal to 20 mb, so that

$$r \approx \sqrt{20 \text{ mb}} = \sqrt{20 \times 10^{-3} \times 10^{-24} \text{ cm}^2} = 1.4 \text{ fm}. \qquad (2.276)$$

Notice that this value of r compares favorably with the value of r_0 in Equation (2.256), which was measured in another way.

The shape of the neutron-proton differential cross section [Figure 2.84(b)] for an incident neutron energy of 90 MeV can be accounted for in the following manner: The differential cross sections for neutron-proton scattering at low energies suggest a hard-sphere interaction, so we must consider an interaction that approximates a hard-sphere interaction for low energies. This is true for the square-well interaction described in Problem 2.4. The square-well interaction is described by the potential energy function $V(r)$ (Figure 2.87), given by

$$V(r) = -V_0, \qquad r < a$$
$$= 0, \qquad r > a, \qquad (2.277)$$

where a is the range of the interaction. Figure 2.88 shows the differential scattering cross section for square-well scattering for the case in which the relative energy E of the two particles is equal to V_0 [see Problem 2.95(f)].

This differential cross section decreases with increasing angle, as does the neutron-proton differential cross section [Figure 2.84(b)] for $\theta < 90°$. However, the latter increases with scattering angle for $\theta > 90°$ and, in fact, appears symmetrical about $90°$. The increase in $\sigma(\theta)$ with θ is the result that would be obtained if we erroneously measured, instead of the recoil angle θ, the angle ϕ at

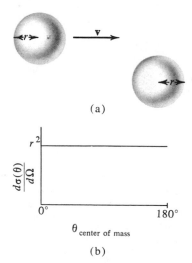

FIGURE 2.86 Hard-sphere scattering as described by newtonian mechanics. (a) One hard sphere incident on an identical sphere. (b) Differential scattering cross-section coefficient [see Problems 2.95(d) and 2.3(c)].

FIGURE 2.87 The square-well potential approximates a strong force at the
surface of the sphere $r = a$. The
potential-energy function $V(r)$ corresponds to a large force around a.

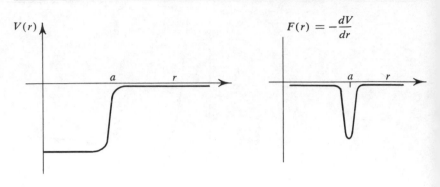

FIGURE 2.88 Differential scattering
cross section for square-well interaction
and relative energy $E = V_0$.

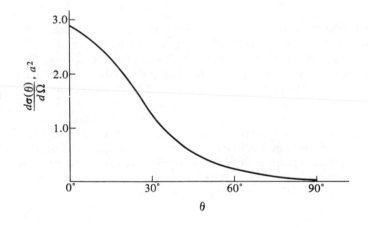

FIGURE 2.89 Differential scattering
cross section as a function of the recoil
angle ϕ. (a) θ is the angle of scatter of
the projectile and ϕ is the angle of
recoil of the target: $\theta + \phi = 180°$: (i)
before the collision; (ii) after the collision. (b) $d\sigma(\theta)/d\Omega$ plotted as a
function of ϕ.

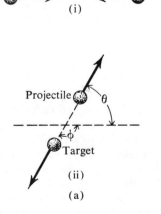

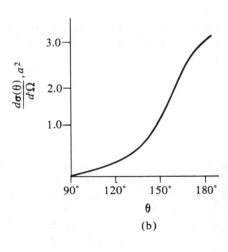

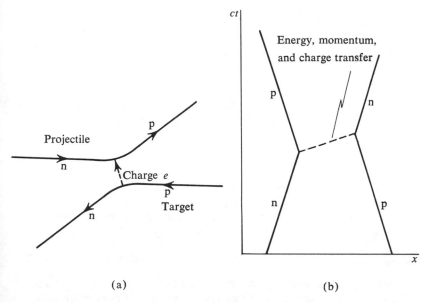

FIGURE 2.90 Diagrams of an interaction with a charge transfer. (a) The proton transfers its charge to the neutron, and the target particle becomes a neutron. (b) Space-time diagram.

(a)   (b)

which the target particle moved off (Figure 2.89). Indeed, suppose that in the collision the electric charge of the proton was transferred to the neutron along with the energy and momentum transfer (Figure 2.90). In this event, the proton and the neutron interchange roles, and by measuring the angle of scatter of the neutral particle, we actually measure the angle ϕ of the recoil particle instead of the angle θ of the scattered projectile. The measured differential cross section would appear, then, as in Figure 2.89(b). This increases with angle for values of $\theta > 90°$ and, by itself, does not show the shape of the 90-MeV neutron-proton differential cross section 2.84(b). However, the cross sections 2.88 and 2.89 can be combined if the charge is exchanged in only one-half the collisions; in this case, the differential cross section appears as shown in Figure 2.91 and is very similar to that shown in Figure 2.84(b).

The symmetry about 90° of these two differential cross sections indicates that charge is exchanged in about one-half the collisions; this suggestion is verified by the quantum-mechanical description of the interaction. The neutron-

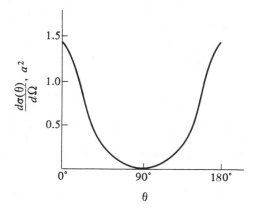

FIGURE 2.91 Differential scattering cross section for square-well scattering if charge is exchanged in one-half the collisions.

proton interaction is said to contain an *exchange force*.* We shall·return shortly to this property of the nucleon-nucleon interaction.

The similarity in the curves 2.84(b) and 2.91 suggests that the nucleon-nucleon interaction has a range a, given approximately by

$$a \approx \sqrt{10 \text{ mb}} = 1 \text{ fm}, \qquad (2.278)$$

in agreement with our previous result. Similarly, the depth of the square well V_0 is approximately equal in that case to the incident relative energy,

$$\tfrac{1}{2}(90 \text{ MeV}) = 45 \text{ MeV}.$$

SUMMARY   The neutral counterpart of the proton is the neutron. A free neutron is not stable, but undergoes β decay.

Protons and neutrons are different aspects of the particle called the nucleon. Nucleons interact strongly over distances of the order of 1 fm. The interaction involves an exchange force.

Problem 2.111

A particle of mass m is incident on a particle of identical mass. The incident particle moves with speed $v \ll c$, and the target particle is at rest relative to the laboratory frame of reference before the force of interaction takes effect. The incident particle is scattered through the angle ϑ in the laboratory frame and through a corresponding angle Θ in the center-of-mass frame.

(a) Show that before the collision the speed of the incident particle was $\tfrac{1}{2}v$ relative to the center-of-mass frame.
(b) Use the conservation law of energy to show that after the collision the speed of the particle is $\tfrac{1}{2}v$ relative to the center-of-mass frame.
(c) Apply the galilean transformation law $\mathbf{v}_{\text{final}} = \mathbf{V}_{\text{CoM}} + \mathbf{v}'_{\text{final}}$ to show that $\vartheta = \Theta/2$.
(d) Show that $d\sigma(\Theta) = d\sigma'(\vartheta)$ (Figure 2.92) and hence that

$$\frac{d\sigma'(\vartheta)}{d\Omega'} = 4 \cos \vartheta \left[\frac{d\sigma(\Theta)}{d\Omega}\right]_{\Theta = 2\vartheta}$$

(e) Find $d\sigma'/d\Omega'$ for the scattering of one hard sphere of radius r by another identical hard sphere.

Problem 2.112

The electrostatic energy between two charged spheres of radii r_1 and r_2, when they are in contact, is given by

$$E = \frac{1}{4\pi\varepsilon_0} \frac{Q_1 Q_2}{r_1 + r_2},$$

where Q_1 and Q_2 are the electric charges on the spheres. Suppose that a nucleus of $_{92}U^{238}$ fissions into two equal parts. Estimate the electrostatic energy of repulsion, in megaelectron volts, when the two parts have just separated.

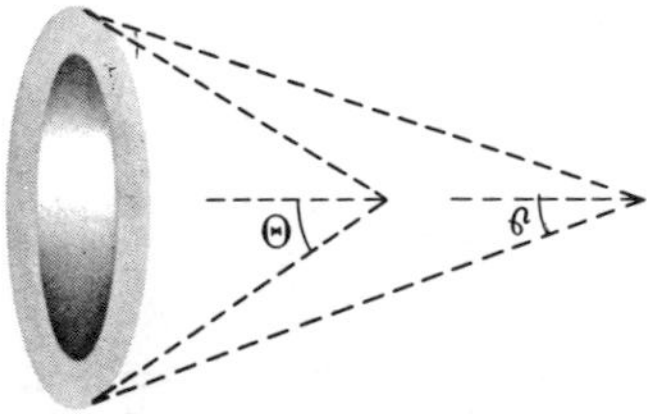

FIGURE 2.92 The particles scattered into the solid angle $2\pi \sin \Theta \, \Delta\Theta$ in the center-of-mass frame are scattered into the solid angle $2\pi \sin \vartheta \, \Delta\vartheta$ in the laboratory frame.

---

* See D. L. Falkoff, "Exchange Forces," *American Journal of Physics*, *18*: 30 (1950).

Problem 2.113

(a) Calculate the energy that would be released if two heavy hydrogen (deuterium) atoms fuse to form one helium atom.

Unified atomic mass units: $_1H^2 - 2.014102$; $_2He^4 - 4.002603$.

(b) 0.015% of the hydrogen found in nature is deuterium. Calculate the energy that could be obtained from the fusion into helium of all the deuterium in 1 cm³ of water.

(c) Estimate the amount of energy available from the oceans from the fusion of deuterium into helium.

(d) Estimate the ratio of the energy available from the fusion into helium of the deuterium in the water that comes into a house daily to the energy supplied by electrical power to the same house daily.

(e) The radiant energy emitted at the surface of the sun is 3.9×10^{26} W. Calculate the amount of deuterium undergoing fusion into helium that would be required to produce this power for 1 day. Express your answer as a percentage of the mass of the sun.

Problem 2.114

Show that the free neutron is unstable against β decay.

Problem 2.115

Calculate the potential energy $(1/4\pi\varepsilon_0)(e^2/r)$ between two protons at a separation distance of $r = 1$ fm and compare the magnitude of this with the estimate given above of 45 MeV for the depth of the square well for the nucleon-nucleon interaction.

Problem 2.116

Use the techniques of Problems 2.4 and 2.95(f) and the estimates of the size of the square well given above to calculate the differential cross section for neutron-proton scattering for a relative energy of 10 MeV. Compare your result with that shown in Figure 2.85.

Problem 2.117

Repeat the calculation of Problem 2.116 for a relative energy of 7 MeV and compare the result with that shown in Figure 2.84(a).

Problem 2.118

The total scattering cross section for neutrons on protons with a relative energy $E \ll 1$ MeV is given by experiments to be $\sigma_{tot} = 20$ barns. Calculate the corresponding total cross section for the square well estimated in the text.

Problem 2.119

Calculate to five figures the mass of the deuteron:

$$B_d = 2.226 \text{ MeV}.$$

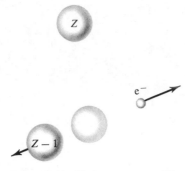

FIGURE 2.93 If the parent nuclide, denoted by Z, decays only into the daughter $Z-1$ and an electron, the energy and momentum of the electron is fixed by the energy-momentum conservation law.

2.4.5 The neutrino

The processes of α and β decay of radioactive nuclides differ in many important ways. In α decay, the constituents of the particle that is emitted, two protons and two neutrons bound together, are present in the nucleus before the decay. In β decay, on the other hand, the electron is not present in the nucleus before the decay but is created at the instant of emission. In α decay, the total energy and momentum of the emitted α particle and the daughter nucleus are equal to the total energy and momentum of the parent nucleus. This is not the case for the sum of the energies of the electron and daughter nucleus in β decay; in general, that sum is less than that of the parent nucleus, nor is their momentum that of the parent nucleus.

Consider a specific β radioactive nuclide. If the emitted electron and the daughter nuclide are the only particles that take part in the process of β decay and if the difference in the energies of the daughter and parent nuclides are the same for all such nuclei, the conservation law of energy and momentum requires that the electron be emitted with the same energy and momentum in every decay (Figure 2.93). Measurements of the energy and momentum of the emitted electrons show, however, that the electrons are ejected with varying amounts of energy and momentum (Figure 2.94). (The maximum kinetic energy T_{max} and momentum P_{max} observed in β decay are those required by 4-momentum conservation for the electron and recoiling nuclide alone.)

Furthermore, the emitted electrons do not lose the difference in energy by collisions with the electrons surrounding the nucleus of the atom. This can be shown by microcalorimeter measurements of the total kinetic energy released in the β decay of a sample of the material; the energy measured in this way agrees with the average energy, as calculated from the energy spectrum for the nuclide such as that shown in Figure 2.94.

This disappearance of energy and momentum is evident also in the β decay of the neutron. The neutron, which has a well-defined energy, disintegrates into an electron plus a proton, and yet the measured values of the energies of the electrons emitted in this process show a broad spectrum (Figure 2.95).

These results suggest that energy and momentum may not be conserved in the β decay process, although they appear to be conserved in every other type of

FIGURE 2.94 Momentum (a) and energy (b) spectra of the electrons in the β decay of $_{83}Bi^{210}$.

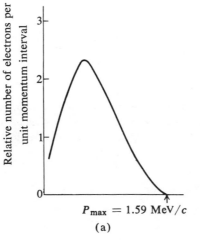

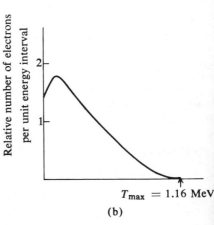

(a)   (b)

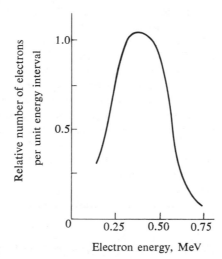

FIGURE 2.95 Spectrum of energies of the electrons emitted in the β decay of neutrons.

process. An explanation of β decay that is consistent with the conservation law of energy-momentum was suggested in 1930 by the Austrian-Swiss theoretical physicist Wolfgang Pauli (1900–1958), who pointed out that energy and momentum could be conserved in the process if the excess energy and momentum were carried off by a particle of very small mass that did not possess an electric charge and hence that would be difficult to detect (Figure 2.96). This suggestion was developed further by the Italian-American physicist Enrico Fermi (1901–1954), who gave the name *neutrino*, meaning little neutral one, to the particle and who formulated a theory of β decay in terms of a point interaction between the neutron, proton, electron, and neutrino (Figure 2.97). This theory and its refinements accounted very well for many of the properties of the β-decay process, even though the neutrino itself was not observed for many years. The neutrino is denoted by the symbol ν.

The difficulty in observing neutrinos results from the fact that they scarcely interact with ordinary matter. Indeed, on the average, a neutrino can travel through approximately 10^2 light-years of solid matter before undergoing an interaction. The fact that neutrinos do not interact very much with ordinary matter can be seen also in the following way: With each type of process, we can associate a typical time that represents the time interval necessary for the process to take place under the best of circumstances. The typical time of 10^{-9} sec can be associated with processes involving neutrinos.* This time can be compared with the typical time for a nucleon-nucleon interaction. Fast nucleons interact with proton targets, so a typical time for a nucleon-nucleon interaction is r/c, where r is of the order of the size of a nucleon; $r/c \gtrsim 10^{-23}$ sec. Both these times appear very short on our scale of things, but a better comparison of these times is given by the ratio of the typical time for β decay to that for a nucleon interaction, $\sim 10^{13}$. This ratio is much larger than the age of the solar system relative to the period of one orbit of the earth, $\sim 10^9$. Thus, in a fundamental particle's scale of things, a typical β decay takes longer than the age of the

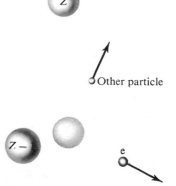

FIGURE 2.96 Pauli's suggestion to save the conservation law of energy-momentum in β decay.

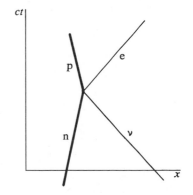

FIGURE 2.97 The Fermi interaction for β decay.

---

* On the other hand, many factors contribute to the size of the half-life of a particular β decay. The β decays of $_{83}Bi^{210}$ and the neutron, described before, have abnormally large half-lives compared to other particles that decay through neutrino interactions.

universe as seen on our scale. From the point of view of a β radioactive nuclide, β decay almost never happens.

In spite of the fact that neutrinos interact so weakly, they have been detected. The effects of an incoming beam of neutrinos were observed in 1956 by the American physicists Frederick Reines (1918–) and Clyde L. Cowan, Jr. (1919–), in an elaborate experiment.* The neutrino beam came from a nuclear reactor, a strong source of neutrinos.

The neutrino can be described by its dynamical variables. In the first place, the neutrino has a zero electric charge. Second, the mass of the neutrino has been shown to be much less than that of the electron, and at present, its mass is believed to be zero. Finally, the spin of the neutrino can be deduced from the conservation law of angular momentum applied to β-decay reactions; the neutrino has a spin of $\frac{1}{2}(h/2\pi)$, equal to that of the electron. Thus, the neutrino appears as a massless, neutral bundle of energy with an intrinsic spin.

$$\text{Neutrino } \nu: \qquad m_\nu = 0, \qquad q_\nu = 0, \qquad S_\nu = \tfrac{1}{2}. \qquad (2.279)$$

The neutrino has not been observed to decay and is believed to be a stable particle.†

Being a zero-mass particle with an intrinsic spin is not the only peculiar property of the neutrino. This particle also exhibits a left handedness‡; the spin of the neutrino, as given by a left-hand rule, lies along the direction of the momentum of the neutrino (Figure 2.98). This striking property of the neutrino was discovered as a result of investigations§ in 1956 by the Chinese–American physicists Tsung Dao Lee (1926–) and Chen Ning Yang (1922–).

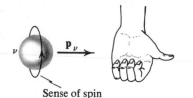

Sense of spin

FIGURE 2.98 The neutrino is left handed!

SUMMARY   The fact that the energy and momentum of the parent nucleus are not equal in general to the energy and momentum of the daughter nucleus and the emitted electron led to the postulation of the neutrino, whose existence has since been verified experimentally. The neutrino is massless, it has zero electric charge, and the direction of its momentum is related to the direction of its spin by a left-hand rule.

Problem 2.120

(a) What is the speed of a neutrino?
(b) What is the magnitude of the momentum, in megaelectron volts per c, of a neutrino of energy 0.93 MeV?

Problem 2.121

(a) Calculate the kinetic energy of the electron if the neutron underwent β decay without the emission of a neutrino particle.

---

* This experiment is described in P. Morrison, "The Neutrino," *Scientific American, 194*: 58, January 1956.
† The stability of zero mass particles is discussed in P. Havas, "Are Zero Mass Particles Necessarily Stable?" *American Journal of Physics, 34*: 753 (1966).
‡ Photons also exist with a handedness, left or right, corresponding to the left- or right-handed circular polarization of the corresponding wave (Problem 1.135). However, both types of photons can be emitted in a general process, so they need not be distinguished as different particles (see Section 2.4.6).
§ See P. Morrison, "The Overthrow of Parity," *Scientific American, 196*: 45, April 1957.

(b) A neutron decays into a proton, and the emitted electron is observed to possess a kinetic energy of 0.50 MeV. Calculate the energy and momentum of the unobserved neutrino particle.

Problem 2.122

The experiment that Lee and Yang suggested involved measuring the relative number of β particles emitted by a nucleus with a nonzero spin, Co^{60}, in the two senses along the spin axis (Figure 2.99). The measured asymmetry in the number emitted in the different directions distinguishes one hand from another or, in other words, destroys the symmetry of mirror images. Explain why a mirror image interchanges left and right but does not interchange up and down.*

Problem 2.123

Suppose that, to produce a neutron in the inverse of β decay, a neutrino had to interact with protons for an average length of time equal to the mean life of the neutron. How much solid matter would a neutrino traverse on the average before interacting with a proton? (Take into account the size of the nucleus as compared to the size of the atom.)

2.4.6 Antiparticles

In 1932, the American physicist Carl David Anderson (1905–) analyzed the cloud-chamber track of a cosmic-ray particle (Figure 2.100) that appeared to be the track of a particle with the mass of the electron but with a positive charge.† Anderson suggested calling this particle the *positron*, as the positive charged equivalent of the negatron, the negatively charged electron. The

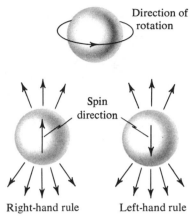

FIGURE 2.99 An asymmetry in the number of β particles emitted up and down distinguishes one hand from the other.

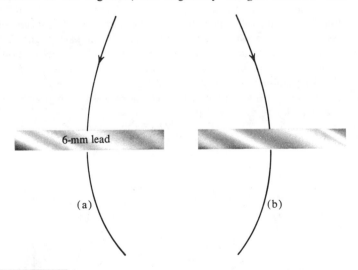

FIGURE 2.100 Anderson observed a track like that illustrated in (a). The track in (b) is that of a 63-MeV electron that loses 40 MeV in the lead plate. The particle of (a) is deviated opposite to that of (b) and thus must be positively charged.

---

* See p. 662 of R. H. Good, Jr., "Massless Particles," *American Journal of Physics*, **28**: 659 (1960).
† See C. D. Anderson, "Early Work on the Positron and Muon," *American Journal of Physics*, **29**: 825 (1961). He described his analysis in "The Positive Electron," *Physical Review*, **43**: 491 (1933), reprinted in *Foundations of Nuclear Physics*, R. T. Beyer (Ed.), Dover, New York, 1949. The discovery of the positron is discussed in greater detail in Chapter 9 of N. R. Hanson, *The Concept of the Positron*, Cambridge Univ. Press, Cambridge, 1963.

existence of this particle opened the door to the discovery, in recent years, of other particles similar in almost all respects to the particles we have discussed above.

The existence of a positively charged counterpart to the electron had been predicted before its discovery by the English theoretical physicist Paul Adrien Maurice Dirac (1902–   ), who holds the same professorship at the University of Cambridge that once was held by Newton. Dirac had developed a theory of the electron in a series of papers beginning in 1928. His electron theory satisfied both the rules of quantum mechanics and those of relativity; it agreed in all respects with the known behavior of electrons except for one point: the equations possessed anomalous solutions corresponding to electrons with negative kinetic energies. These solutions arose from the fact that the equation

$$E^2 = p^2c^2 + m^2c^4 \tag{2.280}$$

is satisfied by two values of E,

$$E = \pm \sqrt{p^2c^2 + m^2c^4}. \tag{2.281}$$

Both these solutions appeared to have their place in Dirac's theory.

We can obtain an interpretation for the negative energy solutions in a manner different from the procedure developed by Dirac; his method depended strongly on the principles of quantum mechanics. Our interpretation follows a theory of electromagnetic phenomena developed largely by the American theoretical physicist Richard P. Feynman (1918–   ).

We have seen that an energy-momentum vector can be associated with the direction of a world line in space-time. Up to the present, we have associated with particles only those energy-momentum vectors whose time-like components E are greater than zero. The corresponding world lines can be marked with an arrow pointing into the future to indicate this fact [Figure 2.101(a)]. The world line of a particle with a negative time-like component E in its energy-momentum vector is marked with an arrow pointing into the past [Figure 2.101(b)].

Let us consider the world lines of two particles that in free motion are described relative to an inertial system S by the 4-momentum $p = (E/c, \mathbf{p})$ with $E > 0$ and by the 4-momentum $p' = (-E/c, -\mathbf{p})$, respectively. The world lines are like those drawn in Figure 2.101. The norms of the energy-momentum vectors are given by

$$p \cdot p = \left(\frac{E}{c}\right)^2 - \mathbf{p} \cdot \mathbf{p} = m^2c^2 \tag{2.282}$$

FIGURE 2.101 The directions associated with world lines. (a) The world line of a particle with $E > 0$. (b) The world line of a particle with $E' < 0$.

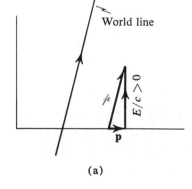

(a)

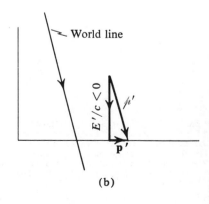

(b)

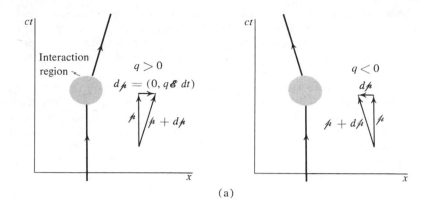

(a)

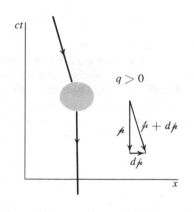

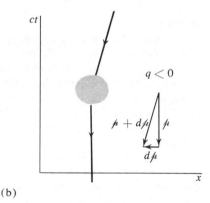

(b)

FIGURE 2.102 The effects of an electric force on positive- and negative-energy particles: (a) $E > 0$. (b) $E < 0$.

and

$$p' \cdot p' = \left(-\frac{E}{c}\right)^2 - (-\mathbf{p}) \cdot (-\mathbf{p}) = \left(\frac{E}{c}\right)^2 - \mathbf{p} \cdot \mathbf{p} = m^2 c^2, \qquad (2.283)$$

so each of the two lines represents the motion of a particle of mass m.

Consider now the effects of an electric force acting on the particles. Let q be the charge of the positive-energy particle and q' that of the negative-energy particle. If the physical electric force relative to S that is experienced by a particle of charge Q is $\mathbf{F} = Q\mathscr{E}$, then the physical forces experienced by the two particles are $q\mathscr{E}$ and $q'\mathscr{E}$, respectively. The relativistic equations of motion for the two particles are

$$\frac{d\mathbf{p}}{dt} = q\mathscr{E}, \qquad \frac{dE}{dt} = q\mathscr{E} \cdot \mathbf{v} = \frac{q}{m_v}\mathscr{E} \cdot \mathbf{p} \qquad (2.284)$$

and

$$\frac{d\mathbf{p}'}{dt} = q'\mathscr{E}, \qquad \frac{dE'}{dt} = \frac{q'}{m_{v'}}\mathscr{E} \cdot \mathbf{p}'. \qquad (2.285)$$

Consider the case in which $\mathscr{E} = |\mathscr{E}|\hat{x}$ and in which the energy-momentum vectors are $(mc, 0)$ and $(-mc, 0)$ before the interaction (Figure 2.102). (The more general case is considered in Problem 2.128.) Then, the physical force $Q\mathscr{E}$ produces in the short time dt changes in the 4-momentums given by

$$dp = (0, q\mathscr{E}\,dt), \qquad dp' = (0, q'\mathscr{E}\,dt). \qquad (2.286)$$

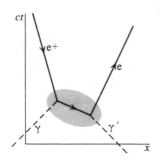

FIGURE 2.103 Production of an electron-positron pair.

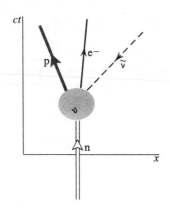

FIGURE 2.104 The β decay of the neutron.

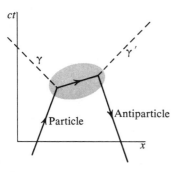

FIGURE 2.105 The annihilation of a charged particle and its antiparticle into two photons, γ and γ'.

These are illustrated in Figure 2.102, from which it can be seen that, in each case, the negative-energy particle of charge q' behaves like a positive-energy particle of charge $q = -q'$. Thus, *a negative-energy electron of 4-momentum p appears as a positively charged particle with 4-momentum $-p$ and hence with positive energy.*

This particle is the *positron* that was discovered by Anderson and is denoted by the symbol e^+. The positron is called the *antiparticle* of the negatively charged electron e (or e^-) and can be also denoted* by $\tilde{e}$. A positron can be produced, in association with the production of an electron, by two photons in the process illustrated in Figure 2.103. [This process with two real photons is extremely improbable; positron-electron pairs are usually produced in the laboratory by the processes described in Problems 2.125(b) and 2.126(b).]

The generality of the arguments given above suggests the existence of an antiparticle to every type of particle, not the electron alone. Indeed, the antiproton, antineutron,† and even the antideuteron (discovered in 1965) have been shown to exist.

The antiparticles of some neutral particles are identical to the particles themselves, whereas there is a difference between other neutral particles and their antiparticles. The antiparticle of a photon with a given handedness is a photon of opposite handedness, but since photons of both polarizations can be emitted in a general process, there is not much point in distinguishing one as the photon and the other as the antiphoton. (This corresponds to calling the particle e^- an electron whether it spins one way or the other.) We say that the antiparticle of a photon is a photon. On the other hand, the neutrino-like particle emitted in a general process always has the same handedness. Therefore, it is convenient to consider the left-handed neutrino-like particle as the neutrino and to distinguish the antineutrino, which is right-handed, from the neutrino itself. The neutrino-like particle that is involved in the β decay of the neutron is, by convention, called the antineutrino. With this convention, the same number of light (e or ν) particles, called leptons, enter a region of interaction as leave that region (Figure 2.104) (with the convention that n antileptons count as $-n$ leptons).

A particle and its antiparticle can annihilate each other; this is illustrated in Figure 2.105 for a charged particle that annihilates with its antiparticle to produce two photons of radiation. The fact that antimatter annihilates with ordinary matter explains why antiparticles are not observed in abundance in our region of the universe. However, antimatter is produced and studied in laboratories. Indeed, even atoms each consisting of an electron and a positron and called *positronium* have been produced and studied in the laboratory.‡

SUMMARY   A negative-energy particle of 4-momentum p' and electric charge q' appears as a positive-energy particle of 4-momentum $p = -p'$ and charge $q = -q'$. The negative-energy counterpart of a particle is called the

---

* The Commission for Symbols, Units and Nomenclature recommends that the tilde ~ be used above the symbol for a particle to denote the corresponding antiparticle, although sometimes an overbar is used instead, for example, ē.
† The discovery of antinucleons is described in E. Segrè and C. E. Wiegand, "The Antiproton," *Scientific American, 194*: 37, June 1956 and E. Segrè, "Antinucleons," *American Journal of Physics, 25*: 363 (1957).
‡ See H. C. Corben and S. De Benedetti, "The Ultimate Atom," *Scientific American, 191*: 88, December 1954.

antiparticle, and a particle-antiparticle pair can be created or annihilated together. For some types of particles, the antiparticle is distinct from the particle, as the positron is from the electron, but for other particles such as the photon, the particle and its antiparticle are the same particle.

Example 2.20

Q. (a) Show that a positron cannot annihilate with an electron alone to produce one photon only.

    (b) A positron and an electron annihilate to produce two photons. Calculate the energies of the photons in the center-of-momentum system of the $e^+ - e^-$ system.

A. (a) Let p_- and p_+ be the energy-momentum vectors of the electron and positron, respectively. The total 4-momentum of the system is

$$\mathcal{P} = p_- + p_+ \tag{2.287}$$

with norm

$$\mathcal{P}^2 = p_-^2 + p_+^2 + 2p_+ \cdot p_-$$

$$= m^2c^2 + m^2c^2 + 2\frac{E_-E_+}{c^2} - 2\mathbf{p}_- \cdot \mathbf{p}_+ \tag{2.288}$$

$$\geq 2(m^2c^2 + \sqrt{|\mathbf{p}_-|^2 + m^2c^2}\,\sqrt{|\mathbf{p}_+|^2 + m^2c^2} - |\mathbf{p}_-|\,|\mathbf{p}_+|)$$

$$> 2m^2c^2.$$

Since the norm of the 4-momentum of a photon is zero, $\mathcal{P}$ cannot be the 4-momentum of a photon; hence, by the conservation law of energy-momentum, an electron and a positron alone cannot annihilate into one photon only.

    (b) Let p_- and p_+ have components $(E_-/c, \mathbf{p}_-)$ and $(E_+/c, \mathbf{p}_+)$ relative to the center-of-momentum system. Then

$$\mathbf{p}_+ + \mathbf{p}_- = 0 \quad \text{or} \quad \mathbf{p}_+ = -\mathbf{p}_-, \tag{2.289}$$

and since the masses of the e^- and the e^+ are equal,

$$\frac{E_+}{c} = \sqrt{|\mathbf{p}_+|^2 + m^2c^2} = \sqrt{|\mathbf{p}_-|^2 + m^2c^2} = \frac{E_-}{c}. \tag{2.290}$$

Let $(E_1/c, \mathbf{p}_1)$ and $(E_2/c, \mathbf{p}_2)$ be the components of the photons' 4-momentums k_1 and k_2, respectively, relative to the center-of-momentum system. The energy-momentum conservation law states that

$$\frac{E_-}{c} + \frac{E_+}{c} = \frac{E_1}{c} + \frac{E_2}{c} \tag{2.291}$$

and

$$\mathbf{p}_- + \mathbf{p}_+ = \mathbf{p}_1 + \mathbf{p}_2. \tag{2.292}$$

Therefore,

$$\mathbf{p}_1 = -\mathbf{p}_2 \tag{2.293}$$

and the two photons travel off in opposite directions (relative to the center-of-momentum system) from the region of annihilation. Also, since $E_1 = |\mathbf{p}_1|c$ and $E_2 = |\mathbf{p}_2|c$, they have equal energies, given by Equations (2.290) and (2.291) as

$$E_1 = E_2 = E_- = E_+ . \tag{2.294}$$

In particular, if the electron and positron are at rest relative to each other ($\mathbf{p}_+ = \mathbf{p}_- = 0$), then

$$E_\gamma = mc^2 = 0.511 \text{ MeV}. \tag{2.295}$$

Example 2.21

Q. A proton of kinetic energy T is incident on a stationary proton, and a proton-antiproton pair is produced in the collision. Calculate the minimum value of T, the *threshold energy*, for which this pair production process is possible.

A. Before the collision, the initial 4-momentum $\mathscr{P}_i$ of the system has components $(2M_p c + T/c, \mathbf{p})$ relative to the laboratory system, where $\mathbf{p}$ is the momentum of the incoming proton with magnitude

$$p = \sqrt{[(M_p c^2 + T)^2/c^2] - M_p^2 c^2} = \sqrt{2M_p T + (T^2/c^2)}. \tag{2.296}$$

Because of conservation laws, the charge $2e$ and mass number 2 of the initial system p + p must equal those of the final system, including an antiproton $\tilde{\text{p}}$. Therefore, the final system must consist of p + p + p + $\tilde{\text{p}}$. Thus, the final system comprises four particles, each of mass M_p. The minimum energy that this system can possess is $4M_p c^2$, and this energy is defined relative to a reference frame in which all the particles are at rest. Therefore, relative to this reference system, the total momentum is zero, and the reference system is the center-of-momentum system. The final 4-momentum $\mathscr{P}_f$ of the system has components $(4M_p c, 0)$ relative to the center-of-momentum system.

The conservation law of energy-momentum gives

$$\mathscr{P}_i = \mathscr{P}_f . \tag{2.297}$$

We know the components of $\mathscr{P}_i$ and $\mathscr{P}_f$ relative to two different reference systems, but we can obtain a relation that does not depend on the reference systems by equating the norms of the two equal vectors:

$$\mathscr{P}_i \cdot \mathscr{P}_i = \mathscr{P}_f \cdot \mathscr{P}_f \tag{2.298}$$

or

$$\left(2M_p c + \frac{T}{c}\right)^2 - p^2 = (4M_p c)^2,$$

which expands to

$$4M_p^2 c^2 + 4M_p T + \frac{T^2}{c^2} - \left(2M_p T + \frac{T^2}{c^2}\right) = 16M_p^2 c^2. \tag{2.299}$$

Hence,

$$T = 6M_p c^2 = 5.63 \text{ GeV}. \tag{2.300}$$

Problem 2.124

A positron of kinetic energy T annihilates with a stationary electron to produce two photons that travel along the line of the incident positron's momentum.

Calculate the energy of the photon that goes in the direction of the positron's momentum and that of the photon that goes in the opposite direction.

Problem 2.125

(a) Calculate the threshold energy for electron pair production by electrons incident on stationary electrons.
(b) Calculate the threshold energy for electron pair production by electrons incident on stationary protons.

Problem 2.126

(a) Show that a photon in an otherwise empty region of space cannot decay into an electron-positron pair. *Hint:* Show that the process violates the energy-momentum conservation law.
(b) Show that a photon can annihilate into an electron-proton pair in the neighborhood of a nucleus. *Hint:* Consider the case in which the energy equivalent of the mass of the nucleus is much larger than the energy of the photon. In this case, the nucleus can absorb momentum while absorbing a negligible amount of energy. Use nonrelativistic mechanics to describe the behavior of the nucleus.
(c) Photons of energy 2.6 MeV are incident on a lead plate to produce electron-positron pairs. The maximum total kinetic energy of each pair has been measured to be 1.6 MeV.* Determine the mass of an electron from these results.
(d) Draw a space-time diagram of the process described in (b).

Problem 2.127

The kinetic energy of the protons bound in a nucleus goes up to about 20 MeV. Find the threshold energy for proton-antiproton production for protons incident on the protons in a stationary nucleus.

Problem 2.128

Show, from a diagram similar to Figure 2.102, that a negative-energy particle of 4-momentum p' and electric charge q' behaves like a positive-energy particle of 4-momentum $p = -p'$ and electric charge $q = -q'$.

2.4.7 Muons and π mesons

Let us now consider some consequences of the existence of nucleon-nucleon interactions. The space-time diagram of the scattering of one nucleon by another is similar to that of the scattering, by electric forces, of one charged particle by another (Figure 2.106). The mechanism for the energy-momentum transfer by which electromagnetic interactions take place is associated with the existence of photons, suggesting that there may also be particles associated with the agency responsible for nucleon-nucleon interactions† (Figure 2.107). We have seen

---

* C. D. Anderson and S. H. Neddermeyer, "Positrons from Gamma-Rays," *Physical Review*, *43*: 1034 (1933).
† See R. J. Blin-Stoyle, "Particles, Field Theory and Force," *Contemporary Physics*, *2*: 325 (1961).

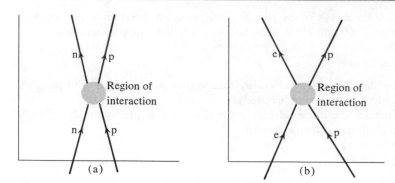

FIGURE 2.106 Space-time diagrams of (a) nucleon-nucleon and (b) charged-particle scattering.

the necessity in nucleon-nucleon interactions of the exchange of charge along with the energy-momentum transfer. This indicates that at least some of the particles associated with the nucleon-nucleon interaction are charged* (Figure 2.108).

The form of the nucleon-nucleon interaction that results in the prediction of the existence of these particles was proposed first by the Japanese theoretical physicist Hideki Yukawa† (1907–   ). Yukawa deduced directly from the principles of quantum mechanics that a consequence of the short range of nucleon-nucleon interactions is that the particle associated with the nucleon-nucleon interaction has a nonzero rest mass; in fact, Yukawa predicted in 1934 that the mass of the particle involved was about $200m_e$. Particles of this mass were called mesotrons, now contracted to *mesons*, from the Greek word μεσος, for middle, and the "tron" of electron.

Particles of about this mass were discovered by C. D. Anderson and his colleague S. H. Neddermeyer almost immediately after Yukawa predicted the existence of the mesotron. The tracks of these particles were observed in cloud chambers exposed to cosmic radiation.‡ At first it was believed that the observed

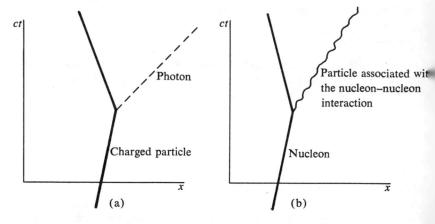

FIGURE 2.107 Production of particles associated with interaction mechanisms. (a) Production of the particle (a photon) associated with charged-particle interactions. (b) Production of the particle associated with the nucleon-nucleon interaction. (The world line of the other charged particle or the other nucleon that plays a part in the production process has been omitted from the diagram.)

---

* See also R. E. Marshak, "Pions," *Scientific American, 196*: 84, January 1957.
† Yukawa described the development of his theory in "The Birth of the Meson Theory," translated by C. Kikuchi, *American Journal of Physics, 18*: 154 (1950).
‡ Such cosmic-ray particles are of interest to other scientists in addition to physicists. For example, their use in finding hidden chambers in pyramids is described in *Physics Today, 19*: 78, September 1966.

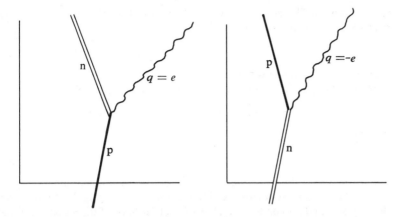

FIGURE 2.108 Some of the particles associated with the nucleon-nucleon interaction are charged.

particles were Yukawa mesotrons. However, it was shown in 1943 that these particles did not interact sufficiently strongly with nuclei to be the particles of the agency responsible for nucleon-nucleon interactions. Physicists were forced to conclude that the observed particles were not the mesotrons of Yukawa but that Anderson had discovered another particle. The particles discovered by Anderson are now known as *muons*,* designated by μ.

The properties† of the muon are so similar to those of the electron that the muon can be considered to be a heavy electron. Like the electrons, muons carry an electric charge of $+e$ or $-e$;‡ furthermore, they have a spin of $\frac{1}{2}(h/2\pi)$. The interactions§ that the muon experiences are identical to those experienced by the electron. Indeed, the only difference between muons and electrons lies in their masses; the muon is much more massive.

Since energy is equivalent to mass, we might expect that the various species of particles differ in mass only because there exist differences in the type of interactions, and thus in their energies of interaction, that these particles experience. However, the muon and the electron appear to undergo identical interactions even though their masses are so different. Why this is so is one of the unsolved problems of physics today.

Muons are not stable particles but decay with a mean life τ_0 of $(2.2001 \pm 0.0008) \times 10^{-6}$ sec relative to their rest frame. This lifetime is sufficiently short that cosmic-ray muons, produced high in the earth's atmosphere by cosmic radiation, cannot traverse the atmosphere in that time interval (measured relative to the earth), even if they travel at the speed c:

$$c\tau_0 \approx 6 \times 10^2 \text{ m}; \quad \text{height in atmosphere of } \mu \text{ production} \sim \tfrac{3}{2} \times 10^4 \text{ m}.$$
$$(2.301)$$

Nevertheless, muons produced high in the atmosphere are observed at the

---

* The term "muon" was introduced as an abbreviated form of μ meson. However, it appears worthwhile to reserve the name meson for those particles of integral spin that interact strongly with nucleons; with this nomenclature, the muon is not a meson-type particle.
† See S. Penman, "The Muon," *Scientific American, 205*: 46, July 1961.
‡ Muons and electrons can interact through their electrical charges; indeed, an atom called muonium, which consists of a μ+ and an e−, has recently been shown to exist. See V. W. Hughes, "The Muonium Atom," *Scientific American, 214*: 93, April 1966.
§ This is true even of the interactions of the electron and muon with their respective weak interaction partners, the neutrinos ν_e and ν_μ [see Equation (2.306)].

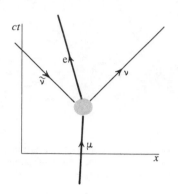

FIGURE 2.109 Space-time diagram of the decay $\mu \rightarrow e + \nu + \tilde{\nu}$.

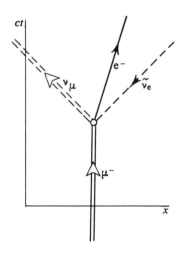

FIGURE 2.110 μ^- decay. The double lines represent μ-family members; the single lines, e-family members.

earth's surface. This results from time dilatation, since the mean life τ measured relative to the earth, given by

$$\tau_0 = \tau \sqrt{1 - (v^2/c^2)}, \tag{2.302}$$

is much larger than τ_0 for a muon speed v near c.

Muons decay into electrons and neutrinos according to the scheme

$$\text{Muon} \rightarrow \text{electron} + 2 \text{ neutrino particles}. \tag{2.303}$$

The decay (Figure 2.109),

$$\mu \rightarrow e + \nu + \tilde{\nu}, \tag{2.304}$$

is consistent with the conservation law of leptons mentioned in Section 2.4.6—the number of leptons entering an interaction region equals the number leaving that region—and the conservation law of electric charge if the muon is considered to be a lepton. However, all the conservation laws that we have encountered so far, in particular that of leptons, are also consistent with the decay

$$\mu \rightarrow e + \gamma, \tag{2.305}$$

although experiments to date have failed to detect this decay mode for muons. The fact that this decay mode does not occur suggests that it violates a conservation law. This conservation law must be satisfied by the decay mode (2.304), so we might expect that whatever is conserved is related to the muon and one of the neutrino particles. If this is true, the neutrino associated with the muon would differ in some respects, presumably its interactions, from the neutrino associated with the electron. The existence of the two types of neutrinos has been demonstrated experimentally,* with the result supporting the conservation law of μ-family members. The neutrinos associated with the electron and the muon are denoted by ν_e and ν_μ, respectively, so that the decay process (2.304), shown in Figure 2.110, can be written

$$\begin{aligned} \mu^- &\rightarrow e^- + \nu_\mu + \tilde{\nu}_e, \\ \mu^+ &\rightarrow e^+ + \tilde{\nu}_\mu + \nu_e. \end{aligned} \tag{2.306}$$

The properties of the muon are given below:

$$m_\mu c^2 = (105.659 \pm 0.002) \text{ MeV}, \qquad q_{\mu\mp} = \mp e, \qquad S_\mu = \tfrac{1}{2}. \tag{2.307}$$

The mesotron of Yukawa was discovered as a component of cosmic rays in 1947 by the English physicist Cecil Frank Powell (1903–) and his colleagues C. M. G. Lattes and G. T. S. Occhialini. These particles are called π mesons, or *pions*. Pions have a mass of about $270m_e$, in agreement with Yukawa's prediction. They exist in three electric charge states—neutral and with charges of $\pm e$. The π^+ and π^- are antiparticles, and the π^0 is its own antiparticle.

Pions decay with a mean life of $(2.55 \pm 0.03) \times 10^{-8}$ sec for the charged pions and $(1.8 \pm 0.3) \times 10^{-16}$ sec for the neutral pion. The difference in lifetimes arises from the fact that the dominant modes of decay for the charged pions,

$$\pi^+ \rightarrow \mu^+ + \nu_\mu, \qquad \pi^- \rightarrow \mu^- + \tilde{\nu}_\mu, \tag{2.308}$$

---

* See L. M. Lederman, "The Two-Neutrino Experiment," *Scientific American*, *208*: 60, March 1963.

involve neutrinos, whereas the dominant decay mode of the neutral pion,

$$\pi^0 \rightarrow 2\gamma, \tag{2.309}$$

proceeds through electromagnetic interactions.

Charged pions at rest usually decay into a muon and a neutrino and, since the masses of these particles are well-defined, they must emerge from the decay process with certain values of energies and momentums (Figure 2.111). It was this characteristic of the dominant decay mode of the charged pions that led to their discovery by Powell.

The masses of the charged pions can be determined from the mass of the muon and the energies of the muons emitted in the decay

$$\pi^\pm \rightarrow \mu^\pm + \nu_\mu \quad (\text{or } \tilde{\nu}_\mu) \tag{2.310}$$

of pions at rest. The components of the 4-momentum of the pion in its rest system are given by

$$\mathscr{P} = (m_\pi c, 0), \tag{2.311}$$

and the 4-momentum after the decay process has the components

$$\left(\frac{E_\mu}{c} + \frac{E_\nu}{c}, \; \mathbf{p}_\mu + \mathbf{p}_\nu \right) \tag{2.312}$$

relative to the same reference frame. Therefore, the magnitudes of the momentums of the muon and the neutrino are equal and, since the neutrino has zero rest mass,

$$|\mathbf{p}_\mu| = |\mathbf{p}_\nu| = \frac{E_\nu}{c}. \tag{2.313}$$

Hence, the conservation law of 4-momentum gives the result

$$m_\pi c = \frac{E_\mu}{c} + |\mathbf{p}_\mu| = \sqrt{\mathbf{p}_\mu^2 + m_\mu^2 c^2} + |\mathbf{p}_\mu|. \tag{2.314}$$

Measurements of the muon momentum give the value 29.8 MeV/c for $|\mathbf{p}_\mu|$; hence,

$$m_\pi c^2 = 139.6 \text{ MeV}. \tag{2.315}$$

The strong nucleon-nucleon interactions not only are associated with the existence of pions but, because they contain a charge-exchange feature, they also contribute to the electric charge structure of nucleons. Just as a neutron experiences the strong interaction at some distance (~ 1 fm) from a nucleon, so also do charged particles experience an electrical interaction as if the charge were distributed over corresponding distances about the nucleon. This charge structure has been observed, and in fact, as we shall see shortly, the study of this charge structure led to the discovery of other particles.

The properties of the pions are listed below:

$$S_\pi = 0, \tag{2.316}$$

$$m_{\pi^\pm} = (139.58 \pm 0.02) \text{ MeV}, \quad \tau_{0_{\pi^\pm}} = (2.55 \pm 0.03) \times 10^{-8} \text{ sec}, \tag{2.317}$$

$$m_{\pi^0} = (134.97 \pm 0.02) \text{ MeV}, \quad \tau_{0_{\pi^0}} = (1.8 \pm 0.3) \times 10^{-16} \text{ sec}. \tag{2.318}$$

The form of the nucleon-nucleon interaction [Figure 2.112(a)] resulted in the successful prediction of the particles called pions [Figure 2.112(b)]. This

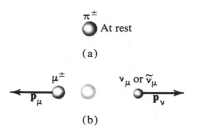

FIGURE 2.111 Decay of a charged pion at rest. The conservation law of 4-momentum gives the result, Equation (2.314), $m_\pi c = \sqrt{\mathbf{p}_\mu^2 + m_\mu^2 c^2} + |\mathbf{p}_\mu|$, so the muon is emitted with a definite momentum. (a) Before decay. (b) After decay.

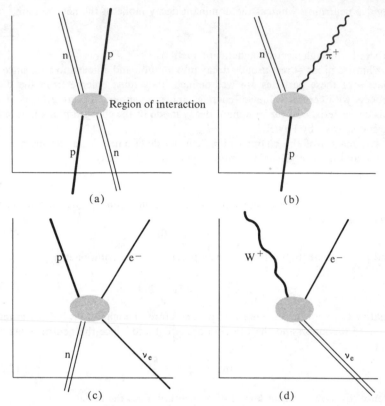

FIGURE 2.112   The interactions and the particles associated with them. (a) The nucleon-nucleon interaction. (b) Production of the particle associated with the nucleon-nucleon interaction. (c) The weak interaction. (d) Production of the particle associated with the weak interaction. [In (b) and (d), the other system, a nucleon that also plays a part in the production mechanism, has been omitted from the diagram.]

suggests that the weak interactions such as β decay [Figure 2.112(c)] may imply the existence of another type of particle, as shown in Figure 2.112(d). The properties of this conjectured particle have been investigated; for example, it can be seen from Figure 2.112(d) that it should carry an electric charge, since the corresponding interaction results in a charge exchange between the nucleon and the lepton. The proposed particle has been called the intermediate boson or, alternatively, the W particle (for weak interaction). Physicists have searched for evidence of the existence of this particle, but so far no such evidence has been found.

SUMMARY   The successful prediction of the existence of particles called photons that are associated with electromagnetic interaction led to the prediction of particles that result from the nucleon-nucleon interaction. These particles are called pions and exist in three charge states. A similar prediction of a particle, the W particle, associated with the weak interaction has not been verified.

There exists another particle, the muon, that at one time was associated erroneously with the nucleon-nucleon interaction but that now appears to be a heavy electron. The muon, as well as the electron, has its own type of neutrino associated with its weak interactions.

Problem 2.129

Calculate the threshold energy for the creation of charged pions by photons incident on stationary protons.

Problem 2.130

Calculate the threshold energy for the creation of neutral pions by protons incident on stationary protons.

Problem 2.131

Calculate the energies of the photons emitted in the decay $\pi^0 \to 2\gamma$ of a neutral pion at rest.

Problem 2.132

The maximum kinetic energy of electrons produced in the decay

$$\mu^- \to e^- + \nu_\mu + \tilde{\nu}_e$$

has been measured to be 55 MeV. Calculate the mass of the μ^-.

Problem 2.133

The mean life of muons is 2.20×10^{-6} sec in their rest system. Consider muons that travel a distance of 6.26×10^3 ft at an average speed of $0.992c$ relative to the earth.

(a) Calculate the ratio of the number of muons that survive to the initial number that were present and at rest in the time required for a particle to travel 6.26×10^3 ft at the speed $0.992c$.
(b) The ratio of the number of muons that survive to those initially present has been measured* to be 0.72. Explain this result.

Problem 2.134

A neutral pion traveling with the speed v decays into two photons that move off at equal angles to the direction of the incident π_0. Let θ be the angle between the directions of the momentums of the π_0 and one of the γ's. Show that $v/c = \cos \theta$.

2.4.8 Strange particles: K mesons and hyperons

In 1947, the compilation of particles appeared to have been completed by the discovery of the pion. It seemed then that the behavior of matter could be explained completely on the basis of the known particles, although the theories were not adequate (and indeed are not yet) to explain why the set of known

---

* The experiment and a thorough analysis of the results are described in D. H. Frisch and J. H. Smith, "Measurement of the Relativistic Time Dilation Using μ-Mesons," *American Journal of Physics, 31*: 342 (1963).

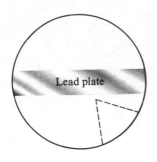

FIGURE 2.113   The characteristic inverted *V* appearance of the tracks of the decay products of a V particle produced in the lead plate.

particles formed the building blocks of matter. The electron and the nuclei were necessary for the explanation of atomic behavior, the nucleons for nuclear behavior, the neutrino for β decay, the pions and photons for the interactions and, finally, the antiparticles were required to satisfy the requirements of relativity theory. Only the muon did not fit into this scheme of natural simplicity.

In 1947,* however, the destruction began of this simple scheme of the rationale for the number of fundamental particles. Cloud-chamber photographs of cosmic-ray particles taken by G. D. Rochester (1908–) and C. C. Butler (1922–) of the University of Manchester showed evidence of the decays of particles with masses of about 500 MeV. Rochester and Butler showed that these particles could not be any of the particles that we have studied in our previous work. One of these particles was neutral and decayed into two charged particles; the tracks appeared as in Figure 2.113, and because of the characteristic pattern of the tracks, the particles were initially called V particles. The study of V particles in the cosmic radiation was supplemented in the early fifties when the Cosmotron at the Brookhaven National Laboratory came into operation, at which time man-made V particles could be studied under laboratory conditions.

These V particles were investigated intensively after it became evident that they indeed were different from the particles known before. It was found that they were produced in abundance in high-energy collisions, for example, of fast negative pions incident on protons. The frequency with which these particles were produced in collisions showed that the interactions responsible for their production were strong or comparable to the nucleon-nucleon forces. Such interactions are called *strong interactions*. The typical time for their production was found to be about 10^{-23} sec, the time required for a fast pion to move past a nucleon. On the other hand, the distance that the V particles traveled in cloud chambers showed that their characteristic decay time ranged from 10^{-8} to 10^{-10} sec, a time typical for *weak interactions* such as those decays involving neutrinos.

This large disparity between the production and decay times appeared strange to scientists at the beginning of the 1950's, as shown by the following example, which is based on the information available on particles at that time.† Suppose that one of the neutral V particles, that called the lambda hyperon Λ^0 with a lifetime of 10^{-10} sec, was produced in the reaction $\pi^- + p \rightarrow \Lambda^0 + \pi^0$ illustrated in Figure 2.114(a). (This assumption is known now to be incorrect.) Then, the Λ^0 should undergo the corresponding decay reaction $\Lambda^0 \rightarrow \pi^- + p + \pi^0$ shown in Figure 2.114(b), or, alternatively, the π^0 might be absorbed by the proton in the reaction to produce the decay $\Lambda^0 \rightarrow \pi^- + p$ [Figure 2.114(c)]. Each of these reactions should take place in a time of about 10^{-23} sec, so the actual decay of this particle is $10^{-10}/10^{-23} \sim 10^{13}$ longer than that predicted by the assumed production mechanism. This puzzling feature of the behavior of these particles led to the V particles being called *strange particles*, a nomenclature that has survived to this day.‡

_____

* A particle with a mass of about 500 MeV had been observed in 1944 by L. LePrince-Ringuet, but the particle did not decay in the cloud chamber and the significance of this single event was not appreciated.

† See also S. B. Treiman, "The Weak Interactions," *Scientific American*, *200*: 72, March 1959.

‡ The choice of this name was premature, for our present understanding of these particles is such that their behavior is not now considered to be any more unusual than is the behavior of any other particle, such as the pion. See the discussion of the choice of this and other names on pp. 5–6 of R. K. Adair and E. C. Fowler, *Strange Particles*, Interscience, New York, 1963.

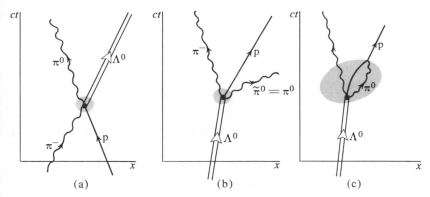

FIGURE 2.114 Argument to suggest that if the Λ° is produced in a fast reaction, such as (a), it should decay swiftly, as shown in (c). This result contradicts the experimental evidence! (a) Assumed production process for the Λ°. (The assumption is incorrect.) (b) How the Λ° could decay if (a) is the production reaction. (c) The decay (b) in which the neutral pion is absorbed by the proton in the decay process.

The resolution of the paradox of the strong production and weak decay of the strange particles was advanced by the Dutch-American theoretical physicist Abraham Pais (1918–). Pais proposed that at least two strange particles must be involved in order that any reaction in which they take part can proceed by a strong interaction. This hypothesis of associated production predicted that strange particles were created in pairs; after the two strange particles are created, they separate, and since each is isolated from the other, they cannot decay in the same manner by which they were produced (Figure 2.115). This hypothesis was verified by experiment shortly after it was proposed.

The principle of associated production was related to a conservation law in 1953 by the American theoretical physicist Murray Gell-Mann* (1929–) and independently by the Japanese physicist Kazuhiko Nishijima (1926–). They postulated that this conservation law, which we shall describe briefly below, was satisfied in processes involving the strong interactions but could be violated in the much slower weak interactions.

The first strange particles observed were the K mesons (or kaons), denoted by K, and the lambda hyperon, denoted by Λ. The neutral states of these particles can be produced by fast negative pions incident on protons:

$$\pi^- + p \to \Lambda^0 + K^0. \tag{2.319}$$

No charged lambda particles have been observed. The Λ^0 was observed to decay via the reaction

$$\Lambda^0 \to p + \pi^-. \tag{2.320}$$

The mass of the Λ^0 is about 10^3 MeV,† and so the Λ^0 was assigned the mass number $A = 1$. In the new terminology, the mass number of a particle with half-integral spin is called its *baryon number* and any particle with nonzero

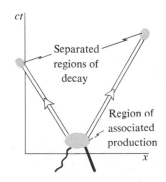

FIGURE 2.115 Associated production: Strange particles experience the strong interactions only when they are together, and thus they cannot decay by those interactions.

---

* Gell-Mann described this phase of his work in M. Gell-Mann and E. P. Rosenbaum, "Elementary Particles," *Scientific American*, *197*: 72, July 1957.
† The Λ particle interacts strongly with nucleons and, in fact, can be absorbed by a nucleus to form a short-lived "hypernucleus." See V. L. Telegdi, "Hypernuclei," *Scientific American*, *206*: 50, January 1962.

baryon number is called a *baryon*. The baryon number of an antiparticle is the negative of that of the corresponding particle.

The neutral K meson was observed to decay weakly through the two processes*

$$K^0 \rightarrow \pi^+ + \pi^- \quad \text{and} \quad K^0 \rightarrow \pi^+ + \pi^- + \pi^0. \tag{2.321}$$

Also, charged kaons were observed that decay through the processes

$$K^\pm \rightarrow \pi^\pm + \pi^0 \quad \text{and} \quad K^\pm \rightarrow \pi^\pm + \pi^+ + \pi^-. \tag{2.322}$$

The mass of the K meson is about 500 MeV, and its spin is zero; thus, the K meson is not a baryon.

The conservation law that governs whether a process will proceed by the fast, strong interactions or by the much slower, weak interactions can be seen from Equations (2.319), (2.320), and (2.321); the fast production process (2.319) involves two strange particles, whereas the slower decay processes (2.320) and (2.321) involve only one. This suggests the introduction of a dynamical variable, somewhat like the electric charge, that has, say, the value 0 for all nonstrange particles, the value -1 for the Λ^0, and the value $+1$ for the K^0; also, this dynamical variable is conserved in strong interactions but not in weak interactions. This dynamical variable is called the *strangeness* and is denoted by S. The strangeness of an antiparticle is the negative of that of the corresponding particle. In the production process (2.319), the initial strangeness is $0 + 0 = 0$, the final strangeness is $1 - 1 = 0$, and strangeness is conserved. On the other hand, in the decay process (2.320), the initial strangeness is -1 and the final strangeness is $0 + 0 = 0$; strangeness is not conserved. This process proceeds via the slow, weak interaction, as the violation of the conservation law of strangeness suggests.

The baryon number A is conserved in all interactions so the conservation law of strangeness can be expressed in terms of the *hypercharge* Y:

$$Y = A + S. \tag{2.323}$$

The hypercharge is conserved in strong interactions but not in weak interactions. The hypercharge of a particle is used to label particles in the recently developed classification schemes for particles.

The hypercharge of a species of particles can be related to the number of differently charged particles in that species (Problem 2.135). If the average electric charge of the species is $\bar{Q}$, then

$$\frac{\bar{Q}}{e} = \frac{Y}{2}. \tag{2.324}$$

This relation is true, for example, for pions $[\bar{Q} = (q_{\pi^+} + q_{\pi^0} + q_{\pi^-})/3 = 0]$, for nucleons $[\bar{Q} = (q_p + q_n)/2 = \frac{1}{2}]$, and for the Λ^0 $(\bar{Q} = q_{\Lambda^0} = 0)$. It is also true for the K mesons, if K^+ and K^0 are the particles and hence $K^- = \tilde{K}^+$ and $\tilde{K}^0$ are the antiparticles. (The neutral particles, the K^0 and the $\tilde{K}^0$, differ by their strangenesses or, in other words, by their production mechanisms.)

Other strange particles were observed that were found to obey the conserva-

---

* It was the study of these two modes, the θ and τ modes, respectively, that led Lee and Yang to postulate that the conservation law of parity may be violated in weak interactions. See P. Morrison, "The Overthrow of Parity," *Scientific American*, *196*: 45, April 1957. The study of the neutral K meson decays has led to a further important discovery, described in E. P. Wigner, "Violations of Symmetry in Physics," *Scientific American*, *213*: 28, December 1965.

tion law of hypercharge. These particles are produced by strong interactions, so it is convenient to introduce the name *hadrons* for all particles, mesons and baryons, that interact strongly. Baryons with nonzero strangeness are called *hyperons*.

A sigma hyperon Σ was found that existed in three charge states: Σ^+, Σ^0, Σ^-. The Σ hyperon has strangeness $S = -1$ and hypercharge $Y = 0$. One further particle, the xi hyperon Ξ^-, was observed to decay weakly via the process

$$\Xi^- \rightarrow \Lambda^0 + \pi^-$$
$$\quad\quad\quad \rightarrow p + \pi^-. \tag{2.325}$$

This decay does not take place sufficiently quickly for the Ξ^- to have $Y = 0$, like $\Lambda^0 + \pi^-$, or $Y = 1$, like $p + \pi^-$, and so Gell-Mann and Nishijima assigned the value $Y = -1$ to the Ξ hyperon. With this assignment, $\bar{Q}/e = -1/2$, which implies that there is another Ξ particle with $Q = 0$, the Ξ^0. This neutral particle was found subsequently and provided confirmation of the assignment and the theory.

The properties of those strange particles described above are listed in Table 2.2.

TABLE 2.2 *Strange Particles*

Particle	Spin	Mass, MeV	Baryon number A	Hyper-charge Y	Mean life, sec
K^+	0	493.78 ± 0.17	0	1	$(1.229 \pm 0.008) \times 10^{-8}$
K^0	0	497.7 ± 0.30	0	1	$\begin{cases}(0.909 \pm 0.015) \times 10^{-10}\\ (5.70 \pm 0.65) \times 10^{-8}\end{cases}$
Λ^0	$\frac{1}{2}$	$1{,}115.44 \pm 0.12$	1	0	$(2.61 \pm 0.02) \times 10^{-10}$
Σ^+	$\frac{1}{2}$	$1{,}189.39 \pm 0.14$	1	0	$(0.794 \pm 0.026) \times 10^{-10}$
Σ^0	$\frac{1}{2}$	$1{,}192.3 \pm 0.2$	1	0	$<1.0 \times 10^{-14}$
Σ^-	$\frac{1}{2}$	$1{,}197.20 \pm 0.14$	1	0	$(1.58 \pm 0.05) \times 10^{-10}$
Ξ^0	$\frac{1}{2}$	$1{,}314.3 \pm 1.0$	1	-1	$(3.05 \pm 0.38) \times 10^{-10}$
Ξ^-	$\frac{1}{2}$	$1{,}320.8 \pm 0.2$	1	-1	$(1.75 \pm 0.05) \times 10^{-10}$

SUMMARY   Strange particles are produced in pairs by the strong interaction, and they decay singly by the weak interaction. These processes can be distinguished if we assign a dynamical variable called strangeness, or one called hypercharge, to each particle. A process can proceed by a strong interaction only if strangeness or hypercharge is conserved, but it can proceed via the weak interaction if these are not conserved.

Problem 2.135

(a) On a graph, mark energies from 0 to 1,500 MeV on the vertical axis and the electric charges of $-e$, 0, and $+e$ on the horizontal axis. Mark the positions with circles on this graph of each of the hadrons that we have discussed so far and mark, with squares, the corresponding positions of each of the antiparticles.

(b) Show from the graph of (a) that the average charge Q of a species of particles is given by $\bar{Q}/e = (A/2 + S/2)$.

Problem 2.136

Determine whether the following reactions proceed via strong interactions or via weak interactions:

(a) $K^- + p \rightarrow \Xi^0 + K^0$.
(b) $\pi^0 + n \rightarrow \Sigma^+ + K^-$.
(c) $K^- + p \rightarrow \Sigma^- + \pi^+$.
(d) $\Xi^- \rightarrow \Lambda^0 + \pi^-$.
(e) $\Sigma^- \rightarrow n + \pi^-$.
(f) $K^+ \rightarrow \mu^+ + \nu_\mu$.

Problem 2.137

A particle of mass M at rest decays into two particles of masses m_1 and m_2. Let $\mathscr{P}$, p_1, and p_2 be the 4-momentums of the particles and let $(E_1/c, \mathbf{p}_1)$ be the components of p_1 in the rest frame of M.

(a) Show that

$$p_2 \cdot p_2 = (\mathscr{P} - p_1) \cdot (\mathscr{P} - p_1)$$

can be evaluated in the rest frame of M to give the relation

$$E_1 = c^2 \frac{M^2 + m_1^2 - m_2^2}{2M}.$$

(b) Let $\Delta M = M - m_1 - m_2$. Show that the kinetic energy of particle 1 relative to the rest frame of M is given by

$$T_1 = \Delta M c^2 \left(1 - \frac{m_1}{M} - \frac{\Delta M}{2M}\right).$$

(c) Calculate the kinetic energies of the proton and the pion in the decay, from rest, of the Λ given by $\Lambda \rightarrow p + \pi^-$.

Problem 2.138

A Λ decays in flight via the scheme $\Lambda \rightarrow p + \pi^-$. Let the 4-momentums p_p and p_π have components $(E_p/c, \mathbf{p}_p)$ and $(E_\pi/c, \mathbf{p}_\pi)$ relative to the laboratory frame. Show that

$$\cos \theta = \frac{(m^2 - m_p^2 - m_\pi^2)c^4 - 2E_p E_\pi}{2|\mathbf{p}_p|\,|\mathbf{p}_\pi|c^2},$$

where θ is the angle between $\mathbf{p}_p$ and $\mathbf{p}_\pi$.

Problem 2.139

Find the threshold energies for the following processes in which mesons are incident on stationary protons:

(a) $\pi^- + p \rightarrow K^0 + \Lambda^0$.
(b) $K^- + p \rightarrow \Sigma^- + \pi^+$.
(c) $K^- + p \rightarrow \Xi^0 + K^0$.
(d) $\pi^- + p \rightarrow \Sigma^- + K^+$.
(e) $K^- + p \rightarrow \Xi^- + K^+$.

2.4.9   *Resonances and other particles*

The discovery of the kaons and the hyperons was followed, in the 1960's, by the discovery of many other particles. The existence of some of these particles had been predicted before their discovery, whereas others were found, without any forewarning, as the result of experiments. In this part, we shall describe illustrative examples of these recently discovered particles.*

An indication of one of these particles had been observed initially in 1952 by E. Fermi and his colleagues, while measuring the cross section for the scattering of pions by protons. This cross section (Figure 2.116) has a strong peak at 196 MeV, showing that the pion and the proton interact particularly strongly in that energy state. (Fermi's measurements extended to 200 MeV only.) The bump in the cross-section curve is interpreted as a resonance in the state of the π-nucleon system that corresponds to the resonances at the proper frequencies of waves in a confined medium.† Further resonances have been observed at higher energies.

Each of these resonances can be interpreted as due to the formation of a particle that quickly decays into a pion and a nucleon (Figure 2.117); they are called *resonance particles*. Consider a particle that decays via the strong interactions in a small multiple of the characteristic time, 10^{-23} sec. The track of such a particle could not be seen in a cloud or bubble chamber because, in that time interval, the particle only travels a few femtometers. However, in a scattering experiment involving the decay products, such a particle would show up as a bump in the cross section at the appropriate energy, owing to the fact that an

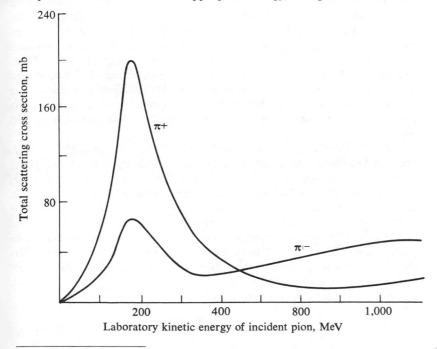

FIGURE 2.116   Cross section for the scattering of pions by protons.

---

* See also R. D. Hill, "Resonance Particles," *Scientific American, 208*: 38, January 1963.
† The mathematics involved in the analysis of the π-nucleon scattering also shows similarities to that used in wave theory.

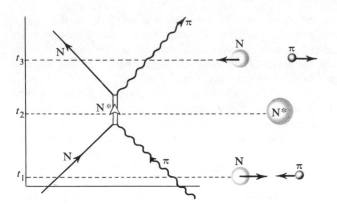

FIGURE 2.117 The formation of a short-lived resonance particle in π–N scattering.

additional mechanism for the interaction—namely, the formation of the unstable particle—is allowed for the scattering of the particles. This interpretation requires only that the definite numbers that are used to describe a particle, such as spin, can be assigned to the resonance. The resonance that appears in pion-proton scattering at $T_\pi = 196$ MeV corresponds to a spin of 3/2, and the resonance particle is denoted by $N^*_{3/2}(1236)$, or Δ_δ.

Two resonances in the interactions of pions with pions were predicted in the late 1950's as being necessary to explain the charge structure of nucleons as measured by electron-scattering experiments. The existence of the strong nucleon-nucleon interaction with its charge-exchange feature implies that, to an incident charged particle, a nucleon appears to possess an electric charge distribution. This comes about in the following manner: The nucleon-nucleon interaction has a nonzero range and also a charge-exchange character. Thus, associated with the interaction, there is an electric charge distribution that extends over the range of the interaction. A charged particle incident on a nucleon will experience the electric force that results from the charge of the nucleon plus that from the charge of the charge-exchange part of its strong interaction.

Analyses of the results of electron scattering from nucleons suggested that, although the charge associated with the charge-exchange part of the strong interaction was usually 1 elem ch, sometimes it was 2 and sometimes 3 elem ch. This, in turn, indicated that the interaction between pions was such that there existed a particle, possibly unstable, that corresponded to two pions and another that corresponded to three pions. However, this prediction could not be tested by π–π scattering experiments since, for example, free pions do not exist sufficiently long that a target of pions can be formed. Another technique was necessary for the examination of the existence of the dipion and tripion particles. One such technique is outlined below.

The 2π resonance was observed in the reactions

$$\pi^- + p \begin{cases} \to \pi^- + \pi^0 + p \\ \to \pi^+ + \pi^- + n. \end{cases} \qquad (2.326)$$

This reaction would proceed as shown in Figure 2.118(a) if there were no dipion resonance. However, if a 2π resonance state exists, the reaction could also proceed as shown in Figure 2.118(b); as shown there, the dipion is sufficiently long lived that it moves intact well away from the region of production before it decays. The characteristic feature of this process is that a single 2π particle exists and possesses a fairly well-defined mass outside the region of production. This feature can be analyzed as follows:

Consider a system that results in the production of three particles, labeled 1, 2, and 3, respectively. These particles are created in some process such as that shown in Figure 2.118. Let $\not{p}_1$, $\not{p}_2$, and $\not{p}_3$ be their respective 4-momentums. The mass M of the *system* consisting of particles 1 and 2 is given by the relation

$$M^2c^2 = (\not{p}_1 + \not{p}_2)^2 = (\mathscr{P} - \not{p}_3)^2, \tag{2.327}$$

where $\mathscr{P}$ is the total 4-momentum of the initial system. We evaluate $(\mathscr{P} - \not{p}_3)^2$ in the center-of-momentum system in which $\mathscr{P}$ has components $(E/c, 0)$ and $\not{p}_3$ has components $(e_3/c, \mathbf{p}_3)$:

$$\begin{aligned} M^2c^2 &= \mathscr{P}^2 + \not{p}_3^2 - 2\mathscr{P}\cdot\not{p}_3 \\ &= \frac{E^2}{c^2} + m_3^2 c^2 - \frac{2Ee_3}{c^2}; \end{aligned} \tag{2.328}$$

E is fixed by the dynamical condition of the initial system, whereas e_3 is not. If all three particles move off separately from the region of production [Figure 2.118(a)], then e_3 can take on all values from m_3c^2 (corresponding to the circumstance in which particle 3 is at rest relative to the center of momentum after creation) up to some maximum value $(e_3)_{\max}$ (corresponding to a circumstance in which both m_1 and m_2 move directly away from m_3). Therefore, M can take on all values from below $[(E/c^2) - (e_3)_{\max}/c^2]$ up to $[(E/c^2) - m_3]$, and for a given production mechanism, there will be a certain mass distribution $P_\Delta(M)$ for the probability that, in any such production reaction, M will lie in the range $[M - (\Delta/2), M + (\Delta/2)]$. Calculations, based on the assumption that there is no correlation between the directions of the three created particles as a result of their interactions, give a curve like that shown in Figure 2.119(a) for $P_\Delta(M)$. However, if m_1 and m_2 always form a particle of mass M^* that does not

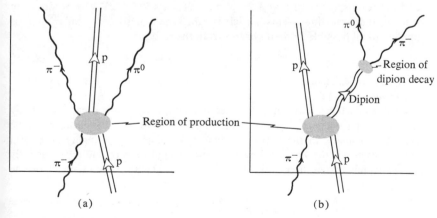

FIGURE 2.118   Two $\pi^- + p \to \pi^- + \pi^0 + p$ processes: (a) $\pi^- + p \to \pi^- + \pi^0 + p$. (b) $\pi^- + p \to$ dipion $+$ p
$\quad\quad\quad\quad\quad\quad\quad\quad \hookrightarrow \pi^0 + \pi^-$.

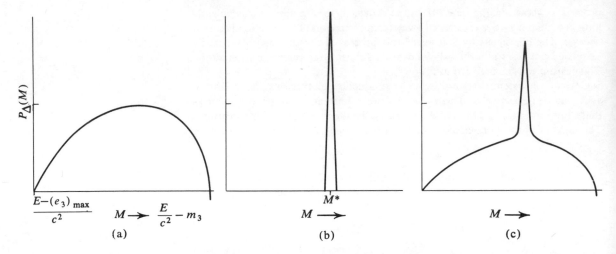

FIGURE 2.119   Mass distribution for two of three particles created in a collision.
(a) No correlation between the separation directions of the three created particles
as a result of their interactions. (b) Particles 1 and 2 always move away from the
region of production as an unstable particle of mass M^*. (c) In some, but not all
the reactions, m_1 and m_2 move off as the unstable particle M^*.

disassociate until that particle is well outside the region of production, then M is
determined uniquely by the conservation law of 4-momentum to be equal to M^*,
and the graph of $P_\Delta(M)$ appears as shown in Figure 2.119(b). If this unstable
particle is found only in some of the reactions, then the mass distribution appears
as in Figure 2.119(c).

An analysis similar to that described above was performed on the 2π products
of the reactions

$$\pi^- + p \quad \begin{aligned} &\to \pi^- + \pi^0 + p \\ &\to \pi^- + \pi^+ + n, \end{aligned} \qquad (2.329)$$

and the results[*] showed that a 2π particle of mass 765 MeV was produced in
some of these reactions. This particle is called the ρ meson. A similar analysis
applied to all possible 3π final systems from the reaction

$$\tilde{p} + p \to \pi^+ + \pi^+ + \pi^- + \pi^- + \pi^0 \qquad (2.330)$$

showed[†] the existence of a $\pi^+ - \pi^- - \pi^0$ resonance state of mass 783 MeV,
called the ω meson.

As these particles were being discovered, attempts were being made to classify
the known particles in a manner somewhat similar to the classification of the
chemical elements in the periodic table of Mendeléyev. A highly successful
classification scheme was developed in 1961 by Gell-Mann[‡] and independently
by the Israeli army officer and physicist Yuval Ne'eman. The existence of one

---

[*] See Figure 2 in A. R. Erwin, R. March, W. D. Walker, and E. West, "Evidence for a π–π Resonance
in the $I = 1, J = 1$ State," *Physical Review Letters, 6*: 628 (1961).
[†] See Figure 1 of B. C. Maglić, L. W. Alvarez, A. H. Rosenfeld, and M. L. Stevenson, "Evidence
for a $T = 0$ Three-Pion Resonance," *Physical Review Letters, 7*: 178 (1961).
[‡] See G. F. Chew, M. Gell-Mann, and A. H. Rosenfeld, "Strongly Interacting Particles," *Scientific
American, 210*: 74, February 1964.

particle, the η meson, was predicted on the basis of the classification scheme independently by Gell-Mann and Ne'eman in advance of its discovery in 1962. Another remarkable success of this theory was the prediction of the Ω particle that was required to fit the scheme after the discovery of some other particles. The previous successes of the classification scheme led to the remarkable effort* that resulted in the discovery of this particle.

These classification schemes for fundamental particles correspond to Mendeléyev's categorization of the chemical elements in the periodic table. And, just as there is more to chemistry than the periodic table, so there is more to particle physics than the classification schemes. Physicists today are studying the patterns that are exhibited in the behavior of these particles and are attempting to provide a rational basis for these relations, in the manner that the structure of atoms provides a basis for Mendeléyev's scheme. These investigations are very demanding on the physicists, and the challenge is made more stimulating by the unexpected discoveries and apparent paradoxes that often emerge from these investigations.

The properties of a few of the recently discovered particles are listed in Table 2.3.†

Particle	Spin	Baryon number A	Hyper-charge Y	Mass, MeV
η	0	0	0	548.8 ± 0.5
ω	1	0	0	782.7 ± 0.5
$\rho^{\pm}$	1	0	0	769 ± 3
ρ^0	1	0	0	769 ± 3
$N^*_{1/2}$ (1,518)	$\frac{3}{2}$	1	1	$1,518 \pm 10$
$N^*_{3/2}$ (1,236)	$\frac{3}{2}$	1	1	1,236
Y^*_0 (1,405)	$\frac{1}{2}$	1	0	1,405
Ω^-	$\frac{3}{2}$	1	-2	$1,675 \pm 3$

TABLE 2.3 *Some Recently Discovered Particles*

SUMMARY   Investigations of pion scattering by nucleons and of the electric charge structure of nucleons led to the discovery of very short-lived particles called resonance particles. With lifetimes of about 10^{-22} sec, they can be detected as peaks in the mass distribution of the system composed of a few of the particles in the final state of a reaction.

A classification scheme for these particles has been developed, which, among other things, has led to the discovery of other particles.

Problem 2.140

A pion of kinetic energy T is incident on a stationary proton, and a resonance baryon particle is produced. Show that the mass M^* of the resonance is given by $M^* = \sqrt{\mathscr{P} \cdot \mathscr{P}}/c^2$, where $\mathscr{P} = \{m_\pi c + (T/c) + m_p c, \ \sqrt{T[2m_\pi + (T/c^2)]}\hat{p}\}$ and $\hat{p}$ is a unit vector in the direction of the initial pion beam.

---

* See W. B. Fowler and N. P. Samios, "The Omega-Minus Experiment," *Scientific American*, *211*: 36, October 1964.
† A more complete table is given in A. H. Rosenfeld, A. Barbaro-Galtieri, W. H. Barkas, P. L. Bastien, J. Kirz, and M. Roos, "Data on Particles and Resonant States," *Reviews of Modern Physics*, *37*: 633 (1965).

Problem 2.141

Resonance baryon particles can be created when pions with the following energies are incident on stationary protons: 196 MeV, 550 MeV, 900 MeV, 1,354 MeV, and 2,349 MeV. Calculate the masses of the corresponding resonance particles.

Problem 2.142

Pions of momentum 1.89 GeV/c are incident on stationary protons and produce the reaction

$$\pi^- + p \rightarrow \rho^- + p$$
$$\quad\quad \rightarrow \pi^- + \pi^0.$$

Calculate the recoil energy and momentum of the proton if the ρ meson travels in the direction of the incoming pion beam.

Problem 2.143

The U.S. Atomic Energy Commission is designing an accelerator that will yield protons with energies of 200 BeV.

(a) The protons will be accelerated from 8 to 200 BeV in 0.8 sec. Calculate the average physical force experienced by the protons in that time interval.
(b) Calculate $(c - v)/c$ for protons of 200 BeV.
(c) Calculate the center-of-momentum energy available from a collision between a 200-BeV proton and a stationary proton.
(d) Consider a 200-BeV proton passing through air with an index of refraction of $n = 1.0060$. Find the angle of emission of the Čerenkov radiation (see Problem A1.3).
(e) (i) What is the radius of the circle upon which a 200-BeV proton travels in a region of constant **B** with $B = 1.5T$ (see Problem 2.76)?
   (ii) What would the radius be for a very intense **B** of $10T$?
(f) How many pions can be produced in the collision of a 200-BeV proton with a stationary proton?
(g) What is the maximum energy of the π^0 produced in the reaction

$$p + p \rightarrow p + p + \pi^0$$

for a 200-BeV proton incident on a stationary proton?
(h) (i) What is the maximum energy of the π^+ produced in the reaction

$$p + p \rightarrow p + n + \pi^+$$

for a 200-BeV proton incident on a stationary proton?
   (ii) What is the maximum energy of the neutrino produced in the resulting decay

$$\pi^+ \rightarrow \mu^+ + \nu_\mu?$$

   (iii) What is the mean distance that the π^+ of (ii) travels before decaying?
   (iv) If the μ^+ and ν_μ are emitted in the π^+ rest system at 90° to the direction of travel of the π^+, what is the angle between the direction of travel of the ν_μ and that of the π^+ in the laboratory system?
(i) It has been proposed that a high-energy proton beam be stored in rings under the action of a guiding magnetic force.

(i) If the magnetic induction is $1.5T$, calculate the radius of the ring for protons of energy (1) 1 BeV, (2) 5 BeV, (3) 10 BeV, (4) 50 BeV, and (5) 200 BeV.

(ii) What energy would be available in the center-of-momentum system if a proton from each of the storage rings of (i) collided head-on with a proton from the 200-BeV accelerator?

Problem 2.144

Scientists have proposed an accelerator that will yield protons of 1 TeV energy. Answer parts (b), (c), (d), (e), (f), (g), (h), and (i) of Problem 2.143 for protons of this energy.

Additional Problems

The data on particles necessary for these problems are given in Section 2.4.

Problem A2.1

The world line of a particle is described, relative to an inertial observer S, by the equation

$$x = \sqrt{1 + (ct)^2} - 1.$$

Draw the world line on the space time diagram of S and draw the 4-momentum vectors on that world line at times $ct = 0$, $ct = 1$, $ct = 2$, and $ct = 3$.

Problem A2.2

A 1-GeV proton incident on a stationary proton is scattered through $10°$. Find the angle of recoil of the target proton.

Problem A2.3

A beam of protons passes through a transparent medium whose index of refraction is 1.9 and emits Čerenkov radiation at $10°$ to the direction of the beam. Calculate the energy of the protons.

Problem A2.4

A 10-GeV proton is incident on a stationary proton.

(a) The incident proton is scattered through $90°$ in the center-of-momentum system. Find the scattering angle of the incident proton and the angle of recoil of the target proton in the laboratory frame.

(b) The incident proton is scattered through $180°$ in the center-of-momentum system. Find the laboratory kinetic energy of each of the protons after the collision.

Problem A2.5

(a) Prove that

$$\frac{dp}{p} = \frac{1}{1 - (v^2/c^2)} \frac{dv}{v}.$$

Use this result to show that m_v is a measure of inertia.

(b) Prove that

$$\frac{dv}{v} = \frac{1}{(E/mc^2)^2 - 1}\frac{dE}{E}.$$

Show from this that, for large E, E can increase appreciably without v undergoing a substantial change.

Problem A2.6

Consider the reaction $1 + 2 \to 3 + 4$, where the particles 1, 2, 3, and 4 have 4-momentums represented by $\not p_1$, $\not p_2$, $\not p_3$, and $\not p_4$, respectively. Choose energy units in which c takes on the value unity.

(a) Let $s = (\not p_1 + \not p_2)^2$. Show that $\sqrt{s}$ is the total energy in the center-of-momentum system.
(b) Show that, if particle 2 is stationary in the laboratory system,

$$E_1^{\text{lab}} = \frac{1}{2m_2}(s - m_1^2 - m_2^2).$$

(c) Let
$$t = (\not p_1 - \not p_3)^2, \qquad u = (\not p_1 - \not p_4)^2.$$

Show that

$$s + t + u = m_1^2 + m_2^2 + m_3^2 + m_4^2.$$

Problem A2.7

Two 3-ton trucks travel toward each other at 90 mi/hr and collide head on. Find the amount of mass of the resulting system above 6 tons.

Problem A2.8

(a) Calculate the kinetic energies of the neutron and the pion in the decay from rest of the Σ^+: $\Sigma^+ \to n + \pi^+$.
(b) Calculate the kinetic energies of the Λ and the pion in the decay from rest of the Ξ^-: $\Xi^- \to \Lambda + \pi^-$.

Problem A2.9

Find the threshold energy for protons incident on stationary protons for the reaction $p + p \to p + n + \pi^+$.

Problem A2.10

(a) Calculate the threshold energy for the process $\pi^- + n \to \Xi^- + K^0 + K^0$ for pions incident on stationary neutrons.
(b) Calculate the threshold energy for the process $K^- + p \to \Lambda + \pi^+ + \pi^-$ for K mesons incident on stationary protons.

Problem A2.11

(a) Find a formula for the threshold energy, for negative pions incident on protons, for the production of N neutral pions.
(b) Calculate the threshold energies for $\pi^- + p \to \pi^- + p + N\pi^0$ for $N = 1$, 2, 3, and 4.

Problem A2.12

Two particles, 1 and 2, initially at rest, interact to produce two other particles, 3 and 4. Show that

$$m_4 = (m_1 + m_2 - m_3)\sqrt{1 - z},$$

$$m_4 - m_2 = (m_1 - m_3)\sqrt{1 - z} - \frac{zm_2}{1 + \sqrt{1 - z}},$$

where

$$z = \frac{2(m_1 + m_2)}{(m_1 + m_2 - m_3)^2} T_3.$$

Problem A2.13

A K^+ meson decays into three charged pions that are emitted symmetrically. Find the magnitude of the momentum of each of the pions.

Problem A2.14

A 160-lb object travels at 60 mi/hr relative to S through empty space. A light of 100 W is attached to the object and shines a beam directly ahead of the object.

(a) Find the speed of the object, after the light is turned on, relative to S at the end of
   (i) 1 hr,
   (ii) 1 day,
   (iii) 1 week,
   (iv) 1 yr.
(b) How long, after the light is turned on, does it take for one-half the mass to be radiated away?

Problem A2.15

(a) A proton of energy 10 MeV experiences the physical force $\mathbf{F} = e\mathbf{v} \times \mathbf{B}$, where $\mathbf{B}$ is a constant magnetic induction vector of magnitude $1.5T$ and is perpendicular to $\mathbf{v}$. Find the radius of the orbit of the proton.
(b) Find the radius of the orbit for a proton with a kinetic energy of
   (i) 100 MeV,
   (ii) 1 GeV,
   (iii) 10 GeV.

Problem A2.16

A beam of negative pions of kinetic energy 100 MeV is produced in a laboratory. How does the relative intensity of the pions vary along the beam?

Problem A2.17

(a) Show that the angular momentum vector $\mathbf{L} = \mathbf{r} \times \mathbf{p}$ about O of a particle traveling with momentum $\mathbf{p}$ at the position $\mathbf{r}$ relative to O is determined by the nonvanishing components of the 3-tensor of the second order $L_{ij} = x_i p_j - x_j p_i$.

(b) The relativistic expression for the angular momentum is given by $L_{\alpha\beta} = x_\alpha p_\beta - x_\beta p_\alpha$. Show that $L_{\alpha\beta}$ is a skew-symmetric tensor of the second order.

(c) Show that $L_{\alpha\beta}$ is conserved (that is, $dL_{\alpha\beta}/dt = 0$) in the motion of a free particle.

(d) The total angular momentum tensor of a system of particles is given by

$$L_{\alpha\beta} = \sum_n L_{\alpha\beta}^{(n)},$$

where $L_{\alpha\beta}^{(n)}$ is the angular momentum tensor belonging to the particle labeled n. Show that $L_{\alpha\beta}$ is conserved if every particle of the system is free.

(e) Show that the vector $\mathbf{L}_0$ defined by

$$\mathbf{L}_0 = ct \sum_n \mathbf{p}^{(n)} - \sum_n \frac{E^{(n)}}{c} \mathbf{r}^{(n)}$$

is conserved for a system of free particles. The bracketed superscripts are labels denoting the various particles.

(f) Show for a system of free particles that the point C specified by

$$\mathbf{R}_C = \frac{\sum_n E^{(n)} \mathbf{r}^{(n)}}{\sum_n E^{(n)}}$$

moves with the constant velocity

$$\mathbf{V}_C = \frac{c^2 \sum_n \mathbf{p}^{(n)}}{\sum_n E^{(n)}},$$

where C is called the *center of inertia*.

(g) Show that the center of inertia coincides with the center of mass in the appropriate nonrelativistic limit.

Problem A2.18

A right-angled lever (shown in Figure A2.1) is pinned at its vertex but is free to rotate about that pin. The arms of the lever are of equal length, $l_{AB} = l_{BC}$, in the rest system S of the lever. The lever is constrained from rotating by two forces F_A and F_C of equal magnitude ($F_A = F_C$) in the rest system of the lever. The forces act in directions perpendicular to the arms of the lever.

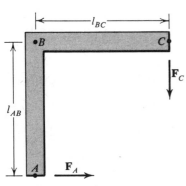

FIGURE A2.1 A right-angled lever under stress.

(a) Show that relative to S the magnitude T_B of the total torque on the system about B is given by

$$T_B = l_{AB}F_A - l_{BC}F_C = 0.$$

(b) Consider the lever from the point of view of an inertial system S' that moves with the constant speed V along the direction from B to C. Show that

$$l'_{AB} = l_{AB}, \qquad l'_{BC} = l_{BC}\sqrt{1 - (V^2/c^2)},$$

$$F'_A = F_A \quad \text{and} \quad F'_C = F_C\sqrt{1 - (V^2/c^2)}.$$

(c) Show that the total torque about B relative to S' is given by

$$T'_B = F'_A l'_{AB} \frac{V^2}{c^2}.$$

Does this torque result in a rotation of the lever relative to S'? State the reasons for your answer.

(d) Show that the force at A does work on the system in the amount $F_A V$ per second relative to S'. Thus, energy enters the system at A and flows out at the point B where a force opposite to $\mathbf{F}_A$ acts. Show that a mass of $F_A V/c^2$ enters A per second and thus results in an increase per unit time of the angular momentum of $(F_A V/c^2) V l_{AB}$. Resolve the paradox encountered in (c).

(e) A bar of length L experiences forces $\mathbf{F}_1$ and $\mathbf{F}_2$ at its ends. The forces are equal in magnitude in the rest system S of the bar and directed away from the center of the bar. An inertial observer travels with the velocity $\mathbf{V}$ relative to S, where $\mathbf{V}$ makes an angle α with the length of the bar. Repeat the problems (a)–(d) above, with suitable modifications for this case.

Problem A2.19

Relative to an inertial frame S, a particle experiences an attractive inverse-square central physical force

$$\mathbf{F} = -\frac{k}{r^2}\hat{r}, \qquad (k > 0).$$

This problem is concerned with the calculation of the orbit of the particle.

(a) Show that the time-like component of the 4-force equation gives the result

$$m_v c^2 - \frac{k}{r} = E, \qquad \left(m_v = \frac{m}{\sqrt{1 - (v^2/c^2)}}\right),$$

where E is a constant.

(b) Show that the spatial component of the 4-force equation gives the result $\mathbf{r} \times m_v \mathbf{v} = \mathbf{L}$, where $\mathbf{L}$ is a constant vector.

(c) Show that $\mathbf{r}$ lies in a plane that contains the force center O.

(d) Let (r, θ) be the polar coordinates of $\mathbf{r}$ relative to O in the plane of the motion. Show that

$$m_v r^2 \frac{d\theta}{dt} = L.$$

(e) Introduce $u[\theta(t)] = 1/r$. Show that

$$\frac{dr}{dt} = -\frac{L}{m_v}\frac{du}{d\theta}.$$

(f) Show that

$$\frac{1}{m^2} - \frac{1}{m_v^2} = \frac{1}{m^2 c^2}\left[\left(\frac{dr}{dt}\right)^2 + r^2\left(\frac{d\theta}{dt}\right)^2\right]$$

and hence that

$$m_v^2 = m^2 + \frac{L^2}{c^2}\left[\left(\frac{du}{d\theta}\right)^2 + u^2\right].$$

(g) Show that

$$m^2 + \frac{L^2}{c^2}\left[\left(\frac{du}{d\theta}\right)^2 + u^2\right] = \frac{1}{c^4}(E + ku)^2.$$

(h) Show that

$$\frac{d^2u}{d\theta^2} + u\left(1 - \frac{k^2}{L^2 c^2}\right) = \frac{Ek}{L^2 c^2}.$$

Assume that the initial conditions are such that $L > k/c$.

(i) Show that the solution of the differential equation of (h) is

$$u = A \cos (\alpha\theta + \beta) + \frac{Ek}{L^2c^2 - k^2} = A \cos (\alpha\theta + \beta) + \frac{E}{Lc} \frac{\sqrt{1 - \alpha^2}}{\alpha^2},$$

where $\alpha = \sqrt{1 - (k^2/L^2c^2)}$ and β and A are arbitrary.

(j) Show that the solution in (i) also satisfies the differential equation of (g) if

$$A^2 = \frac{c^2}{L^2\alpha^4} \left(\frac{E^2}{c^4} - m^2\alpha^2 \right).$$

(k) Show that r, given by

$$\frac{1}{r} = \frac{c}{L\alpha^2} \left[\sqrt{\frac{E^2}{c^4} - m^2\alpha^2} \cos (\alpha\theta + \beta) + \frac{kE}{Lc^3} \right],$$

is bounded if $E/c^2 < m$. Interpret this result.

(l) For $\alpha \approx 1$, show that the orbit approximates a rotating ellipse. Show that, in this case, the advance per revolution in the pericenter is given by

$$\frac{2\pi}{\alpha} - 2\pi \approx \frac{\pi k^2}{L^2c^2}.$$

ADVANCED REFERENCES

1. See R. Penrose and W. Rindler, "Energy Conservation as the Basis of Relativistic Mechanics," *American Journal of Physics*, *33*: 55 (1965), and the references given there, and also J. Ehlers, W. Rindler and R. Penrose, "Energy Conservation as the Basis of Relativistic Dynamics. II," *American Journal of Physics*, *33*: 995 (1965).

2. This point is discussed in the section entitled "The Paradox of Advanced Actions" in J. A. Wheeler and R. P. Feynman, "Classical Electrodynamics in Terms of Direct Interparticle Action," *Reviews of Modern Physics*, *21*: 425 (1949).

3. A translation of this paper (*Annalen der Physik*, *18*: 639, 1905) is given under the title "Does the Inertia of a Body Depend Upon its Energy-Content?" in *The Principle of Relativity*, by H. A. Lorentz, A. Einstein, H. Minkowski, and H. Weyl, Dover, New York, 1924.

General Theory of Relativity

3

No study of the laws of physics would be complete without reference to the general theory of relativity. This is a theory of gravitation that is primarily the result of the brilliant investigations of one man, Albert Einstein.

A detailed discussion of the general theory of relativity requires the introduction of a special branch of mathematics called the tensor calculus [1]. We shall not consider this in any detail, not because it is too difficult for a student at your stage of study, but rather because it would carry us too far afield to discuss it thoroughly here.* In this chapter, we shall outline only the basic features of the theory of general relativity and state some of its consequences.

General relativity is a theory of gravitation that provides, within the appropriate limits, the same description of planetary motions and other mechanical gravitational effects as does Newton's law of gravitation. Therefore it is worthwhile at this point to review some features of the newtonian theory of gravitation before proceeding to study Einstein's more modern theory.

Newton's law states that the gravitational interactions of objects can be described within the framework of newtonian mechanics by the interparticle force law

$$\mathbf{F}_{2(1)} = -G\,\frac{m_1 m_2}{r^2}\,\hat{r}. \tag{3.1}$$

The vector $\mathbf{r}$ represents the position of particle 2 relative to particle 1, $G = 6.670 \times 10^{-11}$ N·m²/kg is a universal constant, and m_1 and m_2 are the masses of the two particles. The resulting description of planetary motions agrees very well with the observations of these motions; there are only very minor discrepancies (of the order of motions of seconds of arc per century) between the theory and the observational data.†

The measure of the strength of the gravitational force exerted or experienced by a particle—a force which corresponds to the electric charge in Coulomb's law—is the inertial mass of the particle. Indeed, we might call that measure the

---

* An elementary introduction to the necessary mathematics and to details of the general theory of relativity can be found in L. R. Lieber and H. G. Lieber, *The Einstein Theory of Relativity*, Holt, Rinehart, & Winston, New York, 1945. More complete treatments of the subject are listed in the advanced references [2].
† It cannot be claimed that Einstein developed general relativity because there was a pressing need to reconcile theory and observation.

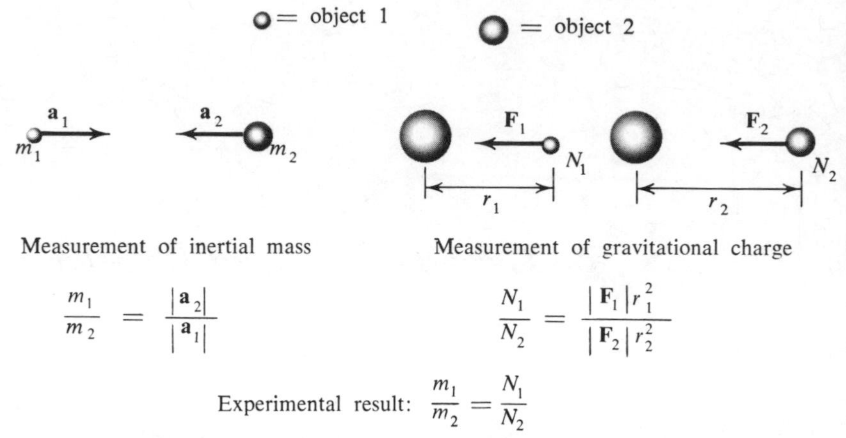

○ = object 1   ● = object 2

Measurement of inertial mass   Measurement of gravitational charge

$$\frac{m_1}{m_2} = \frac{|\mathbf{a}_2|}{|\mathbf{a}_1|}$$

$$\frac{N_1}{N_2} = \frac{|\mathbf{F}_1|\,r_1^2}{|\mathbf{F}_2|\,r_2^2}$$

Experimental result: $\dfrac{m_1}{m_2} = \dfrac{N_1}{N_2}$

FIGURE 3.1 Equality of inertial mass and gravitational charge.

gravitational charge, denote it by N, and wonder at its equality with the inertial mass m—a property of the particle that can be measured in an experiment in which gravitation plays no part (Figure 3.1).

The equality of the gravitational charge and the inertial mass is a coincidence within the newtonian theory of gravitation. Einstein, however, argued that scientists should not be satisfied as long as such a fundamental result appears to be only a coincidence; coincidences should be explainable on the basis of general laws of nature. To explore the possibility of such an explanation, we need to reformulate our description of the coincidence, which up to now states only the equality of the gravitational charge and the inertial mass. The new description must follow in mechanics from this equality (or Newton's law of gravitation), and it must be appropriate for generalization to a law of nature. A new description of the equality may be obtained if we examine the consequences of the force [Equation (3.1)] with N's replacing m's, and distinguish between the gravitational charge and inertial mass until equating these provides distinct new information.

Consider a particle of inertial mass m and gravitational charge N that experiences a gravitational force $\mathbf{F}_{\text{grav}}$ due to the presence of other objects and, in addition, other forces $\mathbf{F}$ that are not gravitational in character. The gravitational force $\mathbf{F}_{\text{grav}}$ is the vector sum of the interparticle gravitational forces given by Equation (3.1), each of which is proportional to N. Hence, the sum is proportional to N:

$$\mathbf{F}_{\text{grav}} = N\mathbf{e}, \tag{3.2}$$

where $\mathbf{e}$ depends only on the gravitational-charge distribution of the other objects relative to the particle under consideration. The equation of motion of the particle is, therefore,

$$ma = \mathbf{F}_{\text{grav}} + \mathbf{F}$$
$$= N\mathbf{e} + \mathbf{F} \tag{3.3}$$

or

$$ma - N\mathbf{e} = \mathbf{F}. \tag{3.4}$$

If we now equate N and m in this equation, it has the form of Newton's equation of motion

$$ma' = \mathbf{F}, \tag{3.5}$$

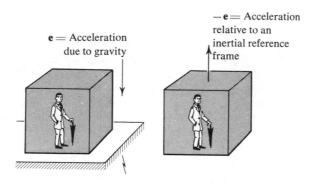

FIGURE 3.2 Two equivalent reference frames. (a) Inertial reference frame and gravitational force. (b) Noninertial reference frame and no gravitational force.

for circumstances in which the gravitational forces do not appear. The acceleration $\mathbf{a}'$ is that of the particle relative to a noninertial frame* experiencing the acceleration $-\mathbf{e}$ with respect to any inertial frame [see Equation (1.10)].

$$\mathbf{a}' = \mathbf{a} - \mathbf{e}. \qquad (3.6)$$

Thus, insofar as mechanical effects are concerned, the statement of the equality of inertial mass and gravitational charge is interchangeable with the statement that the mechanical effects of a gravitational force are equivalent to the apparent mechanical effects of referring motions without gravitational forces to a noninertial reference frame (Figure 3.2). Einstein postulated that this equivalence is a law of nature that applies not just to mechanics but to *all* phenomena. The principle of equivalence thus states, for example, that the path of a ray of light is bent near a massive (gravitating) object (Figure 3.3), a prediction verified by experiment.

The principle of equivalence describes the role of (newtonian) gravitation in other than mechanical phenomena. It extends the range of applicability of Newton's law of gravitation, although it does not replace it since the acceleration $-\mathbf{e}$ is calculated from Equation (3.1).

The use of noninertial reference systems in the principle of equivalence [3] and the intimate connections of space, time, and motion in the special theory

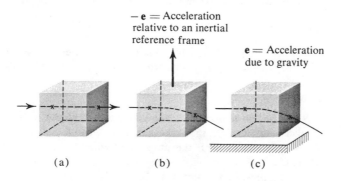

FIGURE 3.3 Path of light ray in (a) an inertial reference frame where no gravitational forces act; (b) a noninertial reference frame; (c) an inertial reference frame where gravitational forces act.

---

* The vector $-m\mathbf{A}$ that appears in the equation of motion $m\mathbf{a}' = \mathbf{F} - m\mathbf{A}$, relative to a frame experiencing the acceleration $\mathbf{A}$ with respect to the fixed stars, is called an inertial or fictitious force, or a pseudoforce, to distinguish it from the true force $\mathbf{F}$ that can be attributed to objects in the neighborhood of the particle.

of relativity provide a hint as to why a study of curved spaces is an important part of a consistent theory of gravitation.

The theory of general relativity involves the concept of curved space-time, and we study this by analogy with curved surfaces, which are curved spaces of two dimensions. These are discussed in Section 3.1. The possibility that gravitational forces reflect a bending or curvature in space-time follows from the principle of equivalence, as described in Section 3.2. The theory of general relativity and its test by experiment and observation are described in Section 3.3.

Problem 3.1

(a) Show that $V(r) = -GmM/r$ is the gravitational potential energy of a particle of mass m at a distance r from a particle of mass M by proving that $V(|\mathbf{r} + d\mathbf{r}|) - V(r) = -\mathbf{F}(\mathbf{r}) \cdot d\mathbf{r}$. $\mathbf{F}(\mathbf{r})$ is the newtonian force of gravity, and the reference point of V is chosen such that $V(r) \to 0$ as $r \to \infty$. *Hint*: Use $(\mathbf{r} + d\mathbf{r}) \cdot (\mathbf{r} + d\mathbf{r}) \approx r^2 + 2\mathbf{r} \cdot d\mathbf{r}$ and $(1 + x)^n \approx 1 + nx$, $x \ll 1$.

(b) Show that, for $|h| \ll R$, $V(R + h) - V(R) = mgh$, where g is the component of the acceleration due to gravity in the direction of the force center.

Problem 3.2

Consider an object suspended by a string at a point on the earth where the object experiences an acceleration $\mathbf{a}_R$ due to the rotation of the earth, and the acceleration due to gravity is $\mathbf{e}$ (Figure 3.4). Show that the direction of the force on the object due to the tension in the string $\mathbf{T}$ is along the direction of $\mathbf{a}_R - N\mathbf{e}/m$, where N is the gravitational charge and m the inertial mass of the object. Explain why the fact that the line of the string is the same for all objects demonstrates the equality of N and m.*

Problem 3.3

A rifle at ground level is aimed at an object balanced precariously on the edge of the wall of a tall building. The rifle is fired at the same instant that the object begins free fall from rest toward the ground. Use the principle of equivalence to show that the bullet will hit the object if the bullet's range is sufficient, and if air resistance can be neglected.

Problem 3.4

Explain, as you would to a high school student, why a light pulse initially directed parallel to the floor across a room is deflected in its motion.

Problem 3.5

Show within a nonrelativistic framework that the principle of equivalence predicts that starlight that approaches to within a distance D of a star of mass M is deflected through the angle Φ (Figure 3.5) given by $\Phi = 2(GM/c^2D)$. *Hint*: Consider the orbit of a particle that travels with a speed approximately equal to c past the gravitational charge of the star (see Problem 2.5).

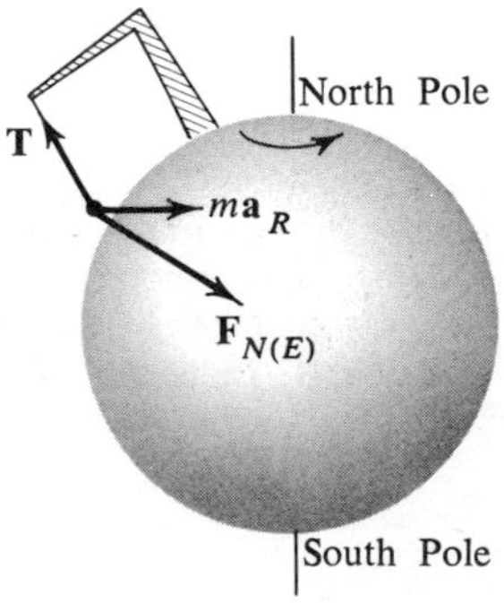

FIGURE 3.4   An experiment to measure $N - m$.

---

* This is the basis of an experiment that shows that the gravitational charge and the inertial mass are equal to a high degree of accuracy. See R. H. Dicke, "The Eötvös Experiment," *Scientific American* 205: 84, December 1961.

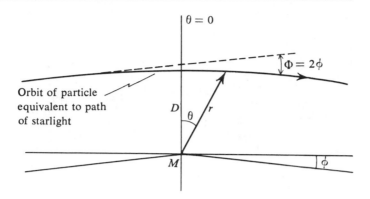

FIGURE 3.5 Deflection of a light ray in the neighborhood of a mass M.

3.1 Geometry and Space

In newtonian mechanics, the motions of objects are described as taking place in a three-dimensional Euclidean space. Within the framework of special relativity, the axioms of Euclidean geometry are valid for the spatial part, relative to each inertial observer, of space-time. Thus, until now we have worked under the assumption that the axioms of Euclidean geometry are valid for space in every inertial system. Prior to the middle of the nineteenth century, this was a necessary assumption for physicists to make, since there was no alternative; they were not aware of the existence of any other form of geometry. Therefore, they assumed that there exists in space a set of elements called points and lines that satisfy the axioms of Euclidean geometry,* such as the axiom that states that any two distinct points are incident with just one line. The particular axiom that will interest us is called the parallel postulate (Figure 3.6): There exists one and only one straight line DE that passes through a given point C and is parallel to a given line AB. The Euclidean axioms imply a set of propositions such as the one stating that "the sum of the angles of a triangle is 180°."

In the nineteenth century, mathematicians discovered consistent sets of axioms specifying properties of elements called points and lines that comprised the axioms of Euclid other than the parallel postulate, and one distinctive axiom replacing the parallel postulate. These systems of axioms and their consequences are called *non-Euclidean geometries*. In the following, we shall consider a two-dimensional non-Euclidean geometry (a curved surface in a three-dimensional Euclidean space) to show how these geometries can be investigated. Also, we shall extend without proof some of the results to three- and four-dimensional non-Euclidean geometries.

3.1.1 *A metric form for the surface of a sphere*

An introduction to non-Euclidean geometries can be provided most simply by a study of the geometry of surfaces, spaces of two dimensions. The relevant Euclidean geometry with which these non-Euclidean geometries can be compared is that of a Euclidean plane. This is the geometry applicable to a sheet of paper or a table top and is the geometry with which you are acquainted from

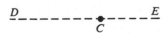

FIGURE 3.6 Euclid's parallel postulate: Through every point C, there is one and only one straight line DE parallel to AB.

---

* See M. Kline, "The Straight Line," *Scientific American*, *194*: 104, March 1956.

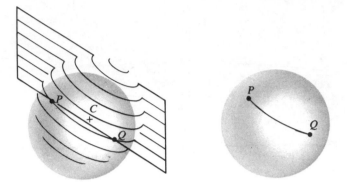

FIGURE 3.7 The line or geodesic joining two points P and Q on a sphere is along the great circle passing through P and Q.

your studies in high school. In the following, we shall compare and contrast this familiar geometry with non-Euclidean geometries.

An example of a corresponding non-Euclidean geometry is the geometry of the surface of a sphere. The points of this geometry are the points on the surface. There are, however, no straight lines on the surface of a sphere, although there do exist lines that satisfy many of Euclid's axioms. Indeed, we live on a spherical surface and often we go from one point to another on the surface by what we consider a "straight path," by which we mean that curve having the shortest distance between those two points. A path of the shortest distance is called a *geodesic* and corresponds to a *line* in the geometry of the sphere. It can be shown that such a line on a sphere is a great circle given by the intersection of the spherical surface with a plane that passes through the center of the sphere (Figure 3.7).

The points and lines in a sufficiently small region on the surface of a sphere cannot be distinguished by their form from the points and lines of a small region of a Euclidean plane. Thus, any small region on the surface of a sphere approximates a portion of a Euclidean plane.* This is why it is possible to consider a small region on the surface of the earth (such as a backyard or a football field) as being flat and to apply there the propositions of Euclidean geometry to measurements of distances and angles.

The geometry of the surface of a sphere is non-Euclidean, as we can show by two simple examples. Consider two lines that are drawn through two distinct points P and Q, respectively, and that are perpendicular to the line joining P and Q (Figure 3.8). On a Euclidean plane, the two lines are parallel and never meet [Figure 3.8(a)], but on the suface of a sphere the two lines meet at a point N [Figure 3.8(b)]. Furthermore, the sum of the angles of a triangle on a non-Euclidean surface is not necessarily 180°; for example, the sum of the angles of triangle PQN in Figure 3.8(b) is greater than 180° by the amount of angle at N subtended by the lines PN and QN.

How can we, who live on a surface, determine whether the surface is flat or curved? We could determine this from a rocket ship in space outside the surface, but we shall restrict our considerations to intrinsic methods that do not involve leaving the surface. Indeed, we are examining the geometry of surfaces only as an

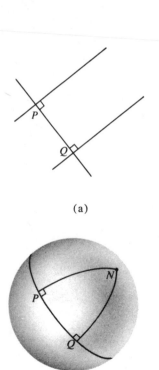

(a)

(b)

FIGURE 3.8 Two lines through P and Q, respectively, and perpendicular to PQ are parallel on a Euclidean plane (a) but meet at a point N on the surface of a sphere (b).

---

* There do exist surfaces with geometries in which it is not possible to approximate even a small portion by a Euclidean plane. We shall not consider these geometries here.

introduction to the possibilities for a geometry for the space in which we live. Therefore, we restrict our considerations to those methods of investigation that can be generalized to apply to the space in which we reside and from which we cannot escape.

One intrinsic method of investigating the features of a surface consists in proceeding along a line from a point P; if we return to P along that line, we have proved that the surface is non-Euclidean. For example, the early explorers demonstrated conclusively, by traveling around it, that the earth is not flat. This is a global method that involves measurements over distances comparable to the size of the surface; this method does not work for non-Euclidean surfaces that are infinite in extent, such as that of the saddle-like surface of which a portion is shown in Figure 3.9. Moreover, we should not have to travel across the universe in order to determine the geometry of space. For our purposes, we want a means of determining properties of a geometry by measurements in a small neighborhood of a point. This can be achieved on a surface in terms of the distances between neighboring points on the surface.

On a Euclidean plane, the distance ds between two points with rectangular coordinates (x, y) and $(x + dx, y + dy)$ is given by the pythagorean theorem (Figure 3.10):

$$ds^2 = dx^2 + dy^2. \tag{3.7}$$

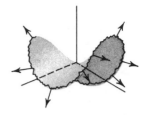

FIGURE 3.9 A portion of a non-Euclidean surface that is infinite in extent.

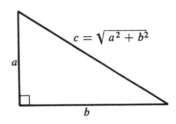

FIGURE 3.10 The pythagorean theorem.

This can be written in a notation that allows for easy generalization to spaces of dimensions higher than two; we set $x = x^1$ and $y = x^2$ (x superscript 2, not x squared) to obtain

$$ds^2 = (dx^1)^2 + (dx^2)^2. \tag{3.8}$$

The formula for the distance ds can be expressed in terms of coordinates other than the cartesian coordinates $x = x^1$ and $y = x^2$. For example, the formula in terms of the polar coordinates $r = x^{1'}$ and $\theta = x^{2'}$ is (Figure 3.11)

$$ds^2 = (dx^{1'})^2 + (x^{1'})^2(dx^{2'})^2; \tag{3.9}$$

however, this form can be reduced by the coordinate transformation

$$x^1 = x^{1'} \cos x^{2'}, \qquad x^2 = x^{1'} \sin x^{2'} \tag{3.10}$$

to the Euclidean form [Equation (3.8)].

An expression similar to Equation (3.8) can be obtained for distances in the

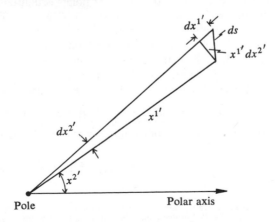

FIGURE 3.11
$$ds^2 = (dx^{1'})^2 + (x^{1'})^2(dx^{2'})^2.$$

neighborhood of a point on any non-Euclidean surface, such as a sphere, that has the property that the surface approximates a portion of a Euclidean plane in sufficiently small regions. However, the characteristic feature of the Euclidean plane is that coordinate systems exist such that Equation (3.8) is valid for distances at all points on the surface and not just those in the small neighborhood about one point, such as the origin of the coordinate system. On a non-Euclidean surface, there are no such coordinate systems; the expression that holds *at every point* on the surface has in any coordinate system the general form

$$
\begin{aligned}
ds^2 &= g_{11}(x^1, x^2)\,(dx^1)^2 + g_{12}(x^1, x^2)\,dx^1\,dx^2 \\
&\quad + g_{21}(x^1, x^2)\,dx^2\,dx^1 + g_{22}(x^1, x^2)\,(dx^2)^2 \\
&= \sum_{i,j=1}^{2} g_{ij}(x^1, x^2)\,dx^i\,dx^j,
\end{aligned}
\tag{3.11}
$$

where the g_{ij}'s depend explicitly on the coordinates of the point (x^1, x^2) in the neighborhood of which the distance ds is being measured; by the definition of a non-Euclidean surface, *it is impossible to choose a coordinate system on a non-Euclidean surface such that*

$$
g_{11}(x^1, x^2) = 1,
$$
$$
g_{12}(x^1, x^2) = g_{21}(x^1, x^2) = 0, \quad \text{and} \quad g_{22}(x^1, x^2) = 1
\tag{3.12}
$$

everywhere.

The form $\sum_{i,j} g_{ij}\,dx^i\,dx^j$ is called the *metric form*, and the set of numbers $g_{ij}(x^1, x^2)$ is called the *metric tensor*. The geometry of a space in which a metric form like Equation (3.11) can be defined is called a *Riemannian geometry*, after the German mathematician Georg Friedrich Bernhard Riemann (1826–1866), who proposed such geometries.*

The surface of a sphere provides an example of the result described above. To examine this, we need to introduce a coordinate system on the sphere, which we do in the following way: Consider any point P on the sphere of radius R (Figure 3.12). In a small neighborhood of P, we set up a rectangular coordinate system. The axes are extended by lines over the sphere. Lines equidistant along the x^1 axis from P are drawn at right angles to the x^1 axis. The x^1 coordinate of any point on one of these lines is equal to the distance from P along the x^1 axis to that line. At points equidistant from P along the x^2 axis, curves are drawn that cut the lines $x^1 = $ constant at right angles. The x^2 coordinate of any point on one of these curves is equal to the distance from P along the x^2 axis to that curve. The resulting coordinate system is shown in Figure 3.12. Note that this coordinate system is not the only one that we could set up on the sphere; however, this one system is sufficient for our purposes.

In the neighborhood of the origin P, the metric takes the form

$$
ds^2 = (dx^1)^2 + (dx^2)^2 \quad [(x^1, x^2)\ \text{sufficiently close to } P],
\tag{3.13}
$$

whereas, about any other point with coordinates (x^1, x^2), the metric form is (Figure 3.13)

$$
ds^2 = \left(\cos\frac{x^2}{R}\right)^2 (dx^1)^2 + (dx^2)^2.
\tag{3.14}
$$

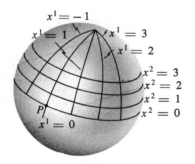

FIGURE 3.12 Coordinate system on the surface of a sphere.

---

* Riemann's investigations followed upon the study by the German mathematician Karl Friedrich Gauss (1777–1855) of two-dimensional surfaces in a Euclidean 3-space.

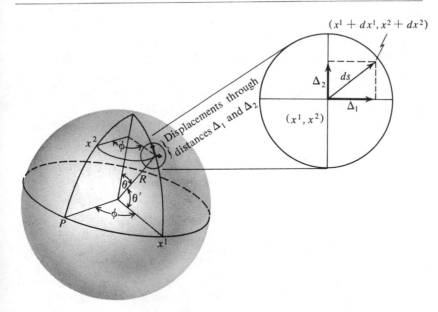

FIGURE 3.13

$ds^2 = [\cos (x^2/R)]^2 \, (dx^1)^2 + (dx^2)^2.$

The distance $\Delta_2 = R \, d\theta' = dx^2$. The distance

$\Delta_1 = R \sin \theta \, d\phi = R \sin (\pi/2 - \theta') \, d\phi$
$\quad = R \cos \theta' \, d\phi = \cos (x^2/R) \, dx^1.$

Therefore,

$ds^2 = \Delta_1^2 + \Delta_2^2$
$\quad = [\cos (x^2/R)]^2 \, (dx^1)^2 + (dx^2)^2.$

This particular form of the metric results from our choice of a coordinate system. *However, there does not exist any choice of coordinates for which this metric form reduces to the expression (3.8) at every point on the surface.* This result can be perceived by noting the following: Suppose we choose such a coordinate system along a small band around a great circle, say $x^2 = 0$, through P. Since the theorem of Pythagoras applies to the coordinates, this coordinate system can be represented by squares along that line. However, it is impossible to carry this hypothesized coordinate system off the equator because we cannot fill succeeding bands with the same number of similar squares (Figure 3.14). The quadrilaterals become pinched in at the side opposite the equator as we go off the equator.

This example illustrates the fact that, even though a small neighborhood of any point on a non-Euclidean surface may be quite similar to a portion of a Euclidean plane, deviations do exist from a Euclidean nature that show up in a variation of the metric tensor across the surface. Thus, *we can detect deviations from a Euclidean character in the behavior of the rate of change or the derivatives of the metric tensor.*

SUMMARY The geometry of the surface of a sphere is non-Euclidean. The nature of the geometry of a surface in which a metric form

$$ds^2 = \sum_{i,j=1}^{2} g_{ij}(x^1, x^2) \, dx^i \, dx^j$$

can be defined is determined by the metric tensor. If it is possible to choose coordinates such that the metric form is

$$ds^2 = (dx^1)^2 + (dx^2)^2$$

everywhere, the surface is a Euclidean plane. Otherwise, the surface is non-Euclidean. The non-Euclidean character of a surface can be detected by an examination of the behavior of the derivatives of the metric tensor.

FIGURE 3.14 We can set up a coordinate system in which there is a band of squares around the equator. However, we cannot introduce a band with the same number of identical squares off the equator.

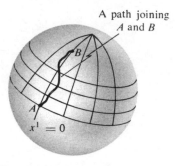

A path joining
A and B

$x^1 = 0$

FIGURE 3.15 The path is described by
the equation $x^1 = f(x^2)$.

Example 3.1

Q. Show that the shortest distance between two points A and B on the surface of a sphere lies along the great circle through those points.

A. The calculation of this problem can be made easier by a suitable choice of coordinates. Let A and B lie along the x^2 axis for which $x^1 = 0$ in a coordinate system like that shown in Figure 3.12.

Consider any path along the surface that joins A and B (Figure 3.15). This path can be specified by an equation of the form

$$x^1 = f(x^2). \tag{3.15}$$

The distance along this path from A to B is given by

$$\int_A^B ds = \int_A^B \sqrt{(dx^2)^2 + \left(\cos\frac{x^2}{R}\right)^2 (dx^1)^2}$$

$$= \int_A^B \sqrt{1 + \left(\cos\frac{x^2}{R}\right)^2 \left(\frac{df}{dx^2}\right)^2}\, dx^2. \tag{3.16}$$

The integrand is always greater than or equal to unity, so the integral has its minimum value for a function $f(x^2)$ such that the integrand always has its smallest value, namely unity. Therefore,

$$\frac{df(x^2)}{dx^2} = 0, \qquad A \le x^2 \le B, \tag{3.17}$$

and since, by our choice of coordinate axes, $f = 0$ at A, $f(x^2) = 0$ for x^2 between A and B.

The line $x^1 = 0$ joining A and B is a great circle; thus, we conclude that the geodesics on a sphere are great circles.

City	Miles
Miami–Moscow	5,731
Miami–Singapore	10,546
Miami–Capetown	7,658
Miami–Cairo	6,484
Moscow–Singapore	5,238
Moscow–Capetown	6,300
Moscow–Cairo	1,803
Singapore–Capetown	6,005
Singapore–Cairo	5,137
Capetown–Cairo	4,500

Problem 3.6

In the table you are given the shortest distances (in miles) between the cities. Prove that the earth is not flat. *Hint*: Try to draw a scale map on a sheet of paper.

Problem 3.7

The components of a vector $\mathbf{V}$ at the point (x^1, x^2) of a spherical surface are (V^1, V^2) in the coordinate system shown in Figure 3.12. Show that the cartesian components of the vector at that point are $[V^1 \cos(x^2/R), V^2]$. The cartesian components $(\mathscr{V}_1, \mathscr{V}_2)$ of the vector are equal to the (directed) lengths of the component vectors $\mathbf{V} = \mathbf{V}_1 + \mathbf{V}_2$ in the appropriate directions, whereas the components (V^1, V^2) are equal to the relative number of coordinate lines that such lengths encompass.

Problem 3.8

Show that the transformation

$$x^1 = x^{1'} \cos x^{2'}, \qquad x^2 = x^{1'} \sin x^{2'}$$

reduces the metric form

$$ds^2 = (dx^{1'})^2 + (x^{1'})^2 (dx^{2'})^2$$

to the form

$$ds^2 = (dx^1)^2 + (dx^2)^2.$$

Problem 3.9

Show, using a rectangular coordinate system, that the path of shortest distance between two points on a Euclidean plane lies along a straight line.

Problem 3.10

Show, using a polar coordinate system, that the path of the shortest distance between two points on a Euclidean plane lies along a straight line.

3.1.2 The curvature tensor

A non-Euclidean surface can be distinguished from a Euclidean plane by the way in which the metric varies from point to point on the surface. This important feature of the metric is illustrated below for the case of a spherical surface. Our concern will be those combinations of the metric tensor and its derivatives that distinguish the spherical surface from a Euclidean plane. The appropriate derivatives of the metric tensor can be determined geometrically in the following manner:

Consider a vector $\mathbf{V}$ defined at some point (x^1, x^2) on the surface; this might represent a velocity, for example. The components (V^1, V^2) of this vector with respect to the given coordinate system represent, as in the Euclidean plane, the relative number of coordinate units required to describe the perpendicular components in the direction of the x^1 and x^2 coordinate curves at that point— that is, as in our previous work on vectors like velocities and forces, a vector $\mathbf{V}$ is represented through a scaling factor, say α, in terms of a displacement $\Delta\mathbf{x}$: $\mathbf{V} = \alpha\,\Delta\mathbf{x}$ or, alternatively, $V^1 = \alpha\,\Delta x^1$, $V^2 = \alpha\,\Delta x^2$. The relevant displacement $\Delta\mathbf{x}$ must be very small in a curved surface, since the length of the coordinate units varies from point to point in a curved space.

Suppose such a vector is assigned at some point, say the origin P, and we wish to compare this vector with another vector assigned at some point on the surface, say that labeled Q on Figure 3.16. The comparison can be made at Q after the vector $\mathbf{V}$, defined at P, is transported to Q. At Q, the vector $\mathbf{V}$ may have components different from (V^1, V^2) because the coordinate axes are not Euclidean rectangular axes. We can proceed naively to obtain the components of $\mathbf{V}$ at the point Q, as follows: Since the neighborhood of a point on the surface is similar to a small portion of a Euclidean plane, we can use the Euclidean concept of parallelism for small displacements. Therefore, we can transport the vector to the point Q along some curve such that, for each small displacement $d\mathbf{x} = (dx^1, dx^2)$ along the curve, the vector at the point $(x^1 + dx^1, x^2 + dx^2)$ is parallel to that at (x^1, x^2) in the Euclidean sense. The change $d\mathbf{V}$ in the components of the vector under such a small displacement depends linearly on the components of the vector and also linearly on the components of the displacement (see Example 3.2). This can be represented as

$$d\mathbf{V} = -\mathbf{\Gamma} \circ (\mathbf{V}; d\mathbf{x}), \tag{3.18}$$

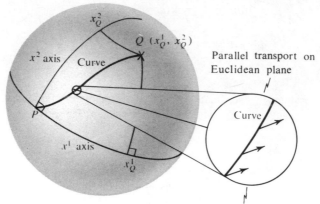

Parallel transport on
Euclidean plane

Portion of Euclidean plane

FIGURE 3.16 A vector **V** given at P
may undergo parallel transport along
the curve to Q.

where the entity **Γ** describes an operation that combines the components of the
vector **V** and the components of the displacement $d\mathbf{x}$ to yield the change $d\mathbf{V}$.
On a spherical surface with the coordinate system of Figure 3.12, the change in
the components of **V** under the displacement $d\mathbf{x}$ can be shown (Problem A3.8)
to be

$$dV^1 = \frac{1}{R}\tan\left(\frac{x^2}{R}\right)(V^1\,dx^2 + V^2\,dx^1),$$

$$dV^2 = -\frac{1}{R}\left(\cos\frac{x^2}{R}\right)\left(\sin\frac{x^2}{R}\right)V^1\,dx^1. \qquad (3.19)$$

By moving the vector parallel to itself through a succession of these small
displacements, we can transport the vector **V** from P to Q along the given curve.

The mathematical entity **Γ**, called the *connection*, that describes the combination of the components of **V** and $d\mathbf{x}$ represents a measure of the rate at which
the metric tensor varies in the neighborhood of the point (x^1, x^2) and involves
the first derivatives of the components of the metric tensor.

Now that we have a method of transporting a vector from P to Q along a
curve, we must inquire next whether the vector transported to Q depends upon
the curve along which it is transported—that is, if we transport the vector to Q
along two different curves, is the result the same? It certainly is for parallel
transport on a Euclidean plane, but the vector transported to Q on a non-
Euclidean surface does depend on the curve, as the following example shows.
Consider the vector $\mathbf{V} = (0, V)$ assigned at P and the triangle on the spherical
surface shown in Figure 3.17. We transport the vector to the point Q along
the curve $x^1 = 0$ and also along the curve PRQ, where PR is the axis $x^2 = 0$,
R has coordinates $(\pi R/2, 0)$, and RQ is the line $x^1 = \pi R/2$. We see from Figure
3.17 that the vectors that undergo parallel transport along the two paths are
perpendicular at the point Q. This result can be stated as follows: *A vector
transported parallel to itself around a closed curve undergoes no change on a
Euclidean plane but undergoes some change on a non-Euclidean surface.* This
result provides us with a means of distinguishing a Euclidean plane from a
non-Euclidean surface, and it also provides a way of determining the warping
or curvature of a surface.

Consider a vector **V** with components (V^1, V^2) assigned at the point P with

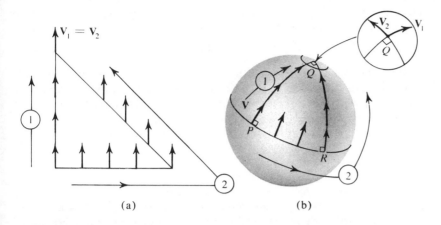

FIGURE 3.17 Parallel transport around a closed curve provides a means of distinguishing (a) a Euclidean surface from (b) a non-Euclidean surface.

(a)   (b)

coordinates (x^1, x^2). Transport this vector parallel to itself around the small area shown in Figure 3.18. We consider the area as a directed entity with the direction being given by the right-hand rule applied to the direction of propagation around the area. In the general case of a multi-dimensional space, the directed area will be represented by an entity that is described by a number of components; let the symbol **A** represent that entity. Then, after being transported parallel to itself around this area, the vector **V** will be changed by Δ**V**, which depends linearly on the components of **V** and on those of the directed area **A** (see Example 3.4). We can represent this symbolically by the form

$$\Delta \mathbf{V} = -\mathbf{R} \circ (\mathbf{V}; \mathbf{A}), \tag{3.20}$$

where the entity **R** describes the operation that combines the components of **V** and those of the directed area **A** that yield the change Δ**V**. On the sphere, this equation has components with the form (see Problem A3.9)

$$\Delta V^1 = -\frac{1}{R^2} V^2 A,$$

$$\Delta V^2 = \frac{1}{R^2} \left(\cos \frac{x^2}{R} \right)^2 V^1 A, \tag{3.21}$$

where A is the magnitude of the area if the area is circumvented in the counterclockwise direction. The entity denoted by **R** depends on the coordinates (x^1, x^2) and is called the *curvature tensor* at the point P. The curvature tensor is a mathematical entity that, like a vector, can be specified by a number of components relative to a given coordinate system but, in fact, is independent of any coordinate system (see Section 1.9). Thus, for example, if all the components of the curvature tensor in a given space vanish everywhere with respect to one coordinate system, they vanish everywhere relative to any other coordinate system in that space. The components of the curvature tensor **R** are written usually as

$$R^i_{jmn}, \quad i = 1 \text{ or } 2, \quad j = 1 \text{ or } 2, \quad m = 1 \text{ or } 2,$$

and
$$\tag{3.22}$$

$$n = 1 \text{ or } 2 \quad \text{(in a two-dimensional space)}.$$

There is no change under parallel transport around a closed curve on a

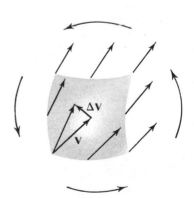

Direction of propagation

FIGURE 3.18 The change in **V** under parallel transport around the area is Δ**V**.

Euclidean plane, so the curvature tensor is zero at every point on a Euclidean plane. The condition for a Euclidean surface, therefore, is the set of equations

$$R^i_{jmn} = 0. \tag{3.23}$$

On the other hand, the curvature tensor does not vanish at every point on a non-Euclidean surface like a sphere; indeed, on a sphere, the curvature tensor is proportional to $1/R^2$ and provides a measure of the bending or warping of the surface.

The curvature tensor **R** involves the rate at which the connection **Γ** varies around the small area; **R** involves the first derivatives of **Γ** and products of the **Γ**'s; **Γ** consists of combinations of the first derivatives of the metric tensor g_{ij}. Therefore, *the curvature tensor contains second derivatives of the components of the metric tensor; furthermore, it is not linear in these components.* These mathematical properties of **R** are important in Einstein's theory of gravitation.

We considered a two-dimensional non-Euclidean geometry above because of the difficulties involved in visualizing a three-dimensional geometry that does not satisfy Euclid's axioms. However, consistent sets of axioms for the geometry of a three-dimensional space can be imagined for which the space does not have a Euclidean character. We can obtain a mental notion of a non-Euclidean space by imagining some sort of a lumpiness or warping in the space. The bending of space can be detected in a manner identical to that used in our analysis of a non-Euclidean surface. We consider a vector **V** designated by the components (V^1, V^2, V^3) at a point P with coordinates (x^1, x^2, x^3). Under parallel transport to the point $\mathbf{x} + d\mathbf{x}$, the components of **V** undergo the change

$$d\mathbf{V} = -\mathbf{\Gamma} \circ (\mathbf{V}; d\mathbf{x}). \tag{3.24}$$

Under parallel transport around a directed area specified by **A**, the vector changes by

$$\Delta\mathbf{V} = -\mathbf{R} \circ (\mathbf{V}; \mathbf{A}), \tag{3.25}$$

where **R** denotes the curvature tensor in the three-dimensional space. The space is Euclidean if the components of the curvature tensor R^i_{jkl} vanish everywhere; it is non-Euclidean otherwise.

Since the possibility of a non-Euclidean space exists, we must inquire as to whether the space in which we live is Euclidean or non-Euclidean. This question can be decided only by experiment. There is no *a priori* reason why space should satisfy one set of axioms, such as those of Euclid, rather than another. However, we can expect, on the basis of our experiences, that space is Euclidean for an observer who is concerned only with a small region of space. We do not expect that deviations from Euclidean geometry will show up to any appreciable extent in the laboratory.

There exists also the possibility that space-time is curved and is not properly described by the *Lorentz metric form*

$$ds^2 = c^2 (dt)^2 - (dx)^2 - (dy)^2 - (dz)^2. \tag{3.26}$$

We shall investigate this possibility in Section 3.2.

SUMMARY   Because a small portion of a spherical surface approximates a region of a Euclidean plane, a vector can be carried by parallel transport step by step along a curve in the surface. The changes in the components of the vector between neighboring points are determined by the connection **Γ**. The change in a vector upon parallel transport around a small area is measured by the curvature tensor **R**. This tensor is zero everywhere on a Euclidean plane but not on a non-Euclidean surface. These results can be generalized to spaces of three or more dimensions.

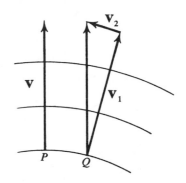

FIGURE 3.19 Under parallel transport, a vector parallel to one of the coordinate axes may acquire an additional component.

Example 3.2

The connection **Γ** at any point on a surface is represented by a set of components

$$\mathbf{\Gamma}: \ \Gamma^i_{jk}, \qquad i = 1 \text{ or } 2, \qquad j = 1 \text{ or } 2, \qquad k = 1 \text{ or } 2. \qquad (3.27)$$

These components and their relation to the changes in the components of vectors under parallel transport arise in the following way:

Consider a vector that has only one nonvanishing component at a point P relative to a curvilinear coordinate system (Figure 3.19). Under parallel transport to a neighboring point Q, this vector may acquire a component along the other coordinate line. Thus, each component V^i of a vector may experience a change that is linear in both the component V^i and the other component.

Similarly, the change from parallel transport through a small displacement dx^j along the x^j coordinate line is proportional to that displacement. Since the components of small displacements are additive, the change from parallel transport in a small displacement $d\mathbf{x} = (dx^1, dx^2)$ is linear in each of the components. Therefore, in general, the change dV^i in the component of a vector under parallel transport through the small displacement (dx^1, dx^2) is linear in the components of the vector (V^1, V^2) and the components of the displacement dx^k:

$$dV^i = -\sum_{j,k=1}^{2} \Gamma^i_{jk} V^j \, dx^k. \qquad (3.28)$$

The components Γ^i_{jk} that represent the connection are called *coefficients of connection*.

The coefficients of connection Γ^i_{jk} at a point depend on the relationship between neighboring coordinate lines and, therefore, on the rate at which the metric tensor varies in the neighborhood of that point. Thus, the Γ^i_{jk} depend on the derivatives of the metric tensor g_{ij}.

Example 3.3

Q. Find the coefficient of connection Γ^2_{21} relative to a polar coordinate system at the point $(x^1, x^2) = (r, \theta)$ on a Euclidean plane.

A. The change dV^i in the component V^i of the vector (V^1, V^2) at (x^1, x^2) under parallel transport through the displacement (dx^1, dx^2) is given by

$$dV^i = -\sum_{j,k=1}^{2} \Gamma^i_{jk} V^j \, dx^k. \qquad (3.29)$$

Our concern here is with Γ^2_{21}. This coefficient appears by itself if we find the change dV^2 in the component V^2 of the vector $(0, V^2)$ under parallel transport through the displacement $(dx^1, 0)$, since in this case,

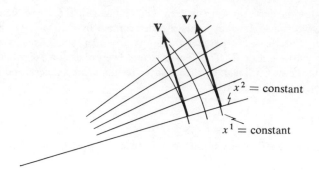

FIGURE 3.20   Under parallel transport, **V** goes over to **V'**.

the sum over j and k reduces to a single term:

$$dV^2 = -\Gamma^2_{21} V^2 \, dx^1. \tag{3.30}$$

The component V^2 represents the relative number of coordinate intervals required to give the length of the vector (Figure 3.20). The distance between the coordinate lines x^2 and $x^2 + dx^2$, which is the length of the coordinate interval, is given by $x^1 \, dx^2 (= r \, d\theta)$, so $V^2 x^1$ is a constant as the component V^2 undergoes parallel transport along the x^1 coordinate line. Therefore,

$$0 = d(V^2 x^1) = (dV^2)x^1 + V^2 \, dx^1 \tag{3.31}$$

or

$$dV^2 = -\frac{1}{x^1} V^2 \, dx^1. \tag{3.32}$$

Thus, the coefficient of connection Γ^2_{21} is given by

$$\Gamma^2_{21} = \frac{1}{x^1}. \tag{3.33}$$

Example 3.4

The components R^i_{jkl} of the curvature tensor arise in the following way:

Consider a vector (V^1, V^2) defined at the point $P(x^1, x^2)$. The changes in the components under parallel transport to $P_1(x^1 + dx^1_{(1)}, x^2 + dx^2_{(1)})$ are given by

$$d_{(1)}V^i = -\sum_{j,k} \Gamma^i_{jk}(x^1, x^2)V^j \, dx^k_{(1)}. \tag{3.34}$$

Under parallel transport from P_1 to P_2 (Figure 3.21), there are additional changes

$$\begin{aligned}
d_{(12)}V^i &= -\sum_{j,k} \Gamma^i_{jk}(x^1 + dx^1_{(1)}, x^2 + dx^2_{(1)})(V^j + d_{(1)}V^j) \, dx^k_{(2)} \\
&= -\sum_{j,k} [\Gamma^i_{jk}(x^1, x^2) + d\Gamma^i_{jk}](V^j + d_{(1)}V^j) \, dx^k_{(2)} \\
&= -\sum_{j,k} (\Gamma^i_{jk}V^j \, dx^k_{(2)} + \Gamma^i_{jk}d_{(1)}V^j \, dx^k_{(2)} + d\Gamma^i_{jk}V^j \, dx^k_{(2)}),
\end{aligned} \tag{3.35}$$

to second order in the dx's; $d_{(1)}V^j$ is given by Equation (3.34) and by

$$d\Gamma^i_{jk} = \frac{\partial \Gamma^i_{jk}}{\partial x^1_{(1)}} dx^1_{(1)} + \frac{\partial \Gamma^i_{jk}}{\partial x^2_{(1)}} dx^2_{(1)}. \tag{3.36}$$

Therefore, under parallel transport around the parallelogram $PP_1P_2P_3$, the changes ΔV^i in the components are given by an expression of the form

$$\Delta V^i = - \sum_{j,k,l=1}^{2} R^i_{jkl} V^j \, dx^k_{(1)} \, dx^l_{(2)}. \tag{3.37}$$

The set of numbers R^i_{jkl} forms the components of the curvature tensor **R** at the point (x^1, x^2). It follows from Equation (3.35) that the components R^i_{jkl} involve the derivatives of the coefficients of connection Γ^i_{jk} and hence the second derivatives of the metric tensor; they also involve products of the Γ^i_{jk} and therefore are nonlinear in the components of the metric tensor.

Problem 3.11

Find the coefficient of connection Γ^2_{12} relative to a polar coordinate system at the point $(x^1, x^2) = (r, \theta)$ on a Euclidean plane.

Problem 3.12

Find the coefficient of connection Γ^1_{22} relative to a polar coordinate system at the point $(x^1, x^2) = (r, \theta)$ on a Euclidean plane.

Problem 3.13

Describe what happens to the components, relative to a polar coordinate system, of a vector under parallel transport around a small quadrilateral at $(x^1, x^2) = (r, \theta)$ on a Euclidean plane. Explain why the coefficients of connection are not zero but the curvature tensor is.

3.2 The Principle of Equivalence and Curved Space-Time

Let us assume for the moment that we are two-dimensional creatures living on the surface of a large sphere. Let us assume further that we had investigated the behavior of objects in a small region on that surface (which we assumed was a portion of a Euclidean plane) and that we had arrived at laws of mechanics

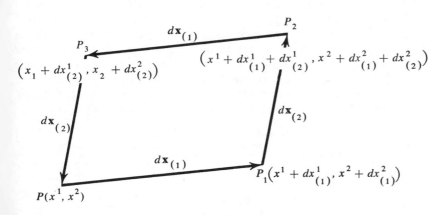

FIGURE 3.21 The parallelogram $PP_1P_2P_3$.

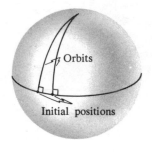

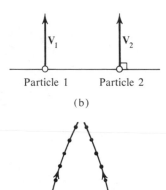

(b)

(c)

FIGURE 3.22 Free motions on a spherical surface. (a) The initial positions and the orbits of the two particles. (b) The initial conditions of the motions. (c) Apparent subsequent motions of the two particles, visualized as taking place on a Euclidean plane.

similar to those of Newton as a result of these investigations. In particular, let us make the assumption that objects in free motion travel along the geodesics, the "straight lines," of our curved space. It is important to note that, at this stage, the "law" of geodesic motion is an assumption that is made in addition to those of the properties assumed for the space.

At this moment, our techniques have developed to the point where we can either make measurements over larger distances or make more precise measurements in our small region; so we perform the following experiment: Two objects are given identical initial velocities along two parallel lines and left alone to travel freely (Figure 3.22). The objects travel along the geodesics of the curved space. If we examined their behavior over distances smaller than $\pi R/2$, we would observe that the objects approach each other as shown in Figure 3.22(c).* We might try to describe this behavior in terms of a force of attraction. Alternatively, we might be more imaginative and attempt to explain this behavior in terms of a curvature in the space in which we live.†

This simple example can provide further insight into the topic of physics in curved spaces. It shows a circumstance in which each particle follows a curve, the geodesic, independent of the properties of that particle. We had studied before one class of real motions that does not depend on the properties of the object undergoing them—namely, motions under gravitational forces. These forces, and the motions they describe, are independent of the properties of the object because of the equality of gravitational charge and inertial mass, and hence may result from the geometry of the space or the space-time in which we exist. In this section, we explore this suggestion further in terms of the principle of equivalence.

3.2.1 Geodesic motion in a region of space devoid of gravitational forces

The simple example given in the introduction indicates that it may be possible to describe motions under a gravitational force as geodesic motions in a curved space. We begin our investigation of this possibility by examining the feasibility of describing free motions in a region devoid of gravitational forces as geodesic motions.

In the example illustrated in Figure 3.22, it was necessary to specify the behavior in time of the objects as well as the geometry of the space. As a result of the assumption of geodesic motions, the geometry of the space determined the orbit of a freely moving particle. However, that assumption of geodesic motions did not determine the time development of the motion; the direction in which the particle moves is given by the geometry, but the rate at which it moves is not. This raises the possibility that we should extend our considerations from space to space-time, since the "direction" of the world line in space-time includes both direction in space and rate of motion (see Figure 1.106). Let us investigate this possibility for the case of a freely moving particle.

* If we followed their motions over larger distances, we would observe that their paths cross and, indeed, return to the initial positions. The point we wish to make is relevant only if we do not observe their motions over such large distances.
† Another such example is described on page 79 of Bertrand Russell, *The ABC of Relativity*, rev. ed., F. Pirani (Ed.), Geo. Allen, London, 1958.

The Lorentz metric form

$$ds^2 = c^2\,dt^2 - dx^2 - dy^2 - dz^2 \qquad (3.38)$$

adequately describes the structure of space-time in regions free of "gravitational forces." For a freely moving particle, ds is the proper time interval along the world line of the particle, and dx, dy, and dz are related to dt by the equation

$$\mathbf{v} = \left(\frac{dx}{dt}, \frac{dy}{dt}, \frac{dz}{dt}\right). \qquad (3.39)$$

Hence

$$ds = c\,dt\sqrt{1 - (v^2/c^2)}. \qquad (3.40)$$

The assumption was made in the simple fictitious example of Figure 3.22 that free particles moved along geodesics. Thus, we could describe the characteristic feature of the orbit of the hypothetical freely moving particle in terms of the metric form. The *characteristic feature of the free motion of a real object is the fact that its velocity $\mathbf{v}$ is a constant.* Can we describe this characteristic feature of real free motions in terms of the Lorentz metric form ds?

Consider the integral

$$\int_1^2 ds = \int_1^2 c\sqrt{1 - [v^2(t)/c^2]}\,dt \qquad (3.41)$$

along any world line between two given events 1 and 2 (Figure 3.23). This integral corresponds to the distance along a possible path of the motion in Figure 3.22. In that example, the orbit corresponded to the shortest path between the endpoints of the motion. Here, we wish to compare the values of the integral $\int_1^2 ds$ over the various possible motions between the two events. Since ds is an invariant, we can evaluate this integral in any reference system, and, in particular, in that inertial system S_0 in which the events 1 and 2 occur at the same place (Figure 3.24). This particular reference system is convenient for our purposes because the free motion of the particle from event 1 to event 2 relative to this frame is described by $\mathbf{v}(t) = 0$. Indeed, the particle cannot go from event 1 to event 2 with a motion described by $\mathbf{v}(t) = 0$ relative to any other reference system. For this reason, the following argument for the result that $\int_1^2 ds$ is a maximum for free motion is the simplest argument by which this result can be obtained, although the same result can be acquired if we refer the motions to any reference system through the use of more complicated reasoning (see Example 3.5).

The integrand $c\sqrt{1 - [v^2(t)/c^2]}$ is always less than c unless $v(t) = 0$. Since the particle can travel between event 1 and event 2 with $v(t) = 0$ relative to S_0, and since $v(t) = 0$ is the condition in S_0 for free motion between these events, the condition in S_0 and, hence, in any inertial reference system for free motion between the two events is that the integral $\int_1^2 ds$ be a maximum.* This corresponds to the condition for a straight line or a geodesic joining events 1 and 2 in space-time (Figure 3.25) relative to any inertial system.

---

* The fact that $\int_1^2 ds$ is a maximum for a geodesic instead of a minimum occurs because dt^2 appears with the opposite sign to that of dx^2, dy^2, and dz^2 in ds^2.

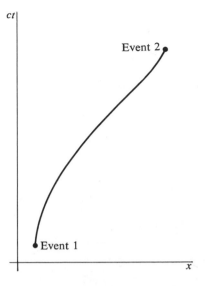

FIGURE 3.23 An arbitrary world line joining events 1 and 2.

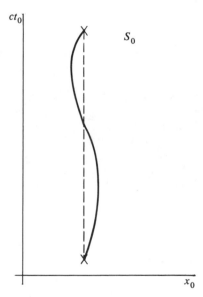

FIGURE 3.24 Arbitrary world line joining events 1 and 2 in that inertial frame S_0 in which events 1 and 2 take place at the same point in space.

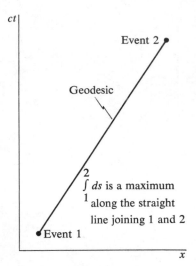

ct

Event 2

Geodesic

$\int_1^2 ds$ is a maximum
along the straight
line joining 1 and 2

Event 1

x

FIGURE 3.25   Condition for free
motion.

Thus, we can describe the characteristic feature of free motion in terms of the Lorentz metric. This indicates the possibility that we can likewise describe motion under a gravitational force in terms of a modified metric form.

SUMMARY The condition for free motion between event 1 and event 2 in a region of space devoid of gravitational forces is that the proper time $\int_1^2 ds$ be a maximum, where

$$ds^2 = c^2\, dt^2 - dx^2 - dy^2 - dz^2$$

is the Lorentz metric.

Example 3.5

The following method can be used to show that $\int_1^2 ds$, evaluated between two fixed events 1 and 2, has a maximum for **v** equal to a constant vector.

We choose our coordinate axes so that the space-time coordinates of the events 1 and 2 are $(0, 0, 0, 0)$ and $(cT, X, 0, 0)$. The motion with constant velocity between events 1 and 2 is described by the (one-dimensional) velocity $V = X/T$, or by the equation

$$x(t) = Vt. \tag{3.42}$$

We consider now other motions between the events 1 and 2. Let $\eta(t)$ be any function that satisfies the relations

$$\eta(0) = \eta(T) = 0. \tag{3.43}$$

Then, the function

$$x_\varepsilon(t) = Vt + \varepsilon\eta(t), \tag{3.44}$$

where ε is a number, describes a motion between the two events 1 and 2 (Figure 3.26). We wish to show that $\int_1^2 ds$ evaluated over this motion is less than that integral evaluated over the motion with constant speed.

Consider the integral

$$I(\varepsilon) = \int_1^2 ds = \int_1^2 \sqrt{c^2 - (dx_\varepsilon/dt)^2}\, dt. \tag{3.45}$$

This is bounded above by $c(t_2 - t_1)$, and thus the integral has a maximum value that occurs for that value of ε for which

$$\frac{dI(\varepsilon)}{d\varepsilon} = 0. \tag{3.46}$$

Since

$$\frac{dx_\varepsilon}{dt} = V + \varepsilon\frac{d\eta}{dt}, \tag{3.47}$$

this condition can be written as

$$0 = \int_1^2 \frac{-(d\eta/dt)[V + \varepsilon(d\eta/dt)]}{\sqrt{c^2 - [V + \varepsilon(d\eta/dt)]^2}}\, dt. \tag{3.48}$$

An integration by parts gives

$$0 = \frac{-\eta[V + \varepsilon(d\eta/dt)]}{\sqrt{c^2 - [V + \varepsilon(d\eta/dt)]^2}}\Bigg|_1^2 + \int_1^2 \eta\frac{d}{dt}\left\{\frac{V + \varepsilon(d\eta/dt)}{\sqrt{c^2 - [V + \varepsilon(d\eta/dt)]^2}}\right\} dt. \tag{3.49}$$

The first term vanishes, since $\eta = 0$ at the events 1 and 2. The variable t appears within the braces in the integrand of the second term only in the terms in $\varepsilon(d\eta/dt)$; so the derivative operator d/dt is equivalent, according to the chain rule, to $d[(\varepsilon\, d\eta/dt)/dt][d/d(\varepsilon\, d\eta/dt)]$ or $\varepsilon(d^2\eta/dt^2)\, d/d(\varepsilon\, d\eta/dt)$. Therefore, the last term is proportional to ε:

$$0 = \int_1^2 \eta \frac{d}{dt} [\cdots]\, dt = \varepsilon \int_1^2 \eta \frac{d^2\eta}{dt^2} \frac{d}{d[\varepsilon(d\eta/dt)]} [\cdots]\, dt. \qquad (3.50)$$

This expression is zero if $\varepsilon = 0$ or $x_\varepsilon(t) = x(t)$. Hence, the integral $\int_1^2 ds$ is a maximum for the motion described by $x = Vt$.

Problem 3.14

Show that $\int_1^2 \frac{1}{2}mv^2\, dt$ is a minimum between the two events 1 and 2 for the nonrelativistic force-free motion between those two events.*

Problem 3.15

Use the following method to show that $\int_1^2 ds$, evaluated between two given events 1 and 2, has a maximum for $\mathbf{v}(t) = d\mathbf{r}/dt = $ a constant vector.

(a) Let $\mathbf{r}_0(t)$ describe that motion from event 1 $(ct_1, \mathbf{r}_1)$ to event 2 $(ct_2, \mathbf{r}_2)$ for which the integral $\int_1^2 ds$ is a maximum. Let $\boldsymbol{\eta}(t)$ be any vector that satisfies $\boldsymbol{\eta}(t_1) = \boldsymbol{\eta}(t_2) = 0$. Show that, for ε a number,

$$\mathbf{r}(t) = \mathbf{r}_0(t) + \varepsilon\boldsymbol{\eta}(t)$$

describes a motion between events 1 and 2.

(b) Explain why the integral

$$\int_{t_1}^{t_2} \sqrt{c^2 - \{[d\mathbf{r}_0(t)/dt] + \varepsilon[d\boldsymbol{\eta}(t)/dt]\}^2}\, dt$$

has a maximum value for $\varepsilon = 0$.

(c) Show that

$$0 = \int_{t_1}^{t_2} \frac{-(d\mathbf{r}_0/dt)\cdot(d\boldsymbol{\eta}/dt)}{\sqrt{c^2 - (d\mathbf{r}_0/dt)^2}}\, dt.$$

(d) Show that

$$0 = \int_{t_1}^{t_2} \boldsymbol{\eta}(t)\cdot\left[\frac{d}{dt}\left(\frac{d\mathbf{r}_0/dt}{\sqrt{c^2 - (d\mathbf{r}_0/dt)^2}}\right)\right]\, dt.$$

(e) Show that the integral of (d) vanishes for every vector function $\boldsymbol{\eta}(t)$ if $\mathbf{r}_0(t)$ satisfies the relativistic equation of motion for a free particle.

3.2.2   The general form of the metric in the neighborhood of a massive object†

We defined inertial reference systems as those systems moving with constant velocities relative to the fixed stars. In the absence of gravitational forces, Newton's laws of motion and other laws of physics take on their simplest forms when referred to these systems; for example, light travels in straight

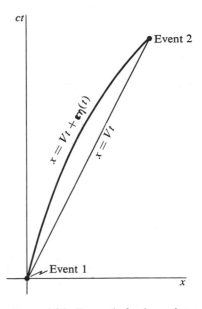

FIGURE 3.26   Two paths for the motion between events 1 and 2.

---

* An extension of this result applicable to motions under a given force is described in Feynman, Leighton, and Sands (vol. 2), Addison-Wesley, Chap. 19, pp. 19–1 to 19–7.
† Taylor and Wheeler, W. H. Freeman, Chap. 3, p. 175.

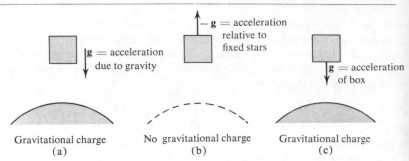

FIGURE 3.27   The principle of equivalence. The laws of physics are the same in the box in (a) as in the box in (b). Alternatively, the box in (c) is equivalent to an inertial reference system in which no gravitational forces act.

lines relative to an inertial reference system, although not in the presence of gravitational forces. The effects of gravitational forces can be taken into account through use of the principle of equivalence. This principle states that the laws of physics that describe the behavior of physical systems relative to an inertial system in which a constant acceleration due to gravity **g** is experienced are equivalent to the laws of physics relative to specific noninertial systems in which no gravitational forces act (Figure 3.27). The equivalent noninertial systems are those that undergo the acceleration $-\mathbf{g}$ relative to the fixed stars.

Indeed, near the surface of the earth, the simplest way to obtain the equivalent to a .local, small region of an inertial reference frame in which objects in undisturbed motions travel with constant velocity and in which light travels in straight lines is to use as the reference frame a box that is falling freely.* There is no need to take account of gravitational forces in such a frame; the effects of the gravitational force appear only upon transformation to a reference system moving with a constant velocity relative to the fixed stars.

The same transformation is not applicable to every point in space, however. It is because of this fact that a gravitational force, which is called a true force, differs from an inertial or fictitious force. An inertial force arises when motions are referred to a reference frame, extending over all space, that is undergoing an acceleration relative to the fixed stars [Figure 3.28(a)]. The inertial force disappears at all points in space under a (uniform) transformation to an inertial system [Figure 3.28(b)]. On the other hand, a gravitational force cannot be transformed away at all points in space. A true gravitational force varies from point to point in space [Figure 3.28(c)], and thus also the corresponding "inertial" reference system varies from point to point [Figure 3.28(d)].

The fact that, in a region of gravitational force, we can transform to the equivalent of an "inertial" system only locally is reminiscent of the circumstances that we encountered before on the surface of a sphere; in that case, we could transform to a Euclidean plane only locally. The fact that we were dealing there with a curved surface manifested itself in the metric form. This suggests that we investigate whether we can replace the effects of gravity by the effects of a metric form that can be reduced locally to that of an inertial system.

* Taylor and Wheeler, W. H. Freeman, Sec. 2, pp. 5–11, define an inertial reference frame to be a freely falling reference frame. This definition of inertial frames is appropriate in the physics of curved space-time, but it is not convenient to use in an introductory study of mechanics. For example, with this definition, familiar motions like those of a baseball or a car are described relative to an inertial frame undergoing an acceleration of 9.8 m/sec² downward with respect to the playing field or the road.

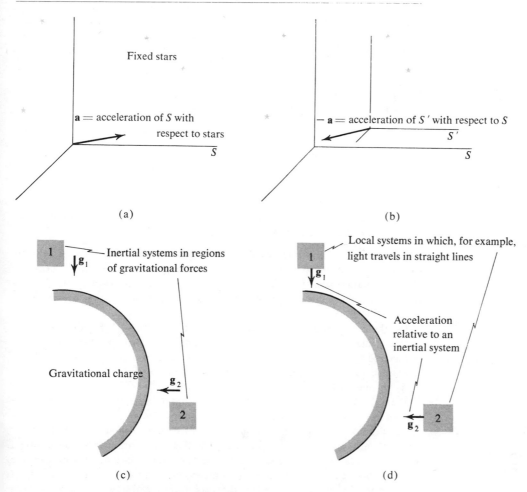

FIGURE 3.28   The difference between an inertial force and a gravitational force. (a) The inertial force experienced relative to S is the same at every point in space. (b) There are no inertial forces anywhere relative to S'. (c) The gravitational force, and hence the gravitational acceleration $\mathbf{g}$, varies from point to point in space. (d) The effects of a gravitational force can be transformed away, but the transformation varies from point to point in space.

The metric form appropriate for an inertial system is the Lorentz metric,

$$ds^2 = c^2\, dt^2 - dx^2 - dy^2 - dz^2$$
$$= (dx^0)^2 - (dx^1)^2 - (dx^2)^2 - (dx^3)^2. \tag{3.51}$$

The world lines of particles undergoing accelerated motions relative to an inertial frame are curved (Figure 3.29); the transformation to an accelerated frame is a transformation among both space and time coordinates. This suggests that the metric form appropriate for the description of the effects of a large gravitational charge has the general form

$$ds^2 = \sum_{\mu,\nu=0}^{3} g_{\mu\nu}(x)\, dx^\mu\, dx^\nu, \tag{3.52}$$

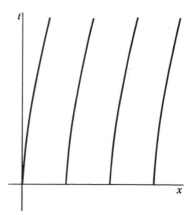

FIGURE 3.29   The world lines of particles undergoing accelerated motions relative to an inertial frame.

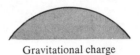

Gravitational charge

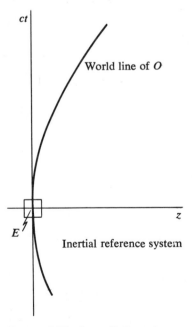

where the components $g_{\mu\nu}$ of the metric tensor are not constants and, indeed (see Figure 3.28), cannot be transformed to those in Equation (3.51) at all events in space-time. The validity of this suggestion is made plausible by arguments in Sections 3.2.3 and 3.2.4.

SUMMARY   The metric that describes the gravitational effects of a massive object has the general form

$$ds^2 = \sum_{\mu,\nu=0}^{3} g_{\mu\nu}(x)\, dx^\mu\, dx^\nu.$$

The metric tensor $g_{\mu\nu}$ cannot be reduced to the Lorentz metric at all points in space-time.

3.2.3 A metric form appropriate for a reference system experiencing a constant acceleration relative to the stars

The principle of equivalence allows us to describe the effects of gravitational forces in terms of behavior relative to a reference system accelerated with respect to the inertial frames. Therefore, we use this principle now to obtain an approximation to a metric form appropriate for a region R of constant gravitational acceleration $\mathbf{g}$.

Let the extent of R in any direction be approximately H (Figure 3.30). We shall assume that g and H are sufficiently small that their product, (a speed)^2, is much less than c^2. The natural measure of such an approximation is gH/c^2, and since we wish to obtain only a first estimate of the effects of g, we shall carry through calculations valid only to the first order in this quantity.

We follow the principle of equivalence and replace the inertial reference system in the region of gravitational acceleration $\mathbf{g}$ by a reference system undergoing the acceleration $-\mathbf{g}$ in a gravitation-free region of space (Figure 3.27). We wish now to incorporate the effects of this acceleration in a metric form. Because of the condition that the acceleration $\mathbf{g}$ is constant, the appropriate metric form is that for a curved coordinate system in a flat space, the Minkowski space-time continuum. However, the procedure we follow is similar to that used in Section 3.1 to determine the metric form on a spherical surface, and this similarity will be exploited below to introduce the metric form appropriate for a region of constant acceleration.

It should be noted that there exist many possible coordinate systems, as was the case with the spherical surface. The particular coordinate system described below is fitting for our work because it allows us to generalize the metric from that applicable to a region of constant gravitational acceleration [Equation (3.58)] to that applicable to a region of variable gravitational acceleration [Equation (3.62)].

In the case of the spherical surface, we first chose one point P on that surface and, in a small neighborhood about that point, we set up a coordinate system relative to which the metric form was Euclidean (Figure 3.31). Here, we choose an event E at one point O in R and, in a small space-time region about that event, we set up a coordinate system relative to which the metric form is Lorentz. The time axis of this coordinate system is along the world line of O at the event E, and one spatial axis, say the z axis, is along the direction of the acceleration $-\mathbf{g}$ (Figure 3.32).

Next, in the case of the spherical surface, we extended the axes through P and we marked off distances along these lines. One of these lines was selected because of some distinctive feature, such as its being the equator in the case of the earth. Coordinates near this line were chosen in such a way that the metric along the line was Euclidean (Figure 3.33). Here, also, we extend the axes of the local coordinate system around E. We will be concerned with the metric form at other points P in R relative to that at O, so we extend the time-like axis along the world line of O and choose coordinates in the neighborhood of that line such that the metric is always Lorentz at O (Figure 3.34):

$$ds^2 = c^2 \, dt^2 - dx^2 - dy^2 - dz^2 \qquad \text{at } O \text{ always.} \qquad (3.53)$$

The fact that the metric is Lorentz along the world line of O restricts our choice of clocks and measuring rods; they must be those appropriate to an inertial system at O at any instant of time. At any time t during its motion, the point O will be instantaneously at rest relative to some inertial frame S_t (Figure 3.35). The clock and the elements of the measuring rods in S_t coincident with O at that instant determine appropriate inertial unit time and distance intervals at O at that time. Thus, for example, the rate of the clock at O is timed in turn by the various inertial clocks that are successively at rest relative to O (at the instants at which they are coincident with O).

Finally, in the case of the spherical surface, we selected the "straight lines" perpendicular to that line along which the metric had the Euclidean form and marked off unit distances along those lines to obtain the completed coordinate system (Figure 3.36). Here also, at an event such as E along the time axis, we select space-like lines perpendicular to the instantaneous direction of the time axis and mark off the unit distances along those lines (Figure 3.37).

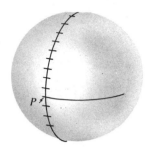

FIGURE 3.33 Extension of the coordinate axes through P and the choice of a line along which the metric is taken to be Euclidean.

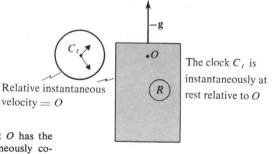

The clock C_t is instantaneously at rest relative to O

FIGURE 3.35 The clock at O has the same rate as the instantaneously coincident clock in the inertial system that is instantaneously at rest relative to O.

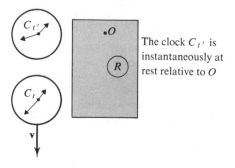

The clock $C_{t'}$ is instantaneously at rest relative to O

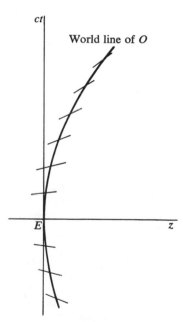

FIGURE 3.34 Extension of the coordinate axes through E and the world line of O along which the metric is taken to be Lorentz. The straight-line segments are along the direction of the spatial axes corresponding to the instantaneous time axes at the points where the straight lines intersect the world line of O.

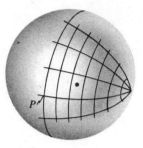

FIGURE 3.36 The completed coordinate system.

This is equivalent to determining distances in R with inertial measuring rods in the following manner: The units of distance are marked off by comparison with the measuring rods of the inertial system S_t that is instantaneously at rest relative to R (Figure 3.38). The inertial system S_t is instantaneously at rest relative to every point in R at the same instant of time as measured by the clocks in S_t. Therefore, the distances in R are marked off simultaneously according to the clocks in the instantaneous rest inertial system. This ensures that the spatial distance between O and P is the same relative to every inertial

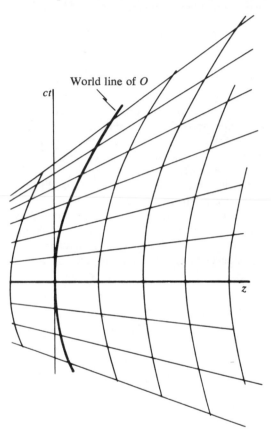

FIGURE 3.37 The completed coordinate system. The straight lines are the spatial axes corresponding to the instantaneous time axes at the points where the straight lines intersect the world line of O.

frame at the instant that the frame is at rest relative to R. Also, because of this, the metric at every point P in R is

$$ds^2 = c^2\, dt_{SP}^2 - dx^2 - dy^2 - dz^2, \tag{3.54}$$

where dt_{SP} is the time interval at P measured relative to the instantaneous rest frame S_t.

The relation between dt_{SP} and the corresponding time interval dt at P that results from our choice of coordinates can be obtained in the following way: Consider the time interval dt between two events A and B on the world line of P (Figure 3.39). Since AB and CD lie along time-like coordinate lines and both lie between two spatial coordinate lines, this time interval has the same length as the time between the two events C and D relative to the inertial frame S_t.

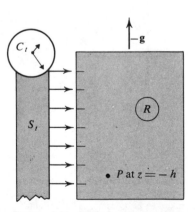

FIGURE 3.38 Distance measurements in R. The distances are marked off by the measuring rods of the inertial system S_t that is instantaneously at rest relative to R.

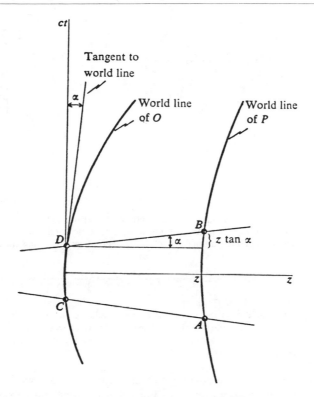

FIGURE 3.39 $AB = CD + 2z \tan \alpha$.

The magnitude of the slope, $\tan \alpha$, of the spatial coordinate lines through A and B on the space-time diagram of the instantaneous rest system is given by v/c, where v is the speed of O at the events C and D. This is the speed acquired under the acceleration $\mathbf{g}$ in the time $dt/2$:

$$\tan \alpha = \frac{v}{c} = \frac{g(dt/2)}{c}. \tag{3.55}$$

The time between the events A and B as measured by the relevant inertial clock in S_t is dt_{SP}, and the size of this time interval is larger than dt by the amount

$$2z \tan \alpha = \frac{zg}{c} \, dt. \tag{3.56}$$

Therefore,

$$c \, dt_{SP} = c\left(1 + \frac{zg}{c^2}\right) dt. \tag{3.57}$$

This result obtains also if we use the same set of clocks that was used at O to provide the rate of the clock at any point P below O in R ($z < 0$). This is shown in Figure 3.40.

The metric at every point P in R is given, to order gH/c^2, in terms of the time and spatial coordinates pertinent to that point by Equation (3.54) with the substitution from Equation (3.57):

$$ds^2 = c^2\left(1 + 2\frac{gz}{c^2}\right) dt^2 - dx^2 - dy^2 - dz^2. \tag{3.58}$$

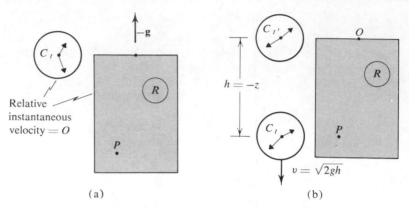

(a)   (b)

FIGURE 3.40   An approximate method for setting the rate of the clock at P, below O in R. The interval of time dt_O at O is related to that dt_P at P by the relation between a single stationary clock in the inertial system S_t and the set of clocks in the inertial systems S_t (see Equation 1.37):

$$dt_P = \frac{dt_O}{\sqrt{1 - (v^2/c^2)}} = \frac{dt_O}{1 - (v^2/2c^2)} = \frac{dt_O}{1 - (gh/c^2)} = \frac{dt_O}{1 + (gz/c^2)}.$$

(a) The clock at O has the same rate as the coincident clock C_t in the inertial system S_t that is instantaneously at rest relative to O. (b) The clock at P is adjusted to have the same rate as the clock C_t, which was used to set the clock at O, when clock C_t is coincident with P. At that event, the clock C_t is moving with speed $v = \sqrt{2gh}$ relative to O.

The geodesics corresponding to this metric form are world lines undergoing the constant acceleration $\mathbf{g}$, as the following argument shows: Relative to an inertial system, the integral $\int_1^2 ds$ has its maximum value for motion with constant velocity between the events 1 and 2. Therefore, relative to R, the integral will have its maximum value with the constant acceleration $\mathbf{g}$, as shown in Figure 3.41.

SUMMARY   The procedure used to introduce a coordinate system on the surface of a sphere can be generalized to provide a metric form for a small spatial region experiencing a constant acceleration relative to an inertial system. This metric form is

$$ds^2 = c^2\left(1 + 2\frac{gz}{c^2}\right) dt^2 - dx^2 - dy^2 - dz^2,$$

where g is the magnitude of the acceleration and $-\hat{z}$ is its direction.

Problem 3.16

Use the procedure given in the text to find a coordinate system on a Euclidean plane that is appropriate for the description of a circular motion in that plane. You should obtain the polar coordinate system.

3.2.4   An approximation to the metric form in the neighborhood of a massive object

The space-time metric form appropriate for a small spatial region experiencing a constant gravitational acceleration $\mathbf{g} = -g\hat{z}$ is given by

$$ds^2 = c^2\left(1 + \frac{2gz}{c^2}\right) dt^2 - dx^2 - dy^2 - dz^2. \tag{3.59}$$

This metric form can be generalized, in the following way, appropriate for any region about a spherical uniform gravitational charge distribution. The additional term that gives the change in the metric from O to P, $2gz/c^2$, depends on the position of the region R relative to the source of gravitational acceleration, since g itself depends on that position. We can write the product gz as $\Delta\phi_0(P)$, the gravitational potential energy per unit mass at the point P relative to that at O [see Problem 3.1(b)].

Suppose now that we consider the change in the metric form between a series of points such that neighboring points are sufficiently close that the acceleration **g** is appreciably constant in the region between them. At some reference point that is a large distance away from the gravitational charge, the metric form is that of Lorentz:

$$ds^2 = c^2\, dt^2 - dx^2 - dy^2 - dz^2. \tag{3.60}$$

As we compare the metric form between one point and its neighbor, starting from the reference point, the metric form changes essentially by the addition of dt^2 times twice the gravitational potential energy per unit mass between the points:

$$\text{The change in the metric form} = 2\,\Delta\phi\, dt^2. \tag{3.61}$$

We can add up these changes to obtain, at the event $(ct, \mathbf{r})$,

$$ds^2 = c^2\!\left(1 + 2\frac{\phi(\mathbf{r})}{c^2}\right) dt^2 - dx^2 - dy^2 - dz^2, \tag{3.62}$$

where $\phi(\mathbf{r})$ is the potential energy per unit mass at $\mathbf{r}$ relative to that at the reference point. For a spherical uniform source of gravitational charge M,

$$\phi(\mathbf{r}) = -\frac{GM}{r} \tag{3.63}$$

[see Problem 3.1(a)], so that*

$$ds^2 = c^2\!\left(1 - \frac{2GM}{rc^2}\right) dt^2 - dx^2 - dy^2 - dz^2. \tag{3.64}$$

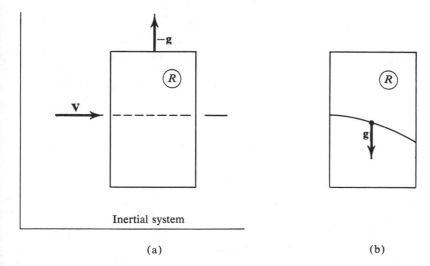

Inertial system

(a)   (b)

FIGURE 3.41 If an object undergoes motion with constant velocity **V** relative to an inertial system, as in (a), it will undergo an acceleration relative to R (b).

---

* See A. Schild, "Equivalence Principle and Red-Shift Measurements," *American Journal of Physics*, *28*: 778 (1960).

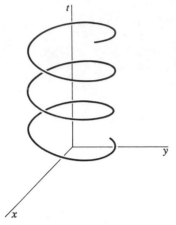

FIGURE 3.42 Geodesic motion for a planet in a circular orbit about the sun, as described by the metric (3.64).

This metric form,* together with the *postulate* that a freely falling object travels along a geodesic, determines the motion of an object in the neighborhood of a gravitational charge M at the point $r = 0$ (Figure 3.42). Thus, *we have replaced the effects of a gravitational force by introducing a curvature in space-time.*

Geodesic motion in the space-time with the metric form (3.64) given by the principle of equivalence also yields other results of physical consequence. We shall discuss these in Section 3.3.

SUMMARY   The metric form in the neighborhood of a mass M is given by the principle of equivalence to be approximately

$$ds^2 = c^2\left(1 - \frac{2GM}{rc^2}\right) dt^2 - dx^2 - dy^2 - dz^2.$$

Problem 3.17

Calculate GM/rc^2 at a point on the earth's surface and also at a point on the sun's surface. Compare your results with unity.

Problem 3.18

What is the density of a spherical uniform star with the mass of the sun and for which $GM/rc^2 = 1$ at the star's surface?

Problem 3.19

Show that the integral $\int_1^2 ds$, evaluated between two fixed events 1 and 2 and with

$$ds^2 = c^2\left(1 + \frac{2gz}{c^2}\right) dt^2 - d\mathbf{r}^2,$$

is a maximum, for $v^2/c^2 \ll 1$, if

$$\frac{d^2\mathbf{r}}{dt^2} = -g\hat{z}.$$

Problem 3.20

Show that $\int_1^2 ds$, evaluated between two fixed events 1 and 2 and with

$$ds^2 = c^2\left(1 + \frac{2\phi(r)}{c^2}\right) dt^2 - d\mathbf{r}^2,$$

is a maximum, for $v^2/c^2 \ll 1$ and $2\phi(r)/c^2 \ll 1$, if

$$m\frac{d^2\mathbf{r}}{dt^2} = \mathbf{F}(\mathbf{r}),$$

where $\phi(|\mathbf{r} + d\mathbf{r}|) - \phi(|\mathbf{r}|) = -\frac{\mathbf{F}(\mathbf{r})}{m} \cdot d\mathbf{r}$

[see Problem 3.1(a)]. What is the equation of motion if $\phi(r) = -GM/r$? Solve this equation for the orbit of a planet (see Problem 2.5).

---

* This metric form is valid only outside a region of gravitational charge. Hence, the singularity in the metric form at $r = 0$ does not appear in the region in which the metric form (3.64) is relevant.

3.3 The General Theory of Relativity

The concept of an inertial system plays an important role in discussions of Newton's laws and the special theory of relativity. In each case, the laws of physics take on their simplest forms relative to an inertial system. In the former, the permissible transformation law—the galilean transformation—between position and time measurements relative to different inertial systems can be deduced from the form of Newton's equation of motion. In this case also, the equation of motion was applicable in a Euclidean 3-space (together with the independent 1-space of time). On the other hand, in the special theory of relativity, we were able to deduce the permissible transformation laws—the Lorentz transformations—from the metric form for an inertial system (Section 1.5). Furthermore, in special relativity, we deduced the possible forms of the equations of motion from these transformation laws and hence indirectly from the form of the metric. The equation of motion of special relativity was applicable in a space-time described by the Lorentz metric. These results must be generalized in the light of the results of the previous section, as the arguments of this section demonstrate. The resulting theory of gravitation is called the general theory of relativity.*

3.3.1 The principle of general relativity

A (localized) inertial system in which a constant gravitational force is acting is equivalent to a noninertial system that experiences a constant acceleration relative to the fixed stars. Therefore, the laws of physics must appear on the same footing relative to a reference system accelerated with respect to the fixed stars as to an inertial system. This result appears more vividly if we consider regions of space-time, such as that around a spherical gravitational charge, in which a variable gravitational force acts. This region can be described with a metric form for a curved space-time. If we interpret an arbitrary gravitational force in an inertial system as representing the effects of a curved space-time, then *the laws of physics must be applicable in a space-time of an arbitrary geometry* (or a geometry subjected to very few restrictions).

The Lorentz transformation was deduced in Chapter 1 from the form of the Lorentz metric and describes the permissible transformations among the co-ordinates—distances in space and intervals of time—under which the forms of the laws of physics remain unchanged in conformity with the special theory of relativity. According to the principle of equivalence, we need not restrict our considerations to the Lorentz metric; a metric form subject to very few restrictions describes an inertial system in which a variable gravitational force acts. Because the space-time metric is not restricted to any extent, we are no longer able to impose restrictions, such as linearity, on the permissible transformations among the coordinates of space and time. Hence, *the laws of physics must take the same form independent of the choice of space-time coordinates.* The choice of coordinates is restricted solely by the condition that neighboring points have coordinates that differ by little or, in other words, that the transformation equations relating different sets of coordinates be continuous. We conclude

---

* An introduction to the theory is given by Einstein himself in A. Einstein, "On the Generalized Theory of Gravitation," *Scientific American, 182*: 13, April 1950. (The last portion of the article contains a generalization of the theory that will not concern us here.)

that the laws of physics must be covariant under an arbitrary (continuous) transformation of the space-time coordinates (see Section 2.1.3).* This is Einstein's *principle of covariance* [4].

The forms of the laws of physics that we have discussed in previous chapters do not satisfy this principle of covariance, since they retain the same form only under a transformation from one inertial reference system to another and not, for example, under a transformation to a noninertial reference system. However, these or any other laws *stated relative to a particular class of coordinate systems* can be expressed in covariant form,† although in general, they would appear in such a form to be so complicated that they would be beyond comprehension and useless for practical applications. Thus, on the one hand, the principle of covariance provides only a guide to a selection between various formulations of a law of physics on the basis of elegance and simplicity. On the other hand, the principle of covariance determines almost uniquely a theory of gravitation if we require that, not only must the equations of the laws of physics have the same form relative to every possible space-time geometry, but also *the laws of physics must determine the geometry appropriate for a particular physical circumstance*. This expanded interpretation of the principle of covariance is called *the principle of general relativity*.

SUMMARY The principle of covariance states that the laws of physics must be covariant with respect to arbitrary coordinate transformations. The principle of general relativity states that, in addition to satisfying the principle of covariance, the laws of physics must determine the geometry of space-time appropriate for a particular physical circumstance.

3.3.2   Einstein's equations for the metric form of space-time

The principle of general relativity provides us with a theory of gravitation, as the following arguments indicate.

The newtonian law of gravity gave a force function that determined the force of gravitational interaction between two elements of gravitational charge in terms of their separation distance in a Euclidean space. From this, we could calculate (at least in principle) the gravitational force experienced by any particle in terms of the observable features of the particle's environment. On the other hand, the introductory discussion of Section 3.2 suggested that the effects of a curved space-time could be only equivalent to those of a gravitational force (because both effects are independent of the properties of a test particle). Therefore, by analogy with the corresponding state of affairs for Newton's law of gravity, we expect that the geometry of space-time for a particular physical circumstance can be deduced in terms of observable features of the environment—namely, the distribution of mass. That is, we expect that there are equations that determine the metric form appropriate to a particular physical arrangement of massive objects and, moreover, that these equations are co-

---

* The Lorentz covariance of special relativity applies to a flat region (or one devoid of gravitational forces) and then only to a restricted class of coordinate systems—namely, those for which the metric takes the form $ds^2 = (dx^0)^2 - (dx^1)^2 - (dx^2)^2 - (dx^3)^2$.

† We carried through a similar procedure in Section 2.3, where we deduced the Lorentz covariant form of the equation of motion from the newtonian form valid in the instantaneous rest system of the particle.

variant with respect to arbitrary continuous transformations of the space-time coordinates. Furthermore, the metric form calculated from these equations describes the effects of what we have called, up to this point, the force of gravity.

Thus, with the additional insight into the problem of gravitation provided by the principles of equivalence and general relativity, we now look for a form for the law of gravity that is sufficiently simple to provide a basis for progress in our understanding of gravitational phenomena. We want a law that appears simple when expressed in a general covariant form and that reduces to the newtonian law in the correct limit. For this purpose, we first examine the conditions on space-time for those particular cases in which there are no gravitational effects present.

A region of space-time devoid of gravitational "forces" appears as a four-dimensional Minkowski space-time, where coordinate systems can be found for which the metric form in that region is that of Lorentz:

$$ds^2 = c^2 (dt)^2 - (dx)^2 - (dy)^2 - (dz)^2. \qquad (3.65)$$

In accordance with the principle of general relativity, the laws of physics must take on the same form when this metric form is expressed in terms of any space-time coordinates $(x^{0'}, x^{1'}, x^{2'}, x^{3'})$ obtained from the above set (x^0, x^1, x^2, x^3) by a continuous transformation law:

$$x^{\mu'} = x^{\mu'}(x^0, x^1, x^2, x^3), \qquad \mu = 0, 1, 2, 3. \qquad (3.66)$$

The metric, expressed in terms of these new coordinates, has the general form

$$ds^2 = \sum_{\mu,\nu=0}^{3} g_{\mu\nu} \, dx^{\mu'} \, dx^{\nu'}, \qquad (3.67)$$

where the $g_{\mu\nu}$'s are functions of the coordinates $x^{0'}, x^{1'}, x^{2'}$, and $x^{3'}$. The characteristic feature that is not altered by the coordinate transformation is that the 4-space is flat. Thus, all the components of the curvature tensor $R^{\mu}{}_{\nu\rho\sigma}$, corresponding to that tensor described in Section 3.1 for a 2-space, are zero everywhere in a Minkowski space-time:

▶ Condition for a region devoid of gravitational "forces":

$$R^{\mu}{}_{\nu\rho\sigma} = 0. \qquad (3.68)$$

These equations provide sufficient restrictions on the components $g_{\mu\nu}$ of the metric tensor that they allow only for those metrics that can be reduced to the Lorentz form [or similar unphysical forms differing from the Lorentz form by the number of plus and minus signs, as in $(dx^0)^2 + (dx^1)^2 - (dx^2)^2 - (dx^3)^2$].

Space-time in the neighborhood of matter—that is, in a region of gravitational force—is not flat like Minkowski space-time but is curved in the sense described in Section 3.2. The metric form

$$ds^2 = \sum_{\mu,\nu=0}^{3} g_{\mu\nu} \, dx^{\mu} \, dx^{\nu} \qquad (3.69)$$

is determined by the particular physical circumstance under consideration, and in any region, this metric form will determine, in turn, what a local observer would call a gravitational force field. Therefore, the theory requires a set of covariant equations that will determine the components $g_{\mu\nu}$ of the metric tensor (as Newton's law of gravitation specifies the force function that determines the motions). There must be sufficient equations to determine the metric tensor. Furthermore, the equations must admit solutions (that satisfy $R^{\mu}{}_{\nu\rho\sigma} = 0$) corresponding to the Lorentz metric, which is applicable in regions devoid of gravitational forces. *These conditions, plus some natural criteria of simplicity, are sufficient to determine uniquely the equations for the metric tensor.* (We state these equations without showing that they satisfy the conditions laid down in the above.) *In the regions of space-time outside the world lines of matter and free of electromagnetic energy,* the equations for the metric tensor have the form *

▶ *Einstein's equations:*

$$R_{\mu\nu} = 0 \qquad (\mu, \nu = 0, 1, 2, 3), \tag{3.70}$$

where

$$R_{\mu\nu} = \sum_{\rho=0}^{3} R^{\rho}{}_{\mu\rho\nu}. \tag{3.71}$$

The set of equations $R_{\mu\nu} = 0$ is a set of *nonlinear* equations in the metric tensor $g_{\alpha\beta}$ and its first and second derivatives. The Einstein equations $R_{\mu\nu} = 0$ are satisfied by the condition $R^{\mu}{}_{\nu\rho\sigma} = 0$ for the Lorentz metric, but they also admit solutions for which $R^{\mu}{}_{\nu\rho\sigma}$ is not zero. The tensor with components $R_{\mu\nu}$ does not vanish at those points in space-time that are not free of both matter and electromagnetic energy.

SUMMARY   The equations that determine the metric form must be covariant and admit the Lorentz metric as a particular solution in a region devoid of gravitational forces. The equations that are determined by natural criteria of simplicity are $R_{\mu\nu} = 0$.

3.3.3 Some theoretical consequences of Einstein's equations

Einstein's equations possess a solution that, to first order, corresponds to the metric form that we deduced in Section 3.2 on the basis of the principle of equivalence.

▶ *Metric form for the principle of equivalence:*

$$ds^2 = c^2\left(1 - \frac{2GM}{rc^2}\right) dt^2 - d\mathbf{r} \cdot d\mathbf{r}. \tag{3.72}$$

The solution of Einstein's equations that corresponds to *the metric form in the neighborhood of a spherically symmetric, time-independent gravitational charge centered at r = 0* is called the *Schwarzschild solution* and is given by the Schwarzschild metric.

▶ *Schwarzschild metric:*

$$ds^2 = c^2\left(1 - \frac{2GM}{rc^2}\right) dt^2 - \frac{dr^2}{1 - (2GM/rc^2)} - d\mathbf{r}_\perp \cdot d\mathbf{r}_\perp, \tag{3.73}$$

where $d\mathbf{r}$ (Figure 3.43) has been written as

FIGURE 3.43 $d\mathbf{r} = (dr)\hat{r} + d\mathbf{r}_\perp.$

---

* These equations are given in the literature also in the form $G_{\mu\nu} = 0$, where the $G_{\mu\nu}$ are linear combinations of the various components of $R_{\mu\nu}$. However, the two sets of equations are completely equivalent, and the $G_{\mu\nu}$ are useful only for reasons that will not enter our discussions here.

$$d\mathbf{r} = (dr)\hat{r} + d\mathbf{r}_\perp, \tag{3.74}$$

and $d\mathbf{r}_\perp$ is perpendicular to $\mathbf{r}$. The coefficient of $dr^2 = (dr\,\hat{r})\cdot(dr\,\hat{r})$ differs from that of $d\mathbf{r}_\perp \cdot d\mathbf{r}_\perp$ because the direction of $\hat{r}$ is distinguished by gravitational effects from that of $d\mathbf{r}_\perp$; the equivalent gravitational force acts along the direction of $-\hat{r}$.

For the motions considered in Section 3.2,

$$d\mathbf{r} \cdot d\mathbf{r} \sim (\text{typical speed})^2\, dt^2; \tag{3.75}$$

hence, in

$$\frac{dr^2}{1 - 2(GM/rc^2)} \approx \left(1 + 2\,\frac{GM}{rc^2}\right) dr^2, \tag{3.76}$$

the term $(GM/rc^2)\, dr^2$ is of the order of the terms that we neglected in our approximation. Thus, in the appropriate limit, the Schwarzschild solution gives the results of Newton's law of gravitation and the principle of equivalence.

The Schwarzschild solution, however, is an exact solution to Einstein's equations; hence, it is worthwhile to compare the predictions that can be made from this solution to those that follow from Newton's law of gravitation and also to those that follow from the metric form (3.72) given to first order by the principle of equivalence. There are three important predictions—the gravitational red shift, the deflection of starlight by the sun, and the precession of the perihelion of the planet Mercury. We shall discuss each of these in turn after we describe the equations that determine the possible motions of particles in a curved space-time.

The motion of an object in the curved space-time with the first-order metric form (3.72) of the principle of equivalence is given by the condition that the integral $\int_1^2 ds$ be a maximum. Within the framework of general relativity, however, the metric form is determined by the distribution of matter, including that of the object whose motion is under consideration. As a consequence of this, Einstein's equations determine the equations of motion in addition to the metric form, as the following arguments suggest.

The Einstein equations (3.70), a set of nonlinear differential equations for the components of the metric tensor, are satisfied in space-time only outside the world lines of the matter present*—for example, consider the motion of a small test particle undergoing motion in the presence of a large gravitational charge that, by itself, generates the Schwarzschild metric form. The metric form of the system is determined not only by the large gravitational charge but also by the object undergoing motion about that charge. Moreover, the metric form for the system of test particle plus large gravitational charge cannot be obtained merely by the addition of the metric forms that each would generate were the other not present; the presence of one disturbs the metric form of the other in a nonlinear manner. This is a result of the fact that the equations are not linear in the $g_{\mu\nu}$. Thus, the behavior of a small test particle in the presence of a large gravitational charge is restricted by Einstein's equations.

These considerations led Einstein and his collaborators, particularly the Polish–Canadian theoretical physicist Leopold Infeld (1898–1968), to examine

---

* The modified equations that hold in the presence of electromagnetic energy are under active investigation today, especially by the American theoretical physicist John Archibald Wheeler (1911–) and his collaborators [5]. See Chapter 3 of E. F. Taylor and J. A. Wheeler, *Spacetime Physics*, W. H. Freeman, San Francisco, 1966.

the motions of a test particle in the presence of a large gravitational charge consistent with Einstein's equations for the metric tensor. They showed that Einstein's equations determine completely the equations of motion and that the law of motion is this: The test particle follows a geodesic of the space-time metric of the large gravitational charge. This *consequence* of Einstein's equations is the result we postulated previously. Thus, *Einstein's equations determine both the metric form associated with a large gravitational charge and the trajectory of a test particle in the presence of that charge.*

SUMMARY   The Schwarzschild solution

$$ds^2 = c^2\left(1 - \frac{2GM}{rc^2}\right) dt^2 - \frac{dr^2}{1 - (2GM/rc^2)} - d\mathbf{r}_\perp \cdot d\mathbf{r}_\perp$$

to Einstein's equations is the metric form in the neighborhood of a spherically symmetric, time-independent mass centered at $r = 0$.

It is a consequence of Einstein's equations, and not an additional postulate, that a small object in free fall in the neighborhood of a large mass moves along a geodesic of the curved space-time resulting from the presence of the large mass.

3.3.4   Tests of Einstein's theory of gravitation

The general theory of relativity is one of the most impressive physical theories that has ever been proposed, but like all physical theories, it must stand or fall upon its agreement with experiment and observation. There are three classical observational tests of general relativity and the principle of equivalence: the gravitational red shift, the deflection of light in the presence of a gravitational charge, and the advance in the perihelion of Mercury. These tests are discussed below. Other tests* have been proposed, some of which are being carried out at the present time.

The tests of general relativity result from a comparison of observation with the solutions for the motions of material objects and also light rays in the curved space-time of the Schwarzschild metric.

Both the Schwarzschild and the first-order metric forms given by the principle of equivalence have the same coefficient for the square of the coordinate time interval dt. Thus, in each, the proper time interval

$$dt_0 = \frac{1}{c} ds \Big|_{d\mathbf{r} = 0} \tag{3.77}$$

of a particle at rest is related to the coordinate time interval by an equation,

$$dt_0 = \left(1 + \frac{2GM}{rc^2}\right)^{1/2} dt, \tag{3.78}$$

that depends on the position. One result of this is that the frequency of a light beam decreases as the light travels away from a gravitational charge (see Problem A1.16). This is called the *gravitational red shift* (Figure 3.44). The red shift has been observed in light emitted by stars and also in terrestrial experiments.

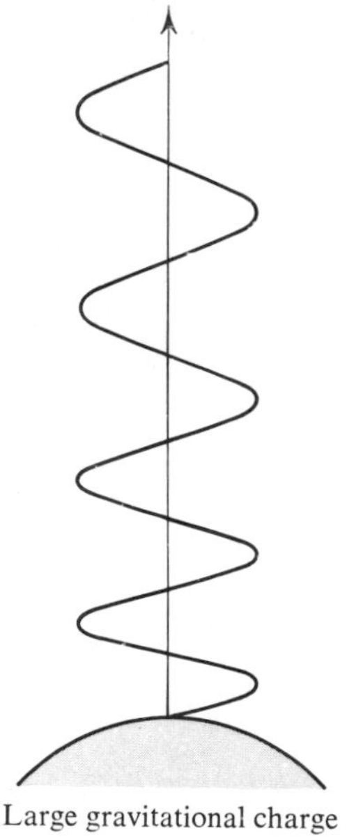

Large gravitational charge

FIGURE 3.44   Schematic picture of the gravitational red shift. The light decreases in frequency and increases in wavelength as it moves away from a gravitational charge.

---

* For example, see I. I. Shapiro, "Fourth Test of General Relativity," *Physical Review Letters*, **13**: 789 (1964) for a discussion of a test involving measurements of the time delays of radar pulses that pass near the sun and are reflected from Venus or Mercury. See also V. L. Ginzburg, "Artificial Satellites and the Theory of Relativity," *Scientific American*, **200**: 149, May 1959.

The results obtained are in good agreement with the predictions of the form of the proper time interval [Equation (3.78)] given by both the principle of equivalence and Einstein's equations.*

We showed in the introduction to this chapter, from the principle of equivalence, that the path of a light ray is bent in a region of gravitational charge; on that basis, we calculated in Problem 3.5 that a light ray that approaches only as close as the distance D to a uniform spherical gravitational charge undergoes a deflection through the angle (Figure 3.45)

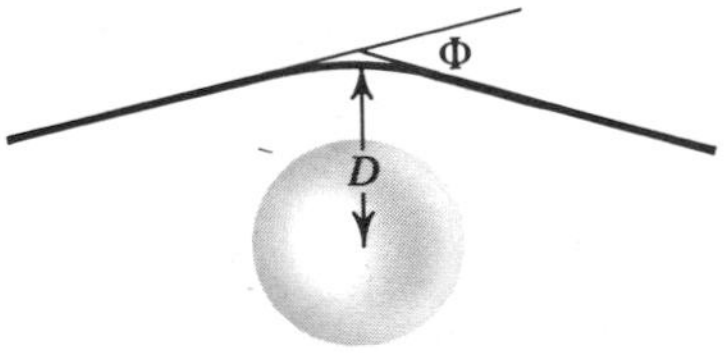

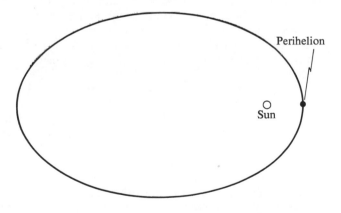

FIGURE 3.45 Deflection of light by a gravitational charge.

$$\Phi_{\substack{\text{equiv} \\ \text{prin}}} = \frac{2GM}{c^2 D}. \tag{3.79}$$

[Also, this result is what we obtain by a calculation with the metric form (3.72) given by the principle of equivalence.] On the other hand, the deflection calculated on the basis of the Schwarzschild metric is given by the formula (which we state without proof)

$$\Phi_{\text{Schwarzschild}} = \frac{4GM}{c^2 D}. \tag{3.80}$$

The results† of a number of observations of the deflection of starlight by the sun during solar eclipses are in agreement with Equation (3.80) but not with Equation (3.79).

The metric form (3.72) derived on the basis of the principle of equivalence leads back to first order in GM/rc^2, to the newtonian equation of motion with Newton's gravitational force (Problem 3.20). Therefore, on the basis of the argument following Equation (3.76), the geodesic motions given by the Schwarzschild metric approximate those given by Newton's law. There are small differences for the motions of the planets of the solar system, however. In particular, general relativity predicts that the position of the perihelion of planetary motion (Figure 3.46) does not remain fixed but advances in time.

Perihelion

O
Sun

FIGURE 3.46 The perihelion of a planetary orbit.

---

* The formula for the red shift is derived, for example, in Section 4.4 of R. Adler, M. Bazin, and M. Schiffer, *Introduction to General Relativity*, McGraw-Hill, New York, 1965. Also see A. J. O'Leary, "Redshift and Deflection of Photons by Gravitation: A Comparison of Relativistic and Newtonian Treatments," *American Journal of Physics*, *32*: 52 (1964). The experimental results are discussed in Section 1–3.1 of the article by B. Bertotti, D. Brill, and R. Krotkov entitled "Experiments on Gravitation," in *Gravitation: An Introduction to Current Research*, L. Witten (Ed.), John Wiley, New York, 1962.
† See Section 1-3.2 of B. Bertotti, D. Brill, and R. Krotkov, *op. cit.*

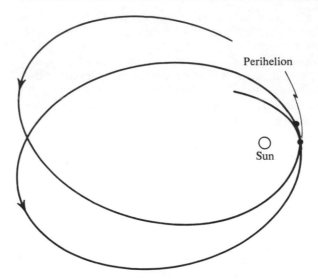

Perihelion

Sun

FIGURE 3.47 Advance in the perihelion of a planet (exaggerated).

This advance in the perihelion is shown in exaggerated form in Figure 3.47.
The advance in the perihelion, $2\pi\delta\tilde{\omega}$, of a planet is given by the fraction of a
revolution (which we state without proof)

$$\delta\tilde{\omega} = \frac{3GM}{c^2 a(1 - e^2)} \tag{3.81}$$

for each revolution of the planet: M is the mass of the sun, a is the semimajor
axis, and e is the eccentricity of the orbit. (An exact treatment with the metric
(3.72) of the principle of equivalence gives the result $\delta\tilde{\omega} = 2GM/[c^2 a(1 - e^2)]$.)
This advance in the perihelion is greatest for the planet Mercury, and the
value 43 sec of arc per century, in perfect agreement with Equation (3.81), had
been determined by astronomers before Einstein proposed his theory. However,
Einstein's theory was proposed independent of this fact, and the resulting agreement between his theory and the astronomers' observations strengthened the
confidence of scientists in the theory of general relativity. Einstein's theory
also is in agreement with the relevant advances in the perihelia of Venus and
Earth.*

The actual observed advance in the perihelion of Mercury is 5,599.7 sec
of arc per century, and various known effects, such as the perturbations by the
other planets, must be taken into account to leave the residual advance of
43 sec of arc per century that is explained by Einstein's general theory of relativity. Recently, the American physicist Robert Henry Dicke† (1916–)
performed an experiment that indicates that the subtracted effects require
correction, so the agreement of general relativity with the additional advance
may have been fortuitous. Dicke measured the shape of the sun and concluded
that the sun may be sufficiently flattened at the poles to contribute about 4 sec
of arc per century to the advance in the perihelion of Mercury. If this contri-

---

* See Section 1-3.3 of B. Bertotti, D. Brill, and R. Krotkov, *op. cit.*
† Just before he announced his results, Dicke described the observational basis for general relativity
and other theories of gravitation in "Gravitational Theory and Observation," *Physics Today*, *20*:
55, January 1967.

bution is proven beyond doubt and the other contributions, which contain uncertainties, are found to be unalterable, it may be necessary to introduce a modification that involves effects other than purely geometrical ones, such as an extra force, into the theory of relativity.

The discovery of the general theory of relativity opened up the way to mathematical studies of cosmology, the theory of the Universe and the general laws that govern it. This is one of the most fascinating studies resulting from physics, but lack of space will not permit a discussion of this topic here.*

SUMMARY The predictions of the general theory of relativity are under active investigation at the present time. There were three classical tests—the gravitational red shift, the deflection of light by massive objects, and the advance in the perihelion of Mercury. Current experimental techniques allow for other tests, some of which are currently under way.

The results of the three classical tests were believed to be in agreement with the predictions until recent evidence appeared that casts this agreement in doubt. Further investigation may support the general theory, or alternatively, it may require a modification in the theory.

Problem 3.21

The measured advances in the perihelia of the three planets Mercury, Venus, and Earth which, until recently,† could not be accounted for by the interactions with other planets are 43.11 ± 0.45, 8.4 ± 4.8, and 5.0 ± 1.2 sec of arc per century, respectively. Compare these results with the predictions of the general theory of relativity.

Planet	Semimajor axis of orbit, km	Eccentricity of orbit	Period, yr
Mercury	5.79×10^7	0.206	0.241
Venus	1.08×10^8	0.00682	0.615
Earth	1.50×10^8	0.0167	1.00

Problem 3.22

(a) Use the techniques of Problem 2.5 to show that the polar equation for the orbit of a particle moving nonrelativistically under the central attractive force with magnitude

$$F(r) = \frac{k}{r^2} - \frac{\alpha}{r^3} \qquad \text{has the form} \qquad \frac{1}{r} = \frac{1 + \varepsilon \cos p\theta}{r_0(1 + \varepsilon)}.$$

(b) Show that for $\varepsilon < 1$ the orbit can be described as an ellipse that precesses.
(c) Find the rate at which the ellipse precesses in terms of $\Delta p = p - 1$ for the case in which $|\Delta p| \ll 1$.

---

* See H. Bondi, *The Universe at Large*, Anchor Books, Garden City, N.Y., 1960, or H. Bondi, *Cosmology*, 2nd ed., Cambridge Univ. Press, Cambridge, 1960.
† See the comments given above on Dicke's work.

(d) Find a value for $\alpha/(kr_0)$ that would account for the 43 sec of arc-per-century advance in the perihelion of Mercury.*

(e) Which explanation of the advance in the perihelion of Mercury do you find most satisfying—that of general relativity or that of part (d) above? Justify your answer.

Problem 3.23

Calculate the advance in the perihelion of Mercury according to the special theory of relativity (see Problem A2.19), and compare your answer with the 43 sec of arc per century, which until recently could not be accounted for by interactions with other planets.

Additional Problems

Problem A3.1

Consider a right circular cylinder of radius R.

(a) Set up a coordinate system on the cylinder with $x^1 = h$, a distance along the axis of the cylinder, and $x^2 = \theta$, an angle about that axis.

(b) Determine the metric on the cylinder in terms of dx^1 and dx^2.

Problem A3.2

Consider a right circular cone with an apex angle of α.

(a) Set up a coordinate system on the cone with $x^1 = h$, a distance from the vertex along the axis of the cone, and $x^2 = \theta$, an angle about that axis.

(b) Determine the metric on the cone in terms of dx^1 and dx^2.

Problem A3.3

Describe qualitatively the connection and the curvature tensor on the surface of a right circular cylinder.

Problem A3.4

Describe qualitatively the connection and the curvature tensor on the surface of a right circular cone.

Problem A3.5

Draw, on the space-time diagram of an inertial system, the coordinate lines for an accelerated observer, as described in the text (Figure 3.37). Use a scale for the z axis such that your diagram covers distances characteristic of a laboratory experiment. For g, use the gravitational acceleration at the surface of the sun.

Problem A3.6

Use the diagram of Problem A3.5 as a basis for a discussion of the statement

---

* See E. Gerjuoy, "Feasibility of a Nonrelativistic Explanation for the Advance of the Perihelion of Mercury," *American Journal of Physics*, 24: 3 (1956).

that the small tube in space-time corresponding to the laboratory is too small to show significant curvature effects.

Problem A3.7

The quantity $r_0 = GM/c^2$ is called the gravitational radius of an object of mass M. Calculate the gravitational radius of

(a) the sun,
(b) the earth,
(c) an 80-kg man.

Problem A3.8

This problem gives a method for calculating the equations for parallel transport [Equation (3.19)] on the surface of a sphere of radius R. Assume that the coordinates (x^1, x^2) determine the position of a point on the sphere as shown in Figure 3.12. Let (V^1, V^2) and (U^1, U^2) be the components of two vectors at the point (x^1, x^2).

(a) Show that the scalar product of the two vectors is given by

$$V^1 U^1 \left(\cos \frac{x^2}{R} \right)^2 + V^2 U^2.$$

(b) Let $(V^1 + \delta V^1, V^2 + \delta V^2)$ and $(U^1 + \delta U^1, U^2 + \delta U^2)$ be the components at $(x^1 + \delta x^1, x^2 + \delta x^2)$ of the vectors obtained by parallel transport of (V^1, V^2) and (U^1, U^2), respectively, from the point (x^1, x^2). Show that, to first order in the δV's, δU's, and δx's,

$$0 = -\frac{2}{R} \cos \frac{x^2}{R} \sin \frac{x^2}{R} V^1 U^1 \delta x^2 + V^1 \left(\cos \frac{x^2}{R} \right)^2 \delta U^1$$

$$+ U^1 \left(\cos \frac{x^2}{R} \right)^2 \delta V^1 + V^2 \delta U^2 + U^2 \delta V^2.$$

(c) Set

$$\delta V^i = -\sum_{m,n=1}^{2} \Gamma^i_{mn} V^m \delta x^n$$

and

$$\delta U^i = -\sum_{m,n=1}^{2} \Gamma^i_{mn} U^m \delta x^n,$$

so that, for example,

$$\delta U^1 = -\Gamma^1_{11} U^1 \delta x^1 - \Gamma^1_{12} U^1 \delta x^2 - \Gamma^1_{21} U^2 \delta x^1 - \Gamma^1_{22} U^2 \delta x^2.$$

Give arguments to show that the Γ's are independent of the vectors (V^1, V^2) and (U^1, U^2) and the small displacement (dx^1, dx^2).

(d) Insert the expressions for δV^i and δU^i of (c) into the equation of (b) and derive from this the following equations:

$$\Gamma^1_{12} = -\frac{1}{R} \tan \frac{x^2}{R}, \qquad \Gamma^1_{11} = 0, \qquad \Gamma^1_{21} \left(\cos \frac{x^2}{R} \right)^2 + \Gamma^2_{11} = 0,$$

$$\Gamma^1_{22} \left(\cos \frac{x^2}{R} \right)^2 + \Gamma^2_{12} = 0, \qquad \Gamma^2_{21} = 0, \qquad \Gamma^2_{22} = 0.$$

(e) Consider a small rectangle with a vertex at (x^1, x^2) and adjoining vertices at $(x^1 + \delta x^1, x^2)$ and $(x^1, x^2 + \delta x^2)$. Show that the other vertex is at the point

$$(x^1 + \delta x^1 - \Gamma^1_{12} \, \delta x^1 \, \delta x^2, \; x^2 + \delta x^2 - \Gamma^2_{12} \, \delta x^1 \, \delta x^2),$$

or, alternatively, at

$$(x^1 + \delta x^1 - \Gamma^1_{21} \, \delta x^2 \, \delta x^1, \; x^2 + \delta x^2 - \Gamma^2_{21} \, \delta x^2 \, \delta x^1).$$

Assume that the rectangle is closed to order $(\delta x)^2$, and hence show that

$$\Gamma^1_{12} = \Gamma^1_{21}, \qquad \Gamma^2_{12} = \Gamma^2_{21}.$$

(f) Solve the equations of (d) and (e) for all the Γ^i_{mn}. Prove Equation (3.19).

Problem A3.9

This problem gives a method for calculating the curvature tensor on the surface of a sphere [Equation (3.21)]. The coordinate system used is that shown in Figure 3.12.

(a) Show that, under the small displacements (δ^1, δ^2) and $(\varepsilon^1, \varepsilon^2)$,

$$V^1(x^1 + \delta^1 + \varepsilon^1, \, x^2 + \delta^2 + \varepsilon^2) - V^1(x^1 + \delta^1, \, x^2 + \delta^2)$$

$$= \frac{1}{R} \tan \frac{x^2}{R} \, [V^1(x^1 + \delta^1, \, x^2 + \delta^2)\varepsilon^2 + V^2(x^1 + \delta^1, \, x^2 + \delta^2)\varepsilon^1]$$

$$+ \frac{1}{R^2} \left(\sec \frac{x^2}{R} \right)^2 \delta^2 [V^1(x^1 + \delta^1, \, x^2 + \delta^2)\varepsilon^2 + V^2(x^1 + \delta^1, \, x^2 + \delta^2)\varepsilon^1]$$

$$= \frac{1}{R} \tan \frac{x^2}{R} \, [V^1(x^1, x^2)\varepsilon^2 + V^2(x^1, x^2)\varepsilon^1]$$

$$+ \frac{1}{R^2} \left(\tan \frac{x^2}{R} \right)^2 [V^1(x^1, x^2) \, \delta^2\varepsilon^2 + V^2(x^1, x^2) \, \delta^1\varepsilon^2]$$

$$- \frac{1}{R^2} \left(\sin \frac{x^2}{R} \right)^2 V^1(x^1, x^2) \, \delta^1\varepsilon^1$$

$$+ \frac{1}{R^2} \left(\sec \frac{x^2}{R} \right)^2 [V^1(x^1, x^2) \, \delta^2\varepsilon^2 + V^2(x^1, x^2) \, \delta^2\varepsilon^1].$$

(b) Show that

$$V^2(x^1 + \delta^1 + \varepsilon^1, \, x^2 + \delta^2 + \varepsilon^2) - V^2(x^1 + \delta^1, \, x^2 + \delta^2)$$

$$= -\frac{1}{R} \cos \frac{x^2}{R} \sin \frac{x^2}{R} V^1(x^1 + \delta^1, \, x^2 + \delta^2)\varepsilon^1$$

$$+ \frac{1}{R^2} \left[\left(\sin \frac{x^2}{R} \right)^2 - \left(\cos \frac{x^2}{R} \right)^2 \right] \delta^2 V^1(x^1 + \delta^1, \, x^2 + \delta^2)\varepsilon^1$$

$$= -\frac{1}{R} \cos \frac{x^2}{R} \sin \frac{x^2}{R} V^1(x^1, x^2)\varepsilon^1$$

$$- \frac{1}{R^2} \left(\sin \frac{x^2}{R} \right)^2 [V^1(x^1, x^2) \, \delta^2\varepsilon^1 + V^2(x^1, x^2) \, \delta^1\varepsilon^1]$$

$$+ \frac{1}{R^2} \left[\left(\sin \frac{x^2}{R} \right)^2 - \left(\cos \frac{x^2}{R} \right)^2 \right] V^1(x^1, x^2) \, \delta^2\varepsilon^1.$$

(c) Consider a small rectangle on the surface of the sphere with a vertex at (x^1, x^2) and adjoining vertices at $[x^1 + dx^1_{(1)}, x^2 + dx^2_{(1)}]$ and $[x^1 + dx^1_{(2)}, x^2 + dx^2_{(2)}]$. Show that the fourth vertex Q is at

$$\left\{ x^1 + dx^1_{(1)} + dx^1_{(2)} + \frac{1}{R} \tan \frac{x^2}{R} [dx^1_{(1)} dx^2_{(2)} + dx^2_{(1)} dx^1_{(2)}], \right.$$

$$\left. x^2 + dx^2_{(1)} + dx^2_{(2)} - \frac{1}{R} \cos \frac{x^2}{R} \sin \frac{x^2}{R} dx^1_{(1)} dx^1_{(2)} \right\}.$$

(d) Show that the components of a vector $\mathbf{V}$ after parallel transport from (x^1, x^2) to $[x^1 + dx^1_{(1)}, x^2 + dx^2_{(1)}]$ to Q are greater than are the components after parallel transport from (x^1, x^2) to $[x^1 + dx^1_{(2)}, x^2 + dx^2_{(2)}]$ to Q in the amount

$$\Delta V^1 = \frac{1}{R^2} V^2(x^1, x^2)[dx^2_{(1)} dx^1_{(2)} - dx^1_{(1)} dx^2_{(2)}],$$

$$\Delta V^2 = -\frac{1}{R^2} \left(\cos \frac{x^2}{R} \right)^2 V^1(x^1, x^2)[dx^2_{(1)} dx^1_{(2)} - dx^1_{(1)} dx^2_{(2)}].$$

Problem A3.10

Let (V^1, V^2) be the components of a vector at the point (x^1, x^2) on a sphere of radius R. The coordinate system used is that shown in Figure 3.12.

(a) Use the results of Problems A3.8 and A3.9 to show that, if the vector undergoes parallel transport around a small area ΔA, the vector is transformed through the angle given by

$$\theta \approx \sin \theta = \frac{\Delta A}{R^2}.$$

Hint: Use cartesian components and the definition of the vector product.

(b) Find the angle through which a vector is transformed under parallel transport around an area, on the surface of the earth, of
  (i) 1 acre,
  (ii) 1 mi².
(c) Around what area must a vector undergo parallel transport on the surface of the earth in order that the vector is transformed by $1°$?

ADVANCED REFERENCES

1. The tensor calculus required for the study of general relativity is presented in Chapters 1, 2, 3, and 8 of J. L. Synge and A. Schild, *Tensor Calculus*, Univ. Toronto Press, Toronto, 1949.
2. A comprehensive treatment of general relativity is given in R. Adler, B. Bazin, and M. Schiffer, *Introduction to General Relativity*, McGraw-Hill, New York, 1965; P. G. Bergmann, *Introduction to the Theory of Relativity*, Prentice-Hall, Englewood Cliffs, N.J., 1942; A. Einstein, *The Meaning of Relativity*, 5th ed., Princeton Univ. Press, Princeton, N.J., 1955.
3. It has been argued that the principle of equivalence served its purpose in providing this hint and should now be discarded. See the preface to J. L. Synge, *Relativity: The General Theory*, North-Holland, Amsterdam, 1960.
4. The principle of covariance is discussed in the article by J. L. Anderson entitled

"Relativity Principles and the Role of Coordinates in Physics," in *Gravitation and Relativity*, H.-Y. Chiu and W. F. Hoffmann (Eds.), W. A. Benjamin, New York, 1964.

5. The work of Wheeler and his colleagues is described in the article by J. G. Fletcher entitled "Geometrodynamics," in *Gravitation: An Introduction to Current Research*, L. Witten (Ed.), John Wiley, New York, 1962.

Answers to Problems

Chapter 1

1.1   (a) It has none.

    (b) Yes. (Consider a "shoe box" with **V** along the diagonal and the edges along the axes.)

    (c) Use **V** − **W**.

    (e) Put two "shoe boxes" corner to corner.

1.2   (a) Use definition of $\cos \theta$.

    (b) Project along the direction of **c**.

    (c) Use $\cos 0 = 1$, $\cos 90° = 0$.

    (d) Expand $(a_x\hat{x} + a_y\hat{y} + a_z\hat{z}) \cdot (b_x\hat{x} + b_y\hat{y} + b_z\hat{z})$ and use (c).

    (e) It is $\mathbf{a} \cdot \mathbf{b}$; see (a).

1.3   (a) $\omega(t + 1) = \omega t + 2\pi \times$ (no. of oscillations per unit time).

    (b) $k(x + \lambda) = kx + 2\pi$.

    (c) $\omega(t + T) = \omega t + 2\pi$.

    (d) Eq. gives $kv = \omega$.

    (e) **k** = magnitude × (unit vector in proper direction). Use ans. to 1.2 (d).

    (f) Use definition of $\mathbf{a} \cdot \mathbf{b}$ in 1.2 (a).

1.4   (a) Use Eq. (1.11) and Prob. 1.3 (d): $\mathbf{A} \sin (kx - \omega t) = \mathbf{A} \sin (kx' - \omega't)$.

    (b) $v_s\lambda = v - V$.

1.5   Expand $(\mathbf{v'} + \mathbf{V}) \cdot (\mathbf{v'} + \mathbf{V})$. $v' + V$, $|v' - V|$.

1.6   Note that t is the same argument in $\mathbf{v'}$ as in $\mathbf{v}$.

1.7 6.71×10^8 mi/hr.

1.8   0.134 sec.

1.9   8.31 min.

1.10 1.28 sec.

1.11 11.0 hr.

1.13 They measure the time to be $2L/c + 2/10$ sec. $L = 9.4\ R_E$.

1.14 (a) $d = 1/2\ gt^2 = 0.31$ m.

    (b) 4×10^{-9}.

1.15 $\nu = [2.998 \times 10^{18}/(\lambda/\text{Å})]$Hz. In fHz: 0.666, 0.666–0.600, 0.600–0.526, 0.526–0.508, 0.508–0.491, 0.491.

1.16 $\omega = n(1/720\ \text{rev})/(2 \times 8{,}633/2.998 \times 10^8\ \text{sec}) = 24.1\ n$ rev/sec, where n is an integer.

1.17 (b) $\omega = n(1/8\ \text{rev})/(2 \times 3.5 \times 10^4/2.998 \times 10^8\ \text{sec}) = 535\ n$ rev/sec, where n is an integer.

1.18 In 10^8 m/sec: 2.997, 2.249, 2.29, 1.93, 1.24.

1.19 In 10^8 m/sec: 1.95, 1.97, 1.99, 2.00.

1.21 $\alpha = 20.5'' \sin (2\pi t/\mathrm{yr})$.

1.23   (a) Use law of cosines, the solution of a quadratic equation and the binomial
       expansion $(1 + x)^n \simeq 1 + nx$, $|x| \ll 1$. $c_E = c - v \cos \alpha - (v^2 \sin^2 \alpha)/(2c)$.
       (b) 0.01%.

1.24 3.40×10^3.

1.25 1.2×10^{-7} m.

1.26   (a) Use answer to 1.23: $c_E = c - v \cos \alpha$. $T_1 = SM\{[c - v \cos(\theta - 45°)]^{-1} +$
       $[c + v \cos(\theta - 45°)]^{-1}\}$. Use $\cos^2 x = (1 + \cos 2x)/2$. $n = [2SMv^2/(\lambda c^2)] \sin (2\theta)$.

1.27 $N \propto v^2 \propto r^2/P^2$. 1.03, 8×10^{-2}, 1×10^{-2}.

1.28   For $v = 500$ mi/hr, ratio = 1.7×10^{-4} %.

1.29   Cf. Fig. 1.29: $c_E \neq c$.

1.30   The speed of sound is isotropic only relative to the rest frame of the trans-
       mitting medium.

1.31   Cf. answer to 1.30.

1.32 2.7×10^{-5} sec before 4:00 P.M.

1.34 $\pm d/2c$.

1.37 $t \neq t'$ so $\mathbf{r}'/t \neq \mathbf{v}'$ (cf. answer to 1.6).

1.38   No. $t \neq t'$.

1.39 $(x' + Vt')^2 + y'^2 + z'^2 = c^2 t'^2$.

1.40   (a) $t = t' = -d/v = -1$ min.
       (b) $x' = 0, x = -1$ mi.
       (c) $x = x' = 1$ mi.
       (d) $x = 2$ mi, $x' = 1$ mi; $x = 1$ mi, $x' = 0$.

1.41 $d_s^2 = d_{s'}^2 + 2V(x_2' - x_1')(t_2' - t_1') + V^2(t_2' - t_1')^2$.

1.42   For $v = 500$ mi/hr, 100,000 yr. $\Delta t = (\Delta t - 1)/(1 - v^2/c^2)^{1/2} = (\Delta t - 1)(1 + v^2/2c^2)$.

1.43   For $v = 500$ mi/hr, -3×10^{-11} %. Use $(1 - v^2/c^2)^{1/2} = 1 - v^2/2c^2$.

1.44   (a) -2.5×10^{-10} %.
       (b) 130 yr.

1.45   (a) 7.1×10^6 m/sec.
       (b) 7.1×10^6 clocks.

1.46 1.3×10^7 m/sec.

1.47   (a) $\mathrm{vol}_s = \mathrm{vol}_{s'} \times (1 - v^2/c^2)^{1/2}$.
       (b) 2.60×10^8 m/sec.

1.48 $l^2 = (1 \text{ m}/2^{1/2})^2[1 + (1 - .99)^2]$. 0.714 m; 82.0°.

1.49 $[x/(1 - v^2/c^2)^{1/2}]^2 + y^2 = 1$ m^2, $y = x \tan \theta$. $l = (x^2 + y^2)^{1/2} = \sec \theta/(5.3 + \tan^2\theta)^{1/2}$.

1.52   For $v = 500$ mi/hr, $(v/c)^2 = 6 \times 10^{-13}$ and $v/c^2 = 2.5 \times 10^{-15}$ sec/m.

1.53 E_1: 0.75 m, 2.5×10^{-9} sec. E_2: 3.33 m, -4.25×10^{-9} sec. E_3: 0.375 m, $7.25 \times$
       10^{-9} sec. E_4: -0.375 m, 8.75×10^{-9} sec.

1.54   (a) 2.58 m.
       (b) 2.96 m.
       (c) 1.13 m.
       The pairs of events are not simultaneous relative to S or S'.

1.55   (a) 6.75×10^{-9} sec.
       (b) 1.15×10^{-8} sec.
       (c) 1.5×10^{-9} sec.
       The pairs of events do not occur at the same position relative to S or S'.

1.56   (a) $t = -(1 \text{ min} - 1.5 \times 10^{-16} \text{ sec})$, $t' = -1$ min; $x = -(1 - 4 \times 10^{-15})$ mi,
       $x' = 0$; $x' = 1$ mi, $x = 1$ mi $- 4 \times 10^{-15}$ mi; $x = (2 - 4 \times 10^{-15}$ mi$)$, $x' = 1$ mi,
       $x = 1$ mi, $x' = 0$.
       (b) 1.5×10^{-14}%, 4×10^{-13}%, 4×10^{-13}%, 2×10^{-13}% (or zero).
       (c) $t = -2.23 \times 10^{-6}$ sec; $t' = -5.36 \times 10^{-6}$ sec; $x = -0.385$ m, $x' = 0$; $x =$
       0.385 m, $x' = 1$ mi; $x = 1.66 \times 10^{10}$ m$(+6 \times 10^2$ m$)$, $x' = 1$ mi, $x = 1.66 \times$
       10^{10} m, $x' = 0$.

1.58   The student uses the same x (i.e., the same clock in the S frame) so the student calculates the time dilatation as observed by S'.

1.59 $\sqrt{1 - v^2/c^2} = 0.57$. 2.46×10^8 m/sec.

1.60   (a) 3.6 m, 4.2 m.
     (b) 0.2 m, -0.6 m.
     (c) -0.24 m, 0.12 m.
     (d) 4.68 m, 5.16 m.

1.61   (a) 92.8 cm.
     (b) 1 hr, 4 min, 4 sec.

1.62   71.1.

1.63   (a) By the formula for the time dilatation of an S clock as observed by S'.

1.64   Use $t = (t' + Vx'/c^2)/\sqrt{1 - V^2/c^2}$: 2.43×10^{-9} sec.

1.65   (a) 1.85×10^8 m.
     (b) 1.08×10^{12} m.

1.67   Set $dx/dt = v_x$ so
$$\frac{dx'}{dt'} = \frac{a_{x'x}dx + \cdots}{a_{t'x}dx + \cdots} = \frac{a_{x'x}v_x + \cdots}{a_{t'x}v_x + \cdots} = \text{constant, etc.}$$

1.68   (b) A change in origin in both space and time.

1.69   (b) (i) Mirror image in the y-z plane.
        (ii) Inversion of the spatial axes through the origin.
        (iii) Time-reversal (e.g., running a movie backwards).
        (iv) Combination of space inversion and time-reversal.

1.70   (b) Equate coefficients of t^2, xt and x^2 (separately) in $c^2t^2 - x^2 = c^2t'^2 - x'^2$.

1.71   (a) Use Eq. (1.103).

1.72 $\phi = \arctan Vt_2/L'$ in the notation of Ex. 1.6. In Eq. (1.115), neglect L'^2 compared to $2LL'$ and then use the binomial expansion $(1 + x)^{1/2} = (1 + x/2)$ for $|x| \ll 1$. (Or just use $OP = c\Delta t$, $OP' = V\Delta t$ in Fig. 1.66.)

1.73   Use Eq. (1.116) as corrected.

1.74 3.3×10^{-9} sec, 0.8 m.

1.75   The length is 1.16 m and the back is inclined at $35°$ to the line of sight.

1.76   Use Eq. (1.116) as corrected, with appropriate interpretations of L', for the appearance as seen by a "pinhole" eye.

1.77   There is no "Lorentz contraction" so the cube appears distorted and rotated.

1.79   (a) $1.98c\hat{x}$. η_0. This is not the speed of an object relative to S' as observed by S'.
     (b) $-0.99995c$.

1.80   Use Eq. (1.119) with $v = -v' = V''$. $c^2[1 - (1 - V^2/c^2)^{1/2}]/V$.

1.82 $(V_{12} + V_{23})/(1 + V_{12}V_{23}/c^2)$.

1.83   Calculate $c^2 - v'^2$ using $v_{x'} = (v_x - V)/(1 - Vv_x/c^2)$, $v_{y'} = v_y/(1 - Vv_x/c^2)$, $v_{z'} = v_z/(1 - Vv_x/c^2)$, with $v_{y'} = dy'/dt'$, for example.

1.84   20 yrs $= 1.89 \times 10^{17}$ m, 2×10^{19} m, 10^{-10} m, 10^{-14} m.

1.85   (a) 1 m/sec $\sim 3 \times 10^{-9}$, 8.9×10^{-8}, 3×10^{-5}.
     (b) 6×10^{-4}, 1.8×10^{-2}, 6 sec of arc.

1.86   (a) Time required for light to travel a distance L.
     (b) 6 ft $= 6 \times 10^{-9}$ sec, 0.133 sec, 8 min 19 sec.

1.87   See Fig. 2.28.

1.88   (b) Use $v = 3c/5$, $V = -4c/5$ to get v'. Then $x' = v't'$: $x' = 35\,ct'/37$.
     (c) $x' = 35\,ct'/37$ and $x' = 4ct'/5$.

1.89   See the world line of M in Fig. 2.28.

1.90   (b) The coordinates are $(2n, 0)$, $(2n + 1, 1)$ relative to S' and $(2 + 3n/2, 2 + 5n/2)$ relative to S, where n is an integer.
     (c) The world lines of the light pulses are inclined at $45°$ to the vertical; the world lines of the mirrors at $\tan^{-1} 3/5$ and cross the x axis at the origin and 4/5 m.

1.91 (a) Axes inclined at $42°43'$ to S axes.
      (b) $OT_1' = OX_1' = 3.54$ m.

1.92 (a) The two events O and X are simultaneous relative to S and each lies on
      one end of the meter stick of S'.
      (b) XX' is on the world line of one end of the meter stick of S' and thus makes
      an angle of ϕ with the vertical.
      (c) OX' makes an angle of ϕ with OX. $(1 - \beta^2)^{1/2}/\sin(\pi/2 - \phi) = OX'/\sin(\pi/2 +
      \phi)$. Use $\cos \phi = (1 + \tan^2 \phi)^{-1/2}$ and $\cos 2\phi = 2 \cos^2 \phi - 1$. $OX' = [(1 +
      \beta^2)/(1 - \beta^2)]^{1/2}$.

1.93 20.5 sec of arc.

1.94 (a) S' axes are inclined at $38°40'$ with the S axes, and are in the second and
      fourth quadrants.
      (b) 5/3, $-4/3$; $-4/3$, 5/3; 1/3, 1/3; 10/3, $-8/3$; $-8/3$, 10/3; 2/3, 2/3; 5, -4;
      -4, 5; 1, 1.
      (c) $OX_1' = OT_1' = 2.13$ m.

1.95 (b) The curves are hyperbolae with asymptotes $x = \pm\tau$, the light cone.
      $(|\tau^2 - x^2|)^{1/2}$ has the same value in all reference systems so $\tau^2 - x^2 = 1$ cali-
      brates the axis $x' = 0$ with $x = 0$ and $\tau^2 - x^2 = -1$ calibrates $\tau' = 0$ with $\tau = 0$
      (at the point where each curve crosses the appropriate axis).

1.96 See Fig. 1.82.

1.98 (a) $-12c/13$.
      (b) 5.
      (c) Slope of τ'-axis $= -12/13$. $OT_1' = 3.54$ m.

1.99 (a) $4c/5$.
      (b) $3/c$.

1.100 See Fig. 1.87.

1.101 See Prob. 1.100.

1.102 (a) $t' =$ time at which the searchlights are directed at $\theta = \omega t'$ above the hori-
      zontal. $t =$ time at which the light beams meet at x. Use $v = (dx/dt')(dt'/dt)$.
      $v = cL\omega/(c \cos^2 \omega t' + L\omega \sin \omega t')$. $v(t' = 0) = L\omega$ and $v(t' = \pi/2\omega) = c$.
      Write $v = 1/f(t')$ and calculate dv/dt'. Then argue that $v = c$ at $t' = \pi/2\omega$
      or $t \to \infty$ for $\omega L < c$ or $\omega L > 2c$ and at $t' = \pi/2\omega$ and one other t' for $c <
      \omega L < 2c$. $v > c$ for $L\omega > c$ to t' such that $\sin \omega t' = L\omega/2c$.
      (b) $x = L \cot \omega t'$, $t = t' + (x^2 + L^2)^{1/2}/c$. $v = c$ at $t' = 0$ and $v = -\omega L$ at $t' = \pi/2\omega$.
      Show $dv/dt' > 0$ so v increases to ∞ then from $-\infty$ to $-\omega L$. $v = c\omega L/
      (\omega L \cos \omega t' - c \sin^2 \omega t')$.

1.103 70 yr $-$ 11 sec.

1.104 $vt = \sqrt{1 - v^2/c^2}\, d$. 2.96×10^8 m/sec.

1.105 (a) $E_1 E_2$, $E_1 E_3$ time-like; $E_2 E_3$ null.
      (c) 0.53 m/c, 0.75 m/c.
      (e) $-3c/4$, $-0.87c$.

1.106 37.5 yr.

1.107 "Sphere" = calibration "curve" (see Prob. 1.95).

1.108 (a) Choose coordinate axes such that OP lies along the x axis, OQ along
      the y axis so $OP = (0, x_1, 0, 0)$ and $OQ = (0, 0, y_2, 0)$.
      (b) Use Eq. (1.60).
      (c) Choose coordinate axes such that OP lies along the τ axis, OQ along the
      x axis.
      (d) $\tau^2 - (x^2 + y^2 + z^2) = 0$ for a null line segment from $(0, 0, 0, 0)$ to (τ, x, y, z).

1.109 (a) -11, -82; -34.
      (b) 19, 5; 19.
      (c) 7, -17; -10.

1.110 (b) (1, 1, 0, 0), (0, 0, 2, 0).
      (c) 0, -4; 0.
      (d) Yes, since $\tau_1 \tau_2 - x_1 x_2 - y_1 y_2 - z_1 z_2 = 0$.

1.111 See Eq. (1.173).

1.112 (a) $a = (a_0, 0)$, $\ell = (b_0, \mathbf{b})$, $a \cdot \ell = a_0 b_0$.
   (b) Choose $a_0 = 0$ so $a \cdot \ell = -\mathbf{a} \cdot \mathbf{b}$.

1.113 (b) $t_0 = 3t/5$.

1.114 (a) $t_0 = 3t/5$.
   (b) Let the plane of the motion determine the x-y plane, $\mathbf{r}(0)$ the x axis. $x = r \cos \theta$ with $\theta r = \omega t$ so $\theta = 2ct/5$.
   (c) $[5ct_0/3, 2 \cos (2ct_0/3), 2 \sin (2ct_0/3), 0]$.
   (d) $dx/dt_0 = [5c/3, -(4c/3) \sin (2ct_0/3), (4c/3) \cos (2ct_0/3), 0]$.

1.115 (a) (c, o), $(2c/3^{1/2}, c/3^{1/2})$, $(10c/19^{1/2}, 9c/19^{1/2})$.
   (b) The length is given by $(\tau^2 + x^2)^{1/2}$ whereas the norm, which is the same for all v's, is $\tau^2 - x^2$.

1.116 Use $d(v \cdot v)/dt_0 = dc^2/dt_0 = 0$. Space-like.

1.117 a is a 4-vector, so use Eq. (1.173) and $a = (0, 32 \text{ ft/sec}^2)_{\text{rest frame}}$. $(-18, 37, 0, 0)$ ft/sec^2.

1.118 The Galilean transformation law for velocities, $\mathbf{v}' = \mathbf{v} - \mathbf{V}$, differs from that, Eq. (1.127), of the Lorentz law.

1.119 $v/c \ll 1$.

1.121 $I = 0$, $P_1 + (-P_1) = 0$.

1.122 $L_{0'0} = 1$, $L_{3'3} = -1$, $L_{1'1} = -\cos \theta$, $L_{1'2} = -\sin \theta$, $L_{2'1} = \sin \theta$, $L_{3'3} = -\cos \theta$, other $L_{\mu'\mu} = 0$.

1.123 (a) $\beta'' = (\beta + \beta')/(1 + \beta\beta')$.
   (b) See Prob. 1.82.

1.124 (a) An entity T described by the 64 components $T_{\alpha\beta\gamma}(\alpha, \beta, \gamma = 0, 1, 2, 3)$ that transform as $T_{\alpha'\beta'\gamma'} = L_{\alpha'\alpha}L_{\beta'\beta}L_{\gamma'\gamma}T_{\alpha\beta\gamma}$.
   (b) A scalar.

1.125 $T_{0'i'} = L_{0'\mu}L_{i'\nu}T_{\mu\nu} = L_{i'\nu}T_{0\nu} = L_{i'i}T_{0i}$, $i = 1, 2, 3$.

1.128 1.70×10^{14} Hz.

1.129 6.14×10^7 m/sec.

1.130 (a) $\mathbf{k} = k\hat{z}$, $ck = \omega$ so $k_{x'} = -\beta\gamma k$, $k_{z'} = k$ and $\tan \alpha = -\beta\gamma$. $\alpha = 20.5''$.
   (b) $\gamma \simeq 1$.

1.131 Use $vk = \omega$, $v' = \omega'/k'$, $(v - V)/(1 - vV/c^2)$.

1.132 (a) $5.38\mu s$, $6.27\mu s$.
   (b) $0.9\mu s$.

1.133 (a) $\theta = 180°$.

1.134 (a) 7.3 Hz.
   (b) 5.1×10^2 Hz.

1.135 (b) $+$.

A1.1 (a) 4.56×10^9 m.
   (b) $\Delta t = 3.55 \times 10^{10} \sin (1.721 \times 10^{-2} \, t/\text{day})/\{c[62.8 - 23.3 \cos (1.721 \times 10^{-2} \, t/\text{day})]^{1/2}\}m$.
   (c) $T = 2r_E/c$.
   (d) 3.00×10^8 m/sec.

A1.2 (a) $p = R/D$ radian.
   (b), (c), (e), (h)

km	mi	AU	ly	pc
8.3×10^{13}	5.2×10^{13}	5.6×10^5	8.8	2.7
2.47×10^{14}	1.53×10^{14}	1.65×10^6	26.1	8.0
4.1×10^{13}	2.6×10^{13}	2.8×10^5	4.3	1.33
2.6×10^{15}	1.6×10^{15}	1.7×10^7	2.7×10^2	8×10
4×10^{15}	2.4×10^{15}	3×10^7	4×10^2	1.2×10^2

A1.3 (a) Use Fig. A1.3. $\cos \theta = c_m/v$.
   (b) $1°24'$, $41°24'$, $40°14'$, $49°49'$, $65°36'$.

A1.4 (a) $c^2 - v^2 = (c + v)(c - v) \simeq 2(c - v)$.
   (b) 1.41×10^{-1}, 7.1; 4.47×10^{-2}, 22.4; 1.41×10^{-2}, 71.
   (c) v differs from c by 1.5×10^2, 1.5×10^{-4} and 1.5×10^{-9} m/sec.

A1.5   (a) $x_2 - x_1 = V(t_2 - t_1)$.

       (b) $t_2' - t_1' = -V(x_2' - x_1')/c^2$.

A1.6   The clocks are $(3/5)$ m apart according to S and the n-th clock reads $ct' = -4n/5$.

A1.7   The sphere appears rotated, i.e., a sphere.

A1.8   (c) $25(x')^2/9 + (y')^2 = 1$.

A1.9   (b) Use $v = (d\alpha/d\tau)(d\tau/dt_0)$ and determine $d\tau/dt_0$ from $v \cdot v = c^2 \cdot v = (\sqrt{4 + \tau^2}, \tau)/2$.

       (c) $t_0 = (2/c) \, ln\{[\tau + (\tau^2 + 4)^{1/2}]/2\}$.

A1.10 See the reference. (At $t' = 0$, $y' = \pm(3/50)$ m, $x' = \pm(2/5)$ m.)

A1.11 See the reference.

A1.12 See the reference.

A1.13 Use the binomial theorem: $(1 \pm vc/nc^2)^{-1} = 1 \mp vc/nc^2$.

A1.14 (a) Use Eq. (1.238) and $\nu = 2\pi/\lambda$.

       (b) $0.11°$, $3.6°$.

       (c) $\cos 0 + \cos 180° = 0$.

A1.15 (a)   (i) No. See (ii).

       (b)   (i) 93 km/sec-Mpc.

           (iii) 4×10^4 km/sec; 4×10^2 Mpc.

           (iv) 5×10^4 km/sec; 1.1×10^5 km/sec; 5×10^2 Mpc; 1.2×10^3 Mpc.

           (vi) 11×10^9 yr.

A1.16 (a) $c \cdot \Delta t = h$. (i) $gh/c^2 = 2.46 \times 10^{-15}$.

       (b) Use $\Delta t = \Delta t'/(1 - v^2/c^2)^{1/2}$, $(1 - v^2/c^2)^{1/2} = 1 - v^2/2c^2$ for $|v| \ll c$ and $\frac{1}{2}v_B^2 + gh = \frac{1}{2}v_A^2$.

A1.16 (b)   (i) $-GM/(R + h) = -(GM/R)(1 + h/R)^{-1} = -(GM/R)(1 - h/R)$.

           (ii) $(1 - gh/c^2)^{-1} = 1 + gh/c^2$.

           (iii) $g(h)dh = d(-GM/R)$.

A1.17 (a)   (i) Since $T_{\mu\nu} + T_{\nu\mu} = 0$, each diagonal component ($\mu = \nu$) is zero.

           (ii) The diagonal components vanish and the off-diagonal components are related in pairs: $(4 \times 4 - 4)/2 = 6$. For example, T_{01}, T_{02}, T_{03}, T_{12}, T_{13}, T_{23}.

       (b)   (iv) Each vector is perpendicular to $\mathbf{c}$ so consider projections of $\mathbf{a}$, $\mathbf{b}$ and $(\mathbf{a} + \mathbf{b})$ in the plane perpendicular to $\mathbf{c}$.

           (v) $\hat{y} \times \hat{z} = \hat{x} = -\hat{z} \times \hat{y}$, $\hat{z} \times \hat{x} = \hat{y} = -\hat{x} \times \hat{z}$, $\hat{y} \times \hat{y} = 0 = \hat{z} \times \hat{z}$.

           (vi) Use $\mathbf{a} \times \mathbf{b} = (a_x\hat{x} + a_y\hat{y} + a_z\hat{z}) \times (b_x\hat{x} + b_y\hat{y} + b_z\hat{z})$.

       (c) (i), (ii) and (d) (iii) and (e). Determine $L_{\mu'\mu}$ and use $T_{\mu'\nu'} = L_{\mu'\mu}L_{\nu'\nu}T_{\mu\nu}$.

       (e) (iii) $V^{1'} = V^1$, $V^{2'} = \gamma(V^2 - \beta P^3)$, $V^{3'} = \gamma(V^3 + \beta P^2)$, $P^{1'} = P^1$, $P^{2'} = \gamma(P^2 + \beta V^3)$, $P^{3'} = \gamma(P^3 - \beta V^2)$.

Chapter 2

2.1   (a) If $\mathbf{v} = \mathbf{v}' + \mathbf{V}$, $\mathbf{V} = $ constant vector, then $\mathbf{a} = \mathbf{a}'$.

       (b) $\mathbf{r}_i = \mathbf{r}_i' + \mathbf{V}t$, so $\mathbf{r}_i - \mathbf{r}_j = \mathbf{r}_i' - \mathbf{r}_j'$.

       (c) $d\mathbf{P}/dt = \sum_i m_i\mathbf{a}_i = \sum_{i,j} \mathbf{F}_{i(j)} = 0$.

2.2   (a) Use Eq. (2.18).

       (b) Use $d(\mathbf{v} \cdot \mathbf{v})/dt = 2\mathbf{a} \cdot \mathbf{v}$.

       (d) 1.6×10^{-5} W.

       (e) 1.88×10^9 m/sec $> c$.

       (f) 10^{-5} Å.

       (g) $1.008u$.

       (h) 4.00×10^{17} kg/m³ $= 0.240u$/fm³, 10^3 kg/m³ $= 6.02 \times 10^{-16}$ u/fm³.

2.4   (a) Use $\Delta V = -\mathbf{F} \cdot \Delta \mathbf{r}$, $\mathbf{F}$ and $\Delta\mathbf{r}$ radial.

       (b) $\sin i/\sin r = v_{inside}/v_{outside}$ and conservation of energy.

       (c) Use $i = \Theta/2 + r$, expand $\sin(\Theta/2 + r)$, set $\cos r = (1 - \sin^2 r)^{1/2}$ and remove surd.

2.5   (a) $d\mathbf{L}/dt = m\mathbf{v} \times \mathbf{v} + m\mathbf{r} \times \mathbf{a}$.

       (b) $|d\mathbf{r}_\perp| = rd\theta$ and use Prob. A1.17 (b) (iii).

       (c) $v^2 = \mathbf{v} \cdot \mathbf{v} = (\mathbf{v}_\parallel + \mathbf{v}_\perp) \cdot (\mathbf{v}_\parallel + \mathbf{v}_\perp) = v_\parallel^2 + v_\perp^2$ and $v_\parallel = dr/dt$.

(d) Replace v^2 in the energy conservation law with the expression for v^2 given in (c), use the hint and replace $d\theta/dt$ using the formula for L in (b).

(e) Use $du/d\theta = -(1/r^2)dr/d\theta = -u^2 dr/d\theta$.

(h) $\cos\theta_i = 1/\varepsilon$, $\Theta = \pi - 2\theta_i$.

2.6 (a) Use $a = v^2/R$ and $|\mathbf{v} \times \mathbf{B}| = vB$.

(b) $d\mathbf{v}_\parallel/dt = 0$ since $q\mathbf{v} \times \mathbf{B} = q\mathbf{v}_\perp \times \mathbf{B}$ is perpendicular to $\mathbf{v}$.

2.7 (a) Use the law of conservation of momentum.

(b) $M(t) - M(O) = \int_0^t (-\mu)dt$.

(c) $v(t) = \int_0^t (dv/dt)dt$.

2.8 (a) $\mathbf{a}' = (d/dt')(d\mathbf{r}'/dt') = (1 - \beta^2)^{1/2}/(1 - \beta v_x/c)d/dt[(v_x - \beta c, (1 - \beta^2)^{1/2}v_y, (1 - \beta^2)^{1/2}v_z)/(1 - \beta v_x/c)]$.

(b) $\mathbf{a}'$ does not depend only on $\mathbf{a}$ and β: it also depends on $\mathbf{v}$.

2.9 Newton's laws are much easier to apply to everyday motions and give very accurate descriptions of these.

2.10 The relationship between inertial frames.

2.11 No, since $T_{\mu'3'} = L_{\mu'\mu}L_{3'\nu}T_{\mu\nu}$ whereas $b_{\mu'} = L_{\mu'\mu}b_\mu$.

2.12 (a) Unlike x, $\mathbf{r}$ cannot be defined independent of a reference system.

(b) Show that $v = v_0$ is equivalent to $\mathbf{v} = \mathbf{v}_0$.

2.13 One needs to know more than ma and β to determine ma'.

2.14 (a) 8.0×10^{14} MeV.

(b) 2.7×10^{23} MeV/c.

2.16 (b) $E \simeq T$, $[1 - (E_0/E)^2]^{1/2} = 1 - \frac{1}{2}(E_0/T)^2$.

2.17 (a) 0.511.

(b) 0.511, 0.512, 1.511 MeV, 1 GeV.

(c) 1.00, 1.002, 2.96, 1.96×10^3.

(d) 1.00, 0.937, 5.88×10^{-2}, 1.30×10^{-7}.

2.18 (a) 78 MeV.

(b) 77.5 MeV.

(c) $c(1 - 2 \times 10^{-5})$.

2.19 (a) Use $v_x = p_x c^2/E$.

(b) Use (a) and $E = m_v c^2$.

2.20 (b) The energy-momentum vectors are tangent to the world line and have norm $m^2 c^2$.

2.21 (a) 2.488×10^{-28} kg.

(b) 139.6 MeV, 1.396 MeV/c; 140.3 MeV, 14.03 MeV/c; 161.2 MeV, 80.6 MeV/c; 320 MeV, 288 MeV/c; 990 GeV, 980 MeV/c; 3.12 GeV, 3.12 GeV/c; 9.87 GeV, 9.87 GeV/c.

2.22 $v = pc^2/E$ and $(E/c)^2 - p^2 = m^2 c^2$.

2.23 (a) $\mathbf{k} = \alpha\mathbf{p}$, α a scalar.

(b) $\mathbf{k} \cdot \mathbf{k} = 0$ since $\mathbf{k}$ is null.

2.25 (a) $v = \left\{1 - [m/(m + T)]^2\right\}^{1/2} = 1 - \frac{1}{2}[m/(m + T)]^2 = 1 - 3.1 \times 10^{-4}$.

(b) Use Eq. (2.65). 37 GeV.

(c) $c(1 - 3.1 \times 10^{-4})$.

(d) Use Eq. (2.66). 958 MeV.

(e) Use Example 2.4. 6.30×10^6 m/sec.

(f) $1 - 3 \times 10^{-4}$.

2.26 (a) $1 - 1.3 \times 10^{-9}$.

(b) 1.8×10^4 GeV.

(c) $1 - 1.3 \times 10^{-9}$.

(d) 4.4 GeV.

(e) 0.91.

(f) $1 - 1.3 \times 10^{-9}$.

2.27 (a) Use conservation of $e + E$ with $\mathbf{p} + \mathbf{P} = 0$ as well as $E^2/c^2 = P^2 + M^2$. Then show that the resulting equation has, with $p_f^2 = p_i^2 + \varepsilon$, only the solution $\varepsilon = 0$.

(b) Calculate V from Eq. (2.68), then use Lorentz transformation on $\not{k}^* =$

$(20, 0, 0, 20)$ MeV/c to get $\tan \phi = (1 - V^2)^{1/2}/V$, $\phi = 88.8°$.

2.28 (a) 11.4 MeV/c $\hat{x}$, 11.4 MeV; 8.07 $(\hat{x} + \hat{y})$ MeV/c, 11.4 MeV; $(0, 3.37\hat{x} - 8.07\hat{y})$MeV/c; -76(MeV/c)^2.

2.29 (a) Each $p^{(a)}$ lies inside the future light cone so a diagram shows the vector sum cannot point out of that cone.
(c) $E_0 = (E^2 - \mathbf{P}^2 c^2)^{1/2}$.
(d) Use (a).
(e) Use the Lorentz transformation.
(f) Transform to the center-of-momentum system. Each $p^{(a)}$ has fourth component $\geq m^{(a)}c$, the equality holding only if $\mathbf{p}^{(a)}$ lines along $\mathbf{V}$.

2.30 (a) Take norm of $p + \mathscr{P} - p' = \mathscr{P}'$.
(b) Use $p_0 + \mathscr{P}_0 = p'_0 + \mathscr{P}'_0$.
(c) Use $q = |\mathbf{P}'|$ and (b).
(d) Use $q^2 = (p - p')^2$.
(e) Use $(p + \mathscr{P} - \mathscr{P}')^2 = p'^2$.

2.31 Use Prob. 2.30(b).
(i) 0.16 MeV.
(ii) 9.6 MeV.
(iii) 17.6 MeV.

2.34 (a) The speeds are the same.
(d) Use $v'_x = dx'/dt'$.
(e) Substitute for $(1 - v_i'^2)^{1/2}$ from (d).

2.35 Use Eq. (2.119) to calculate ϕ from β_1 and θ.

2.36 Consider particles of unit mass colliding and then S'' is the center-of-momentum system.
$\pm \mathbf{V}/(1 + [1 - V^2/c^2]^{1/2})$.

2.37 (a) $E' = (c/X'_0)(EX_0/c - \mathbf{p} \cdot \mathbf{X})$.
(b) $p' = (E'^2/c^2 - m^2c^2)^{1/2}$.
(c) $v' = p'c^2/E'$.

2.38 (a) $p_1 \cdot p_2/m_2$, $[(p_1 \cdot p_2/m_2 c)^2 - m_1^2 c^2]^{1/2}$, $c[1 - (m_1 c)^2 (m_2 c)^2/(p_1 \cdot p_2)^2]^{1/2}$.
(b) $cp_1 \cdot (p_1 + p_2)/[(p_1 + p_2) \cdot (p_1 + p_2)]^{1/2}$, $[\{p_1 \cdot (p_1 + p_2)\}^2/\{(p_1 + p_2) \cdot (p_1 + p_2)\}$
$m_1^2 c^2]^{1/2}$, $c[1 - (m_1 c)^2 (p_1 + p_2) \cdot (p_1 + p_2)/\{p_1 \cdot (p_1 + p_2)\}^2]^{1/2}$.
(c) 3.9 GeV, 1.01 GeV, 2.6×10^8 m/sec.
(d) 4.3 GeV, 8.1 GeV, 2.96×10^8 m/sec.
(e) 3.5 GeV, 4.7 GeV, 2.98×10^8 m/sec.

2.39 (a) 2×10^{17} for $\Delta v = 1$ mm/sec.
(b) 10^{14} for $\Delta v = 1$ mm/sec.

2.40 (a) $2\text{-}3 \times 10^{-10}$.
(b) See (a): $\Delta m \ll m$.

2.41 (a) 3.8×10^9 J/ton.
(b) (i) 7.6×10^{13} J. (ii) 1.9×10^{17} J.
(c) (i) 0.8 g. (ii) 2 kg.
(d) 8 kg.

2.42 (a) $v^2/c^2 = 1 - 1/(1 + T/mc^2)^2$.
(b) $n_e = 3.8 \times 10^{13}$, 1.64 MeV, 4.80 MeV.

2.43 3.5×10^{-6} kg.

2.44 10^{-12}.

2.45 (a) 1.4×10^3 J/m^2-sec.
(b) Solar constant $\times$ area/c $= 10^{-6}\%$.

2.46 7×10^{-12}.

2.47 Use Eq. (2.174). 1.0×10^{-13}.

2.48 (a) $W(1 + \rho)/c$.
(b) $W \cos \theta (1 + \rho)/c$.

2.49 (a) 7×10^5 kg. No.
(b) 4×10^9 kg. No.

2.50 2×10^{-2} μg.

2.51 1.25×10^6.

2.52   The energy comes from the drop in the height of the mass E/c^2.

2.53 $F_{\text{radiation}} = (3.9 \times 10^{26} \ W/c)(\pi r^2/4\pi R^2)$.
        $r < (6 \times 10^{-4} \ \text{kg}/m^2)/\rho \sim 1\mu$.

2.54   (a) Solve the first equation for m/M, substitute in the second and then find γ^{-2}.
        (b) Solve for E'/Mc^2 from each equation and then use the calculation of (a).
        (d) The photon rocket uses less mass to attain a given speed.
        (e) For the photon rocket, $M_{\text{final}}/M_{\text{initial}} = \{[(c - V)/(c + V)]^{1/2}\}^4 = 4 \times 10^{-7}$.

2.55   Use Eq. (2.162).

2.56   (a) $\cos \theta = c/v$.
        (b) $\cos \theta = |E/\mu c^2|/[1 + (E/\mu c)^2]^{1/2}$.

2.57   Calculate $\mathbf{v}' \cdot \mathbf{v}'$ and expand the numerator in powers of $\mathbf{v} \cdot \mathbf{V}$.

2.58 dv/dt_0 reduces, for $v \ll c$, to $\mathbf{a} = 0$ and Prob. 2.8 shows that if $\mathbf{a} = 0$ relative to S, $\mathbf{a}' = 0$ relative to S'.

2.60   (b) $v = c^2 t/(c^2 t^2 + \alpha^2)^{1/2}$.
        (c) $\alpha dt/(\alpha^2 + c^2 t^2)^{1/2} = \alpha dt/x$.
        (d) Use $d/dt_0 = (\alpha/x)d/dt$.

2.62   Eq. (2.187) with subscripts x and y interchanged.

2.65   (a) 2.15×10^8 m/sec.
        (b) 0.70.
        (c) 0.87 yr.
        (d) $4.0 \times 10^{15} \ m = 0.42 \ ly$.

2.66   (a) (i) 4.3×10^6 sec $= 0.14$ yr. (ii) 1.4×10^7 sec $= 0.43$ yr.
        (b) (i) 6.1×10^6 sec. (ii) 1.9×10^7 sec.

2.67   Use $g' = M'g/M_E$. $M' \sim 10^{35}$ kg.

2.68   (a) $v = c/[1 + (c/gt)^2]^{1/2}$.
        (b) Multiply Eq. (2.218) by m.
        (c) $v \to c$ while $p - mgt \to \infty$, whereas $p_{NR} = mv_{NR}$.

2.69   The force needed to give the same acceleration to the particle increases with speed so $\int F v dt$ is greater. T increases indefinitely whereas v never reaches c.

2.70   (a) $gt/c \ll 1$ so $v = gt$; $T = \frac{1}{2}mv^2$ for $v/c \ll 1$.
        (b) $v \simeq c$ so, by Eq. (2.218), $(1 - v^2/c^2)^{-1/2} = gt/c$. $mg =$ force and $ct =$ distance.

2.71   (a) $v = gt$. $F = md/dt\{gt/[1 - (gt/c)^2]^{1/2}\} = mg/[1 - (gt/c)^2]^{3/2}$.
        (b) $(mg^2t/\{c[1 - (gt/c)^2]^2\}, mg/[1 - (gt/c)^2]^2)$.

2.72   (a) $c^2 dt_0^2 = c^2 dt^2 - dx^2$. Differentiate with respect to t_0.

2.73   (a) Use $d/dt = (dx/dt)(d/dx)$.
        (b) Multiply by dx and integrate between $t = 0$ and t.
        (c) Use $dt = dx/v$. x goes from 0 to a in 1/4 oscillation.
        (d), (e) Use (b) to get $\gamma(x)$; expand with the binomial theorem.

2.74   (a) Multiply the equation of motion with dv/dt_0 after expanding $d(\mu v)/dt_0$. Use $v \cdot (dv/dt_0) = 0$.
        (b) $d\mu/dt_0 = b$, by the procedure given for (a).
        (c) Use $t_0 = (1 - v^2/c^2)^{1/2} t$.

2.75   (a) Use the procedure given above for Prob. 2.74 (a).
        (c) $(1/c)(T - vX/c)/(1 - v^2/c^2)^{1/2}$.

2.76   (a) $\mathbf{F} \cdot \mathbf{v} = 0$ so m_v and p are constant.
        (b) See Prob. 2.6.

2.77   (a) See Prob. 1.57 (a).
        (b) See Prob. 1.57 (b).
        (c) Use $\mathbf{f} = \gamma \mathbf{F}$.
        (d) S' is the rest system.

2.78   (a) Use $\mathbf{f} = \gamma \mathbf{F}$ in Prob. 2.77 (a).
        (b) Use $\mathbf{v} \cdot \mathbf{F}_0 = v F_x$.

2.79   (a) $d\mathbf{p}/dt = \mathbf{F}$.

(c) Use $v^i = p^i c^2/E$, $(v^0)^2 - \Sigma(v^i)^2 = c^2$ and $v^0 = c/(1 - v^2/c^2)^{1/2}$ to get $v_x = v^i c/(\Sigma(v^i)^2 + c^2)^{1/2}$. $v_x = p_0^i c/(2p_0^2 + 2F^2 t^2 + m^2 c^2)^{1/2}$, $v_y = Ftc/(2p_0^2 + 2F^2 t^2 + m^2 c^2)^{1/2}$, $v_z = 0$.

(d) $p_x = $ constant $= m_v v_x$ and $m_v = m_v(t)$.

(e) $(x, y, z) = (x_0, y_0, z_0) + \int_0^t (v_x, v_y, v_z)dt = (x_0, y_0, z_0) + ((p_0 c/2^{1/2}F) \ln\{[2^{1/2}Ft + (2F^2 t^2 + 2p_0^2 + m^2 c^2)^{1/2}]/(2p_0^2 + m^2 c^2)^{1/2}\}, (c/2F)[(2F^2 t^2 + 2p_0^2 + m^2 c^2)^{1/2} - (2p_0^2 + m^2 c^2)^{1/2}], 0)$.

2.80   (a) Use $d\mathbf{p}/dt = \mathbf{F}$.

(b) $t_{10} \neq t_{20}$ in general.

2.81   (a) $ma = eE$, $t = l/v$, $d = \frac{1}{2}at^2$ and $\theta = d/l$.

(b) $ma = evB$.

(c) 1.4×10^{11} C/kg, 2.7×10^7 m/sec.

2.82   (a) $F = mg$; $m = $ volume $\times$ density.

(b) $F = mg - $ (weight of displaced air).

(c) $F_{total} = F_{effective\ gravity} - 6\pi\eta av = 0$ when acceleration $= 0$.

(d) $F_{new\ total} = F_{electric} - (F_{total}$ of $c))$.

(e) Calculate $a = 2.04 \times 10^{-6}$ m. $q = 1.13 \times 10^{-18}$ C.

(g) 1.642×10^{-19} C.

2.83   (a) $eVJ = V$ electron volts.

(b) $(1.602 \times 10^{-19} V)$J.

(c) Use the binomial theorem: $(1 - v/c)^{-1/2} = 1 + \frac{1}{2}v^2/c^2 + (3/8)(v^4/c^4) \ldots$ so want v such that $[(3/8)(v^4/c^4)]/(v^2/2c^2) = .01$. $v = 3.5 \times 10^7$ m/sec.

2.84 2.60×10^8 m/sec.

2.85   Calculate $(-q_e/m_v)[1/(1 - v^2/c^2)^{1/2}]$.

2.86   (a) 16J.

(b) Use Prob. 2.16 (b). 1.3×10^{-29}, 5.1×10^{-15}.

2.87   (a) 3.6×10^{-19} J.

(b) 140, 14.

2.88   (a) $v' = v[(1 - V/c)/(1 + V/c)]^{1/2}$.

(b) Use $V/c = (v^2 - v'^2)/(v^2 + v'^2)$.

2.89   0.718Å, 0.735Å, 0.752Å.

2.90 $\lambda' = \lambda + [h/(10^5 m_e c)](1 - \cos \theta)$.

2.92   (b) Plot $\lambda' - \lambda$.

2.93   (a) $\not{p}_e = \not{p}_v + \not{p}_e'$. $\not{p}_e' \cdot \not{p}_v = 0$ only if $v = 0$.

(b) $\not{p}_e + \mathcal{P}_0 = \not{p}_e' + \mathcal{P}' + \not{p}_v$ with $\mathcal{P}' = (Mc, M\mathbf{v})$.

(c) $hv = eV$.

(d) (i) 1.74×10^4 V, (ii) 5.82×10^4 V.

2.94   (b) $T_{max} = 0$. $v_0 = \phi/h$.

(c) 5.95×10^{14} Hz.

2.95   (b) $dA = (2\pi R \sin \Theta)(R\Delta\Theta)$.

(c) The number scattered into the ring between Θ and $d\Theta$ is $|$(incident intensity)$(d\sigma/d\Omega)(d\Omega)|$ and equals the number incident through the ring between S and dS, $|$(incident intensity)$(2\pi SdS)|$. Also $dS/d\Theta < 0$.

(d) Use $2 \sin (\Theta/2) \cos (\Theta/2) = \sin \Theta$.

(e) Use $E = \frac{1}{2}mv_i^2$.

2.96   (a) Use $V = Kq_1 q_2/r$ with $q_1 = 92e$, $q_2 = 2e$. $r = 6.3 \times 10^{-14}$ m.

(b) $V = 29$ MeV.

2.97   0.2 mm, 0.02μ.

2.98   (a) 2.3×10^{17} kg/m³.

(b) 2.5×10^8 tons.

(c) From 5×10^{-4} to 2×10^{-4}.

2.99   (a) 17.3 MeV.

(b) 22.4 MeV.

(c) $(2.99 \times 10^8 \pm 5 \times 10^5)$m/sec, $(2.975 \times 10^8 \pm 5 \times 10^5)$m/sec.

2.100 (a) $(Mc + \Delta E/c, 0) = (E/c + hv/c, hv/c - P)$. $E/c = [(hv/c)^2 + M^2 c^2]^{1/2}$. Solve conservation of mass-energy equation for $h\nu$ and use the binomial theorem to simplify the result.

(b) 1.07×10^{21} Hz.

2.101 (a) Use the conservation laws of charge and mass number.

2.102 (a) Use the formulae of Prob. 2.101.

(b) 5.25, 4.58, 4.08.

2.103 $\exp[(\lambda_{238} - \lambda_{235})t] = 0.007/0.993$. 6.0×10^9 yr.

2.104 (a) 6.5×10^9 yr, 3.6×10^5 yr.

(b) 46%, $10^{-6 \times 10^8}$%.

2.105 (d) (i) $33d$.

(ii) Neglect $\exp(-\lambda_A t)$. $145d$.

(iii) $925d$.

(iv) In units of N_0/day: 1.2×10^{-3}, 1.1×10^{-1}; 3.1×10^{-3}, 5.3×10^{-2}; 3.2×10^{-3}, 1.3×10^{-7}; 4.2×10^{-4}, 1.1×10^{-31}.

2.106 (a) 20.94 GeV, 20 GeV/c, 2.87×10^8 m/sec.

(b) Use Eq. (2.66). 6.20 GeV.

2.107 (a) $c(1 - 1.4 \times 10^{-13})$.

(b) 9.8×10^5 MeV.

2.108 (a) $E^* = [(E + M)^2 - (E^2 - M^2)]^{1/2}$.

2.109 (a) 2.6×10^{-5}, 8 cm.

(b) 1.0×10^{-12} N.

(c) 4.1 cm.

(d) 2 mi.

2.110 (a) 4.4×10^{-23}.

(b) 9.4×10^{-12}.

(c) 30 sec.

2.111 (a) Use the symmetry.

(b) $|\mathbf{V}_{\text{CoM}}| = |\mathbf{v}'_{\text{final}}|$.

(d) $d\Omega \propto \sin\Theta d\Theta = 2\sin(\Theta/2)\cos(\Theta/2)d[2(\Theta/2)]$.

(e) Use Prob. 2.95 (d) and Fig. 2.5.

2.112 260 MeV.

2.113 (a) 23.8 MeV.

(b) 8.6×10^7 J.

(c) $\sim 10^{31}$ J.

(d) $\sim 10^5$.

(e) 3×10^{-12}%.

2.114 $m_n > m_p$.

2.115 1.4 MeV.

2.118 Use Prob. 2.95: $\int d\sigma = \frac{1}{2}S^2|_0^a = a^2/2 = 5\ mb$.

2.119 3.3434×10^{-27} kg.

2.120 (a) c.

(b) 0.93 MeV/c.

2.121 (a) Use $\mathscr{P}_p = \mathscr{P}_n - \mathscr{P}_e$. 0.78 MeV.

(b) Use $E_n = E_p + E_e + E_\nu$. 0.28 MeV, 0.28 MeV/c.

2.123 3×10^{16} m.

2.124 $mc^2 + T/2 \pm (2Tmc^2 + T^2)^{1/2}/2$.

2.125 (a) 3.07 MeV.

(b) $T = 1.023$ MeV.

2.126 (c) 0.5 MeV.

2.127 20 MeV/980 MeV $\ll 1$ so use Newtonian mechanics. 4.44 MeV.

2.129 150 MeV.

2.130 280 MeV.

2.131 67.5 MeV.

2.132 The neutrinos are emitted in the direction opposite to the electron's. Also

$E_v = p_v c.$ 111 MeV.

2.133 (a) 0.054.

2.136 (a) Strong.
(b) Weak.
(c) Strong.
(d) Weak.
(e) Weak.
(f) Weak.

2.137 (b) Expand the formula given for T_1 and compare that with $T_1 = E_1 - m_1 c^2$ with E_1 given in (a).
(c) 5.4 MeV, 32.5 MeV.

2.139 (a) 767 MeV.
(b) 0.
(c) 657 MeV.
(d) 912 MeV.
(e) 663 MeV.

2.141 Use Prob. 2.140. 1.24 GeV, 1.48 GeV, 1.69 GeV, 1.92 GeV, 2.36 GeV.

2.142 Use $\not{p}_\pi + \mathscr{P}_p - \mathscr{P}_p' = \not{p}_\rho$. 16.3 MeV, 176 MeV/$c$.

2.143 (a) 1.6×10^{-16} N.
(b) 1.1×10^{-5}.
(c) 17.5 BeV.
(d) 6.3°.
(e) (i) 4.4×10^2 m, 67 m.
(f) 130.
(g) The maximum occurs when the two final protons move along the direction of the incident proton, and their energy-momentum vectors are equal. 200 BeV.
(h) (i) 200 BeV.
(ii) 200 BeV.
(iii) $1 - 1 \times 10^4$ m.
(iv) 2.4′.
(i) (i) Use Prob. 2.76. (1) 2.2 m. (2) 11.1 m. (3) 22.2 m. (4) 111 m. (5) 445 m.
(ii) (1) 21.4 BeV. (2) 61.1 BeV. (3) 87.5 BeV. (4) 198 BeV. (5) 398 BeV.

2.144 (b) 4.4×10^{-7}.
(c) 41.5 BeV.
(d) 6.3°.
(e) (i) 2.2×10^3 m.
(ii) 3.3×10^2 m.
(f) 302.
(g) 1 TeV.
(h) (i) 1 Tev.
(ii) 1 TeV.
(iii) 5×10^4 m.
(iv) 0.48′.
(i) (ii) (1) 50 BeV. (2) 139 BeV. (3) 198 BeV. (4) 445 BeV. (5) 893 BeV.

A2.1 $v = (c[1 + c^2 t^2]^{1/2}, c^2 t)$.

A2.2 Use Eq. (2.107), 74.9°.

A2.3 See Prob. A1.3. $E = 1.11$ BeV, $T = 172$ MeV.

A2.4 (a) Use Eq. (2.107) with $\theta = \phi$. 22.5°.
(b) 0, 9.1 GeV.

A2.5 (a) Differentiate $p = mv/(1 - v^2/c^2)^{1/2}$.
(b) Differentiate $E = mc^2/(1 - v^2/c^2)^{1/2}$. Integrate the equation for large E to get $v = c \exp(1/E^2)$.

A2.6 (a) See Ex. 2.66.

A2.7 5×10^{-11} kg.

A2.8 Use $(\not{p}_1 - \not{p}_2)^2 = \not{p}_3^2$.

    (a) 18.2 MeV, 92.6 MeV.

    (b) 8.6 MeV, 57.2 MeV.

A2.9 293 MeV.

A2.10 (a) 2.24 GeV.

      (b) 0.

A2.11 (a) $[(m_p + m_{\pi^-} + Nm_{\pi^\circ})^2 - (m_p + m_{\pi^-})^2]/(2m_p) = T$.

      (b) 165 MeV, 349 MeV, 553 MeV, 776 MeV.

A2.12 Use $(\not{p}_1 + \not{p}_2 - \not{p}_3)^2 = \not{p}_4^2$. Multiply the zm_2 term by $[1 - (1 - z)^{1/2}]/[1 - (1 - z)^{1/2}]$.

A2.13 87 MeV/c.

A2.14 $M(t) = 160 \text{ lb} - 100 \text{ W}t/c^2$, $F = -100 \text{ W}/c$.

      Use $v = 60 \text{ mi/hr} + \int_0^t [F/M(t)]dt$.

      (a) Less than 60 mi/hr by

          (i) 1.7×10^{-5} m/sec.

          (ii) 4.0×10^{-4} m/sec.

          (iii) 2.8×10^{-3} m/sec.

          (iv) 1.4×10^{-1} m/sec.

      (b) 1 billion years.

A2.15 (a) 31 cm.

      (b) (i) 1.0 m.

          (ii) 3.8 m.

          (iii) 24 m.

A2.16 $\exp(-0.094 \ x/\text{m})$.

A2.19 (a) $-(k/r^2)\hat{r} \cdot \mathbf{v} = -(k/r^2)(dr/dt) = d(k/r)/dt$.

      (d) See Prob. 2.5 (b).

      (e) $dr/dt = (dr/du)(du/d\theta)(d\theta/dt)$.

      (f) $v^2 = (dr/dt)^2 + r^2(d\theta/dt)^2$.

      (k) Energy must be supplied to free the particle with energy $\geq mc^2$.

Chapter 3

3.1   (a) Use $|\mathbf{r} + d\mathbf{r}| = [(\mathbf{r} + d\mathbf{r}) \cdot (\mathbf{r} + d\mathbf{r})]^{1/2}$ and $r = (\mathbf{r} \cdot \mathbf{r})^{1/2}$.

      (b) $g = GM/R^2$.

3.5   Find θ such that $1/r \to 0$ and from that calculate the scattering angle. Use the fact that $v \simeq c$ and $D \simeq S$.

3.6   Draw a single line segment representing the distance between two of the cities and then use a compass to find the position of a third city. Continue with a fourth, etc.

3.7   The prototype of a vector is $d\mathbf{r}$ and every vector can be represented by a prototype vector.

3.9   The calculation is simplified by choosing coordinate axes such that the two points lie along one axis.

3.10 Choose the polar axis to contain the two points.

3.11 $dV^i = -\Sigma\Gamma^i_{jk}V^j dx^k$ so take $\mathbf{V} = (V_1, 0)$, $d\mathbf{r} = (0, dx^2)$. $x^1 dV^2 = -V^1 dx^2$ so $\Gamma^2_{12} = 1/x^1$.

3.12 Take $\mathbf{V} = (0, V_2)$, $d\mathbf{r} = (0, dx^2)$. $dV^1 = -(x^1 V^2)dx^2$ so $\Gamma^1_{22} = -x^1$.

3.13 Draw a diagram, like Fig. 3.20, of the coordinate lines in the neighborhood of (r, θ) and note what happens to the vectors $\mathbf{V}_1 = (V_1, 0)$ and $\mathbf{V}_2 = (0, V_2)$, keeping in mind that rV_2 is the cartesian component, as they are transported around the quadrilateral.

3.14 $\int_1^2 \frac{1}{2}mv^2 dt$ is a minimum if $m\ddot{x} = 0$.

3.17 7×10^{-10}, 2×10^{-6}.

3.18 $r = 1.5$ km. 1.5×10^{20} kg/m³.

3.21 42.9″, 8.6″, 3.8″.

3.22 (c) The fraction Δp of one revolution per revolution.

(d) 1.6×10^{-7}.

(e) In this problem, one parameter, α, is introduced to explain one result, 43″/century.

3.23 7″/century.

A3.1 (b) $ds^2 = (dx^1)^2 + R^2(dx^2)^2$.

A3.2 (b) $ds^2 = \sec^2\alpha(dx^1)^2 + (x^1)^2\tan^2\alpha(dx^2)^2$.

A3.3 The cylinder is flat, i.e., has zero curvature, as can be seen by cutting it along a line parallel to the axis and spreading it out on a plane: the intrinsic geometry is not changed (except at the cut).

A3.4 The curvature is zero.

A3.7 1.47 km, 4.4 mm, 6×10^{-26} m.

A3.8 (b) The scalar product is conserved under parallel transport.

(e) Go to the final vertex by parallel propagation along two different paths.

(f) $\Gamma^1_{21} = \Gamma^1_{12} = -(1/R)\tan(x^2/R)$, $\Gamma^2_{11} = (1/R)\sin(x^2/R)\cos(x^2/R)$, all other components are zero.

A3.9 $\delta_1 V^i = -\Sigma\Gamma^i_{jk}(x)V^j\delta x^k_1$

$\delta_2 V^i = -\Sigma(\Gamma^i_{jk} + \Gamma^i_{jk,l}\,x^l_1)(V^j - \Gamma^j_{mn}V^m\delta x^n_1)\delta x^k_2$.

(d) Use $\Delta V^i = (\delta_1 V^i + \delta_2 V^i)_{0\to1\to2} - (\delta_1 V^i + \delta_2 V^i)_{0\to2'\to1'}$.

A3.10 (a) Use $x' = x\cos\theta + y\sin\theta$, $y' = -x\sin\theta + y\cos\theta$ for small θ: $x' = x + \theta y$, $y' = y - \theta x$.

(b) (i) 2.1×10^{-5}″.

(ii) 1.3×10^{-2}″.

(iii) $7.1 \times 10^{11} m^2 = (840\ \text{km})^2$.

Author Index

Subject Index

A CATALOGUE OF
SELECTED DOVER BOOKS
IN ALL FIELDS OF INTEREST

A CATALOGUE OF SELECTED DOVER
BOOKS IN ALL FIELDS OF INTEREST

CELESTIAL OBJECTS FOR COMMON TELESCOPES, T. W. Webb. The most used book in amateur astronomy: inestimable aid for locating and identifying nearly 4,000 celestial objects. Edited, updated by Margaret W. Mayall. 77 illustrations. Total of 645pp. 5⅜ x 8½.
20917-2, 20918-0 Pa., Two-vol. set $9.00

HISTORICAL STUDIES IN THE LANGUAGE OF CHEMISTRY, M. P. Crosland. The important part language has played in the development of chemistry from the symbolism of alchemy to the adoption of systematic nomenclature in 1892. ". . . wholeheartedly recommended,"—Science. 15 illustrations. 416pp. of text. 5⅝ x 8¼.   63702-6 Pa. $6.00

BURNHAM'S CELESTIAL HANDBOOK, Robert Burnham, Jr. Thorough, readable guide to the stars beyond our solar system. Exhaustive treatment, fully illustrated. Breakdown is alphabetical by constellation: Andromeda to Cetus in Vol. 1; Chamaeleon to Orion in Vol. 2; and Pavo to Vulpecula in Vol. 3. Hundreds of illustrations. Total of about 2000pp. 6⅛ x 9¼.
23567-X, 23568-8, 23673-0 Pa., Three-vol. set $27.95

THEORY OF WING SECTIONS: INCLUDING A SUMMARY OF AIR-FOIL DATA, Ira H. Abbott and A. E. von Doenhoff. Concise compilation of subatomic aerodynamic characteristics of modern NASA wing sections, plus description of theory. 350pp. of tables. 693pp. 5⅜ x 8½.
60586-8 Pa. $8.50

DE RE METALLICA, Georgius Agricola. Translated by Herbert C. Hoover and Lou H. Hoover. The famous Hoover translation of greatest treatise on technological chemistry, engineering, geology, mining of early modern times (1556). All 289 original woodcuts. 638pp. 6¾ x 11.
60006-8 Clothbd. $17.95

THE ORIGIN OF CONTINENTS AND OCEANS, Alfred Wegener. One of the most influential, most controversial books in science, the classic statement for continental drift. Full 1966 translation of Wegener's final (1929) version. 64 illustrations. 246pp. 5⅜ x 8½.   61708-4 Pa. $4.50

THE PRINCIPLES OF PSYCHOLOGY, William James. Famous long course complete, unabridged. Stream of thought, time perception, memory, experimental methods; great work decades ahead of its time. Still valid, useful; read in many classes. 94 figures. Total of 1391pp. 5⅜ x 8½.
20381-6, 20382-4 Pa., Two-vol. set $13.00

GEOMETRY, RELATIVITY AND THE FOURTH DIMENSION, Rudolf Rucker. Exposition of fourth dimension, means of visualization, concepts of relativity as Flatland characters continue adventures. Popular, easily followed yet accurate, profound. 141 illustrations. 133pp. 5⅜ x 8½.
23400-2 Pa. $2.75

THE ORIGIN OF LIFE, A. I. Oparin. Modern classic in biochemistry, the first rigorous examination of possible evolution of life from nitrocarbon compounds. Non-technical, easily followed. Total of 295pp. 5⅜ x 8½.
60213-3 Pa. $4.00

PLANETS, STARS AND GALAXIES, A. E. Fanning. Comprehensive introductory survey: the sun, solar system, stars, galaxies, universe, cosmology; quasars, radio stars, etc. 24pp. of photographs. 189pp. 5⅜ x 8½. (Available in U.S. only)
21680-2 Pa. $3.75

THE THIRTEEN BOOKS OF EUCLID'S ELEMENTS, translated with introduction and commentary by Sir Thomas L. Heath. Definitive edition. Textual and linguistic notes, mathematical analysis, 2500 years of critical commentary. Do not confuse with abridged school editions. Total of 1414pp. 5⅜ x 8½.
60088-2, 60089-0, 60090-4 Pa., Three-vol. set $18.50

Prices subject to change without notice.

Available at your book dealer or write for free catalogue to Dept. GI, Dover Publications, Inc., 180 Varick St., N.Y., N.Y. 10014. Dover publishes more than 175 books each year on science, elementary and advanced mathematics, biology, music, art, literary history, social sciences and other areas.